Upstream Bioprocesses Biomass-Based Platforı and Derivatives

Upstream Bioprocesses to Biomass-Based Platform Chemicals and Derivatives

Editors

Miguel Ladero
Victoria E. Santos

MDPI

Basel • Beijing • Wuhan • Barcelona • Belgrade • Novi Sad • Cluj • Manchester

Editors

Miguel Ladero
Complutense University of
Madrid
Madrid
Spain

Victoria E. Santos
Complutense University of
Madrid
Madrid
Spain

Editorial Office
MDPI AG
Grosspeteranlage 5
4052 Basel, Switzerland

This is a reprint of articles from the Special Issue published online in the open access journal *Fermentation* (ISSN 2311-5637) (available at: https://www.mdpi.com/journal/fermentation/special_issues/JAUTCRK4XP).

For citation purposes, cite each article independently as indicated on the article page online and as indicated below:

Lastname, A.A.; Lastname, B.B. Article Title. *Journal Name* **Year**, *Volume Number*, Page Range.

ISBN 978-3-7258-2377-2 (Hbk)
ISBN 978-3-7258-2378-9 (PDF)
doi.org/10.3390/books978-3-7258-2378-9

Cover image courtesy of Miguel Ladero

Contents

About the Editors

Miguel Ladero

Ph.D. in Industrial Chemistry interested in the development and optimization of bioprocesses and catalytic green chemical processes within the biorefinery concept and under batch and intensified flow conditions. He is Full Professor at Department of Chemical and Materials Engineering of UCM. He got his M.Sc. in chemistry in 1992, focused on cellulase production by *Neurospora crassa* on wheat straw. His PhD thesis was focused on the hydrolysis of lactose based on the use of free and immobilized β-galactosidases of several origins, applying new mathematical techniques and kinetic models to reflect the effect of main phenomena taking place in reacting systems, including enzyme deactivation. His research has covered different aspects (process development, kinetic modelling, phenomenological studies) studying several microbial (fungi, bacteria), enzymatic and catalytic systems: diesel biodesulfurization, anaerobic and microaerobic bioprocesses to 2,3-butanediol, D-lactic, fumaric, succinic acids, sunscreen ingredients by thermal and catalytic procedures, plasticizers and prodrugs with industrial lipases, derivatives of rosin by esterification and disproportionation, stabilization of enzymes by immobilization and reaction media engineering, development of thermal and catalytic green processes to obtain carbonates and ketals from glycerol, and development of physicochemical, enzymatic and microbial processes in lignocellulosic/oleaginous integrated biorefineries, among other research lines. He has been involved, since 1993, in 23 research projects (Regional, National and International), being PI in 5 of them and general coordinator of a recent M.ERAnet (2021) focused on Retinal Ganglion Cell production from Induced Pluripotent Stem Cells differentiated on dedicated biomaterials based on exopolysaccharides (BioMAT4EYE). He has been also involved in 8 contracts with chemical companies of the energy and chemistry sectors, which resulted in 16 private reports.

Victoria E. Santos

Her research career began in 1997 with an M.Sc. thesis on chemical reaction engineering computational methods for application in the kinetic study of complex chemical reactions. In the spring of 1988, she started working on the biotechnological process to produce xanthan (under a research contract with Repsol S.A.); in October of that year, she got a position as an Assistant Laboratory Professor, starting her doctoral thesis combining research and teaching. In June 1993, she presented her doctoral thesis on xanthan gum production. Dr. Santos has focused her research work on the study, development, and modeling of microbial bioprocesses from a chemical engineering point of view, identifying phenomena and carrying out their experimental study and modeling (addressing the kinetic modeling of microbial bioprocesses in a prominent way, as well as the study of the oxygen transfer, uptake rates, and cell damage due to the hydrodynamics necessary to achieve the correct transfer of oxygen). In 2000, she spent 40 weeks at the CIB-CSIC in the group of Professor José Luis García López. In 2008, she began to co-direct the research group with Professor García-Ochoa Soria. Thereafter, she focused her work on collaborations with international groups, obtaining funding as a PI for three projects within the ERA-Nets ERA-IB and SUSFOOD2 (being the main coordinator of the consortium in this project) involving European (EVONIK, SOLVAY, ASA SPECIALENZYME, EKODENGE, ELYS Conseil SASU, SUDZUCKER, BIOPOLIS, PLANTRESPONSE, and PAGO de CARRAOVEJAS) companies and public research centers in several European countries (France, Norway, Polland, Turkey, and Germany). Around 2012, she started working on microbial bioprocesses within the 2G biorefinery approach. In 2023, she began directing the FQPIMA research group, a role that, since September 2024, she has shared with Prof. Miguel Ladero.

fermentation

MDPI

Editorial

Upstream Bioprocesses to Biomass-Based Platform Chemicals and Derivatives

Miguel Ladero Galán

FQPIMA Group, Department of Chemical Engineering and Materials, Faculty of Chemical Sciences, Complutense University of Madrid, Avenida Complutense s/n, 28040 Madrid, Spain; mladerog@ucm.es

Over the past few decades, the need for new, more accessible and renewable raw materials has become evident. The biomass or matter contained and created by living beings through their metabolisms is created annually in quantities of more than 180 billion tons, an amount far greater than the 8 billion tons of fossil resources mobilized each year [1]. Most of this biomass is lignocellulosic (more than 80%) and is not directly edible for humans, which avoids tensions in the food market. Combining the plentiful availability of biomass and the knowledge and, in part, the technology gained during the last 150 years in crude refining and transformation, biorefineries are thought as complex, multifeedstock, multiprocess, integrated facilities, transforming diverse biomasses in a wide range of products via diverse platform chemicals: bio-oils, biochars, biogas, sugars, fatty acids, proteins and more [2].

Originally, biorefineries have been inspired by processes in the food and pulp and paper industries [3,4]. The use of food resources is typical in the so-called first-generation biorefineries, initially meant to produce huge amounts of bioethanol by yeast fermentation of sucrose derived from corn via enzymatic hydrolysis (in the USA) or from sugar factories in Brazil transforming enormous quantities of sugarcane—which is the bioethanol mixed with gasoline in various amounts, up to 85% v/v [5]. In parallel, most biodiesel is obtained from food-grade oil that is rich in triglycerides extracted from oil palm nuts, soya beans, rapeseed, sunflower seeds, etc. [6]. The extraction operations are typical in the food industry, and only the transesterification of triglycerides with low-molecular-weight alcohols can be envisaged as a different type of process; even the hydrogenation of triglycerides to HVO—a biomass-based substitute of diesel—is based on oil hydrogenations common in margarine manufacturing. Therefore, as biofuels are increasingly being blended with gasoline and diesel (up to 85% in gasoline; up to 20% in diesel) and most bioethanol is still produced via the aforementioned first-generation biorefining, a notable stress in the food sector has emerged. HVO is emerging as a supplement for diesel, together with biodiesel from transesterification; however, to reiterate, most oil transformed is from an edible source [7].

The second- and third-generation biorefineries emerge as a solution to the stress posed by first-generation biorefinery processes and expand the biorefinery solution by including the majority of biomass created annually at a global scale [8]. The huge cultivation of macro- and microalgae needed for third-generation biorefineries is still in its infancy, though rapidly progressing. In contrast, lignocellulosic biomass is widely available and can originate from the agriculture, forestry and food sectors, either directly or as residues. Lignocellulose transformation can be performed by combining physical, chemical, biocatalytic and microbial processes via a thermochemical approach [9], a fractionation strategy (as shown in Figure 1) [10] or combining them. However, the high oxygen content of this feedstock, the even higher resistance to physical and chemical change and the chemical diversity of lignocellulosic biomass pose notable technical and scientific challenges.

Citation: Ladero Galán, M. Upstream Bioprocesses to Biomass-Based Platform Chemicals and Derivatives. *Fermentation* **2024**, *10*, 59. https://doi.org/10.3390/fermentation10010059

Received: 9 January 2024
Accepted: 11 January 2024
Published: 15 January 2024

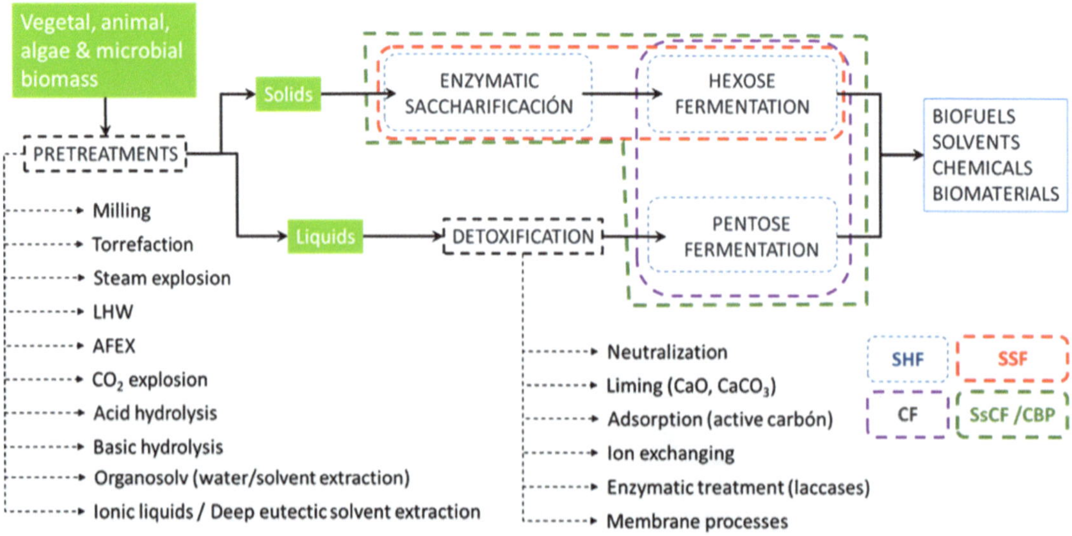

Figure 1. The general fractionation biological biorefinery scheme: From biomass to chemicals and materials via bioprocessing with enzymes and microorganisms. Pretreatments are needed to render the most recalcitrant biomasses amenable for bioprocessing, together, sometimes, with detoxification processes to remove acids, aldehydes, lipopolysaccharides and phenols (including diverse lignin types and lignin oligosaccharides). Enzymatic and microbial bioprocessing can be applied in-series (SHF: separated hydrolysis and fermentation) or together (SSF: simultaneous saccharification and fermentation). Some microorganisms are capable of using pentoses from hemicelluloses and pectin and hexoses from all polysaccharides at the same time (CF: co-fermentation) and can be combined with enzyme cocktails working at the same bioreactor in a one-pot strategy to obtain the targeted compounds (SsCF: simultaneous saccharification and co-fermentation). The ultimate approach is to create genetically engineered microorganisms that can synthetize enzymes, which are able to degrade the biomass to low-molecular-weight compounds that can be transported within the cell and metabolize to products of interest (CB: consolidated bioprocessing).

In this Special Issue, several of the key processes in the upstream section of biorefineries, devoted to the creation of platform chemicals and industrial intermediates, key ingredients or reagents for the final market products, are tackled. It is a collection of notable contributions showcasing diverse key aspects of the first processes of a biorefinery: the upstream section.

Contribution 1 focused on the statistical optimization of steam explosion [11] as a pretreatment that renders a lignocellulose biomass suitable for the next transformation here studied: the production of bioethanol at high-solid load after the enzymatic transformation of the pretreated biomass into a glucose-rich hydrolysate. To this end, the authors employed both horizontal and vertical fermenters which had suitable propellers for operation, with large solid concentrations. In contribution 8, the authors applied an empirical modeling of the response surface, severity factor and multi-response desirability function methodology to explore the feasibility of soybean hull, a major lignocellulosic residue, as a source of bioethanol. To this end, a thermochemical pretreatment based on H_2SO_4 was optimized and supported the SHF process to bioethanol with *Saccharomyces cerevisiae* as the biocatalyst in good mass yields.

Pretreatments open the door to economically feasible productions of bioethanol, lactic acid, succinic acid, fumaric acid, microbial alginate—a most promising exopolysaccharide—and bioplastics, all of which are potential substitutes, at least partially, of petrochemical-based polymers [12–14], to name but a few of the large diversity of potential biorefinery platform chemicals and material ingredients. The importance of bioethanol as co-fuel has been commented before; even if it is not a perfect substitute for gasoline, its power as a polar solvent is of utmost importance in the extraction of bioactive, antimicrobial, antioxidant mixtures out of several lignocellulose biomasses. Lactic, succinic and fumaric acids act as acidulants and flavors in the food and feed industry, but they are also considered promising biomonomers to synthetize biobased polyesters and polyamides.

Lactic acid production is the target of contribution 3, where the authors start from steam-exploded sugarcane bagasse that is hydrolyzed through a mixture of Cellic CTec3 and Cellic HTec3 enzymatic cocktails and further transformed into the acid by *Bacillus coagulans* DSM 2314 with high yields and productivities. Furthermore, the importance of inhibitor elimination and its relation to pretreatment conditions was evidenced. The authors in contribution 5 explore the possibility of using a lignocellulose-rich source, cow manure, as feedstock for lactic acid production, showing that the proper application of a SHF strategy resulted in promising results for lactic acid production with *Bacillus coagulans* DSM 2314 regarding yield, productivity and titer. Contribution 2 highlights the importance of novel and cost-effective membranes for microbial fuel cells able to treat animal slurry, thus reducing the polluting potential of these concentrated wastewater streams, while producing electricity and struvite, a fertilizer rich in nitrogen, magnesium and phosphate.

The development of efficient operations and biocatalysts and of empirical, simple kinetic models aid their future implementation at an industrial scale. In this sense, authors of contributions 4 and 11 explore diverse operation types for the production of succinic acid from glucose and from xylose using *Actinobacillus succinogenes* DSM 22257, proposing and fitting non-segregated unstructured empirical kinetic models suitable for data retrieved in a wide variety of conditions, including nitrogen stress.

Fumaric acid is another dicarboxylic acid and a top biorefinery platform of interest in the food and the chemical industries. Its production via biotechnological strategies using *Rhizopus arrhizus* NRRL 1526, glucose as the carbon source and batch and fed-batch processing as operational approaches is the subject of contribution 6. Here, the authors show how, under adequate conditions and, as a consequence, morphology, the fungus is able to produce up to 195 g/L of the target biomonomers, reaching industrial-relevant concentrations of the acid within 16 days under closely controlled pH by $CaCO_3$ addition.

Moreover, the need to more efficiently create enzymes and microorganisms, from agricultural and food waste (such as lactose in whey), as well as efficient biocatalysts to transform biomass is presented. In this regard, the obtention of thermostable enzymes, which are key for cellulose saccharification, with a natural higher stability at higher process temperatures shows the need for novel saccharification processes under more efficient conditions [15]. This issue is addressed in contribution 7, where a thermostable endoglucanase (MtEG5-1) from *Myceliophthora thermophila* is overexpressed in *Pichia pastoris* GS115 and is fully characterized. The authors observe that the enzyme expressed in this new host had an even higher thermostability than when expressed in others.

Likewise, the use of food residues to obtain these critical enzymes alleviates the production costs, one of the main economical hindrances in the implementation of second-generation biorefineries [16]. In contribution 9, *Trichoderma guizhouense* NJAU4742 is shown to produce high levels of cellulase under lactose, but not with other carbon sources such as glucose, galactose and sucrose. Additionally, the combination of lactose and wheat straw resulted in higher activities of the targeted enzymes. Moreover, the authors performed a detailed study to show that TgRas family genes are critical to the growth of fungi, whereby this family's proteins play key effects on cell wall integrity, growth site selection, polarity establishment and maintenance. In fact, this work sheds more light on the evidence that TgRas family genes play an indispensable role in fungal survival and lactose metabolism.

Furthermore, in this Special Issue, the production of bioplastics and biomonomers from food wastes is also presented. The authors of the review presented in contribution 12 dwell on the use of sugar beet pulp—a plentiful waste carbon source—for the production of two key precursors of bioplastics: lactic acid and polyhydroxyalkanoates (PHAs). Microbial exopolysaccharides can also be used as ingredients for biomaterials, in addition to notable applications in the food and fine chemistry sectors. In this regard, in contribution 10, the authors set the conditions for the continuous production of alginate from *Azotobacter vinelandii* under diazotrophic and nondiazotrophic conditions. The results of this work show the feasibility of enhancing alginate production (yields and specific productivity rates) and quality (molecular weight) under nitrogen-fixing conditions.

If carbon sources in bioprocesses present an important raw material cost, complex nitrogen sources such as peptone, tryptone or yeast extract is an even higher economical expenditure [17]. The proper consideration of all operational costs (OPEX) should be carefully performed to ensure the economic feasibility of industrial-scale lignocellulose-based biorefineries.

Funding: This editorial has received funding from the European Union Horizon 2020 Research and Innovation Program under grant agreement no. 958174. In particular, I wish to express my gratitude to the Spanish "Agencia Estatal de Investigación", which granted me a national subproject within the M.ERA-Net Call 2021, project 9147 "BioMAT4EYE": PCI2022-132971. Furthermore, this work is supported by the Spanish Science and Innovation Ministry-State Research Agency through research project PID2020-114365RB-C21.

Acknowledgments: As Guest Editor of this Special Issue, "Upstream Bioprocesses to Biomass-Based Platform Chemicals and Derivatives", devoted to biorefineries, I would like to show my deep gratitude to the authors that have submitted their valuable contributions published in this Special Issue, thus contributing to its success. Likewise, I would like to thank editorial office for their support during this rewarding experience.

Conflicts of Interest: The author declares no conflicts of interest.

List of Contributors:

1. Troncoso-Ortega, E.; Valenzuela, R.; Reyes-Contreras, P.; Castaño-Rivera, P.; Schiappacasse, L.N.; Parra, C. Maximizing Bioethanol Production from Eucalyptus globulus Using Steam Explosion Pretreatment: A Multifactorial Design and Fermenter Development for High Solid Loads. *Fermentation* **2023**, *9*, 965. https://doi.org/10.3390/fermentation9110965.
2. Iniesta-López, E.; Hernández-Fernández, A.; Garrido, Y.; Ieropoulos, I.A.; Hernández-Fernández, F.J. Microbial Fuel Cell Using a Novel Ionic Liquid-Type Membrane–Cathode Assembly for Animal Slurry Treatment and Fertilizer Production. *Fermentation* **2023**, *9*, 844. https://doi.org/10.3390/fermentation9090844.
3. Alves, W.R.; Da Silva, T.A.; Zandoná Filho, A.; Pereira Ramos, L. Lactic Acid Production from Steam-Exploded Sugarcane Bagasse Using *Bacillus coagulans* DSM2314. *Fermentation* **2023**, *9*, 789. https://doi.org/10.3390/fermentation9090789.
4. Escanciano, I.A.; Ripoll, V.; Ladero, M.; Santos, V.E. Study on the Operational Modes Using Both Growing and Resting Cells for Succinic Acid Production from Xylose Kinetic Modelling. *Fermentation* **2023**, *9*, 663. https://doi.org/10.3390/fermentation9070663.
5. Garrido, R.; Falguera, V.; Pérez Navarro, O.; Acosta Solares, A.; Cabeza, L.F. Lactic acid production from cow manure: Experimental process conditions analysis. *Fermentation* **2023**, *9*, 604. https://doi.org/10.3390/fermentation9070604.
6. Kuenz, A.; Eidt, L.; Prüße, U. Biotechnological Production of Fumaric Acid by *Rhizopus arrhizus*—Reaching Industrially Relevant Final Titers. *Fermentation* **2023**, *9*, 588. https://doi.org/10.3390/fermentation9070588.
7. Zhou, W.; Tong, S.; Amin, F.R.; Chen, W.; Cai, J.; Li, D. Heterologous Expression and Biochemical Characterization of a Thermostable Endoglucanase (Mt EG5-1) from *Myceliophthora thermophila*. *Fermentation* **2023**, *9*, 462. https://doi.org/10.3390/fermentation9050462.
8. Gil Rolón, M.; Leonardi, R.J.; Bolzico, B.C.; Seluy, L.G.; Benzzo, M.T.; Comelli, R.N. Multi-Response Optimization of Thermochemical Pretreatment of Soybean Hulls for 2G-Bioethanol Production. *Fermentation* **2023**, *9*, 454. https://doi.org/10.3390/fermentation9050454.
9. Miao, J.; Chen, C.; Gu, Y.; Zhu, H.; Guo, H.; Liu, D.; Shen, Q. The TgRas1 Gene Affects the Lactose Metabolism of *Trichoderma guizhouense* NJAU4742. *Fermentation* **2023**, *9*, 440. https://doi.org/10.3390/fermentation9050440.
10. Contreras-Abara, P.; Castillo, T.; Ponce, B.; Urtuvia, V.; Peña, C.; Díaz-Barrera, A. Continuous Bioproduction of Alginate Bacterial under Nitrogen Fixation and Nonfixation Conditions. *Fermentation* **2023**, *9*, 426. https://doi.org/10.3390/fermentation9050426.

11. Escanciano, I.A.; Ladero, M.; Santos, V.E.; Blanco, Á. Development of a Simple and Robust Kinetic Model for the Production of Succinic Acid from Glucose Depending on Different Operating Conditions. *Fermentation* **2023**, *9*, 222. https://doi.org/10.3390/fermentation9030222.
12. Marzo-Gago, C.; Díaz, A.B.; Blandino, A. Sugar Beet Pulp as Raw Material for the Production of Bioplastics. *Fermentation* **2023**, *9*, 655. https://doi.org/10.3390/fermentation9070655.

References

1. Deng, W.; Feng, Y.; Fu, J.; Guo, H.; Guo, Y.; Han, B.; Jiang, Z.; Kong, L.; Li, C.; Liu, H.; et al. Catalytic conversion of lignocellulosic biomass into chemicals and fuels. *Green Energy Environ.* **2023**, *8*, 10–114. [CrossRef]
2. Garcia-Ochoa, F.; Vergara, P.; Wojtusik, M.; Gutiérrez, S.; Santos, V.E.; Ladero, M.; Villar, J.C. Multi-feedstock lignocellulosic biorefineries based on biological processes: An overview. *Ind. Crops Prod.* **2021**, *172*, 114062. [CrossRef]
3. Dahiya, S.; Kumar, A.N.; Sravan, J.S.; Chatterjee, S.; Sarkar, O.; Mohan, S.V. Food waste biorefinery: Sustainable strategy for circular bioeconomy. *Bioresour. Technol.* **2018**, *248*, 2–12. [CrossRef] [PubMed]
4. Gupta, G.K.; Shukla, P. Insights into the resources generation from pulp and paper industry wastes: Challenges, perspectives and innovations. *Bioresour. Technol.* **2020**, *297*, 122496. [CrossRef]
5. Ayodele, B.V.; Alsaffar, M.A.; Mustapa, S.I. An overview of integration opportunities for sustainable bioethanol production from first-and second-generation sugar-based feedstocks. *J. Clean. Prod.* **2020**, *245*, 118857. [CrossRef]
6. Callegari, A.; Bolognesi, S.; Cecconet, D.; Capodaglio, A.G. Production technologies, current role, and future prospects of biofuels feedstocks: A state-of-the-art review. *Crit. Rev. Environ. Sci. Technol.* **2020**, *50*, 384–436. [CrossRef]
7. No, S.Y. Application of hydrotreated vegetable oil from triglyceride based biomass to CI engines—A review. *Fuel* **2014**, *115*, 88–96. [CrossRef]
8. Goswami, L.; Kayalvizhi, R.; Dikshit, P.K.; Sherpa, K.C.; Roy, S.; Kushwaha, A.; Kim, B.S.; Banerjee, R.; Jacob, S.; Rajak, R.C. A critical review on prospects of bio-refinery products from second and third generation biomasses. *Chem. Eng. J.* **2022**, *448*, 137677. [CrossRef]
9. Seo, M.W.; Lee, S.H.; Nam, H.; Lee, D.; Tokmurzin, D.; Wang, S.; Park, Y.K. Recent advances of thermochemical conversion processes for biorefinery. *Bioresour. Technol.* **2022**, *343*, 126109. [CrossRef] [PubMed]
10. Rodionova, M.V.; Bozieva, A.M.; Zharmukhamedov, S.K.; Leong, Y.K.; Lan, J.C.; Veziroglu, A.; Veziroglu, T.N.; Tomo, T.; Chang, J.S.; Allakhverdiev, S.I. A comprehensive review on lignocellulosic biomass biorefinery for sustainable biofuel production. *Int. J. Hydrogen Energy* **2022**, *47*, 1481–1498. [CrossRef]
11. Tuan Hoang, A.; Phuong Nguyen, X.; Quang Duong, X.; Ağbulut, Ü.; Len, C.; Quy Phong Nguyen, P.; Kchaou, M.; Chen, W.-H. Steam explosion as sustainable biomass pretreatment technique for biofuel production: Characteristics and challenges. *Bioresour. Technol.* **2023**, *385*, 129398. [CrossRef]
12. Kohli, K.; Prajapati, R.; Sharma, B.K. Bio-based chemicals from renewable biomass for integrated biorefineries. *Energies* **2019**, *12*, 233. [CrossRef]
13. Talan, A.; Pokhrel, S.; Tyagi, R.D.; Drogui, P. Biorefinery strategies for microbial bioplastics production: Sustainable pathway towards Circular Bioeconomy. *Bioresour. Technol. Rep.* **2022**, *17*, 100875. [CrossRef]
14. Schmid, J. Recent insights in microbial exopolysaccharide biosynthesis and engineering strategies. *Curr. Opin. Biotechnol.* **2018**, *53*, 130–136. [CrossRef]
15. Patel, A.K.; Singhania, R.R.; Sim, S.J.; Pandey, A. Thermostable cellulases: Current status and perspectives. *Bioresour. Technol.* **2019**, *279*, 385–392. [CrossRef] [PubMed]
16. Kosamia, N.M.; Samavi, M.; Piok, K.; Rakshit, S.K. Perspectives for scale up of biorefineries using biochemical conversion pathways: Technology status, techno-economic, and sustainable approaches. *Fuel* **2022**, *324*, 124532. [CrossRef]
17. De la Torre, I.; Ladero, M.; Santos, V. Production of d-lactic acid by *Lactobacillus delbrueckii* ssp. delbrueckii from orange peel waste: Techno-economical assessment of nitrogen sources. *Appl. Microbiol. Biotechnol.* **2018**, *102*, 10511–10521. [CrossRef]

fermentation

MDPI

Article

Development of a Simple and Robust Kinetic Model for the Production of Succinic Acid from Glucose Depending on Different Operating Conditions

Itziar A. Escanciano, Miguel Ladero *, Victoria E. Santos and Ángeles Blanco

Department of Chemical Engineering, Faculty of Chemical Sciences, Complutense University of Madrid, Avenida Complutense s/n, 28040 Madrid, Spain
* Correspondence: mladerog@ucm.es

Abstract: Succinic acid (SA) is one of the main identified biomass-derived chemical building blocks. In this work we approach the study of its production by *Actinobacillus succinogenes* DSM 22257 from glucose, focusing on the development and application of a simple kinetic model capable of representing the evolution of the process over time for a great diversity of process variables key to the production of this platform bio-based chemical: initial biomass concentration, yeast extract concentration, agitation speed, and carbon dioxide flow rate. All these variables were studied experimentally, determining the values of key fermentation parameters: titer (23.8–39.7 $g \cdot L^{-1}$), yield (0.59–0.72 $g_{SA} \cdot g_{glu}{}^{-1}$), productivity (0.48–0.96 $g_{SA} \cdot L^{-1} \cdot h^{-1}$), and selectivity (0.61–0.69 $g_{SA} \cdot g_{glu}{}^{-1}$). Even with this wide diversity of operational conditions, a non-structured and non-segregated kinetic model was suitable for fitting to experimental data with high accuracy, considering the values of the goodness-of-fit statistical parameters. This model is based on the logistic equation for biomass growth and on potential kinetic equations to describe the evolution of SA and the sum of by-products as production events that are not associated with biomass growth. The application of the kinetic model to diverse operational conditions sheds light on their effect on SA production. It seems that nitrogen stress is a good condition for SA titer and selectivity, there is an optimal inoculum mass for this purpose, and hydrodynamic stress starts at 300 r.p.m. in the experimental set-up employed. Due to its practical importance, and to validate the developed kinetic model, a fed-batch fermentation was also carried out, verifying the goodness of the model proposed via the process simulation (stage or cycle 1) and application to further cycles of the fed-batch operation. The results showed that biomass inactivation started at cycle 3 after a grace period in cycle 2.

Keywords: succinic acid; fermentation; kinetic model; operational conditions; carbon dioxide

Citation: Escanciano, I.A.; Ladero, M.; Santos, V.E.; Blanco, Á. Development of a Simple and Robust Kinetic Model for the Production of Succinic Acid from Glucose Depending on Different Operating Conditions. *Fermentation* **2023**, *9*, 222. https://doi.org/10.3390/fermentation9030222

Academic Editor: Diomi Mamma

Received: 1 February 2023
Revised: 18 February 2023
Accepted: 22 February 2023
Published: 25 February 2023

1. Introduction

The growing concern about the effects of climate change and the depletion of fossil resource reserves have driven a movement focused on the use of new renewable energy sources and bio-based chemicals [1]. One of the most promising alternatives is the development of biorefineries, where biomass will be sustainably processed into bioproducts (materials and chemicals) and bioenergy (biofuels, electricity, and heat) [2].

Within the top 12 high-value-added chemicals from biomass, according to the US Department of Energy (DOE) [3], succinic acid is considered one of the key carboxylic acids for the bioeconomy era. Although this compound is widely used in the production of polyesters, polyols, resins, coatings, and pigments, and in the pharmaceutical and food industries, in the context of biorefineries, its main applications are the generation of intermediate chemical products such as 1,4-butanediol, tetrahydrofuran, 2-pyrrolidine, or maleic acid. Furthermore, the potential application of succinic acid and its derivatives for the manufacture of biodegradable polymers should also be mentioned [4–6].

Currently, the production volume of succinic acid through the biological route has already exceeded that generated by traditional chemical production. In addition, this growth is expected to continue in the coming years, so that the biotechnological production of this compound will go from generating a market value of USD 170 million in 2020 up to USD 2.22 billion by 2026 [7].

Among the microorganisms that produce succinic acid, it is worth highlighting *Actinobacillus succinogenes*, isolated from the rumen, as one of the most promising bacteria, since this acid is a final product during its anaerobic metabolism [8,9].

Numerous studies have been carried out focused on determining the influence of different variables on the production process of succinic acid using this microorganism as a biocatalyst. Despite the initial concentration of substrate being one of the most studied operating conditions, there is not yet a scientific consensus on its optimal value. For example, in the experiments carried out by Luthfi et al. [10] and Salvachúa et al. [11], the highest yields were achieved in fermentations starting from a concentration of 60 g L^{-1} of glucose, while Ferone et al. [12] achieved the best results when this concentration was reduced to 40 g L^{-1}. These latter authors also observed that low concentrations of xylose, around 5 g L^{-1}, increased the yield of succinic generation. However, Pateraki et al. [13], when using a mixture of sugars rich in xylose as a carbon source, achieved better results at a total sugar concentration of 32.5 g L^{-1}.

Another of the most studied parameters is the CO_2 source. Diverse authors have employed different carbonates ($NaHCO_3$, K_2CO_3, $CaCO_3$, or $MgCO_3$) at several concentrations [14,15], comparing them with pure gaseous CO_2 or biogas at different partial pressures [16–19], and even maximizing their solubility by increasing the pressure in the reactor [20,21]. Other studies have focused on cell status and on the mode of operation, performing experiments with free, immobilized cells or resting cells in batch, repeated batch, fed-batch, and continuous mode operations [22–28]. It is worth noting the work of Kim et al. [23], who carried out a continuous fermentation of recycled cells to maximize the biocatalytic activity, achieving a productivity of 3.86 g L^{-1} h^{-1} of succinic acid. However, the yields and productivities are especially high when working with immobilized cells, operating both in repeated batch. For example, Cao et al. [27] increased the production of succinic acid by approximately 50% compared to the batch operation. Furthermore, in continuous mode, Ercole et al. [25] produced 36.5 g L^{-1} h^{-1} of succinic acid with cells entrapped in alginate beads.

However, there is a lack of knowledge on other conditions that also affect the fermentation process with this microorganism, such as the initial biomass concentration or the agitation speed. In addition, until now, studies focused on the nitrogen source are scarce and the impact that this variable may have on production has not been fully explored [16,29,30]. It is worth noting the work of Tan et al. [31], who compared the use of 15 g L^{-1} of yeast extract (YE) and corn steep liquor (CSL), achieving a succinic acid amount 3.7% lower than using yeast extract as the nitrogen source and reducing by a fifth the costs associated with the nitrogen source. Xi et al. [32] studied the effect of using CSL at different initial concentrations, obtaining similar yields in fermentations with CSL and YE as long as the amount of CSL doubled that of YE. Jiang et al. [33] successfully replaced the yeast extract with a spent brewer's yeast hydrolysate with vitamin supplements.

Besides the many studies and proofs of concept published for the development of biorefinery processes, the main drawback of most of the proposed concepts is their viability at industrial scale. To carry out the scaling of a fermentation process, it is essential to develop kinetic models with equations capable of predicting the behavior of the species involved throughout production. These models allow an adequate selection of the type of operation, as well as the optimal design and operation of the bioreactor. In addition, they are very useful for the implementation of the control system and the performance of techno-economic studies [7,34].

However, the state-of-the-art research shows that the kinetic models developed on succinic acid production are based on results of fermentations carried out at different initial concentrations of substrate. Therefore, as these studies are empirical in nature, they are only valid for the particular experimental conditions tested. Furthermore, there are no kinetic studies taking into account the effects of other operating conditions that are as influential as the concentration of the main carbon source. For these reasons, this work goes beyond the state-of-the-art research, using a novel approach to accomplish an exhaustive study of the influence exerted by diverse key operation variables—the initial concentration of biomass, the speed of agitation, and the concentration of yeast extract—in a batch production of succinic acid from glucose through the action of *A. succinogenes*. The developed new kinetic model is simple but capable of predicting the evolution of the concentration of biomass, glucose, succinic acid, and by-products for each of the experiments, and it was also applied to the production stages of a fed-batch fermentation.

2. Materials and Methods

2.1. Microorganism

Actinobacillus succinogenes DSM 22257, supplied by the German Collection of Microorganisms and Cell Cultures GmbH, was used in all experiments.

2.2. Culture Media

For storage, a 1:1 v/v glycerol/Tryptic Soy Broth (TSB) mixture was used [15,23,35] and for its subsequent reactivation, only TSB was used. TSB composition was (in grams per liter): 17 tryptone, 3 soytone, 2.5 glucose, 5 NaCl, 2.5 K_2HPO_4.

The medium for inoculum preparation as well as for the production reactor was the same [36] (in grams per liter): 3 K_2HPO_4, 0.43 $MgCl_2$.6H2O, 0.2 $CaCl_2$, 1 NaCl, 40 glucose, 2.5/5/7.5/10 yeast extract. In the case of the inoculum, 40 g L^{-1} of $NaHCO_3$ was added as a CO_2 source and for pH control. Carbon and nitrogen sources were autoclaved separately.

2.3. Cultivation Conditions

First, the stored cells were thawed at −80 °C and injected into bottles with 60 mL of TSB, which were previously purged with N_2 for 2 min at a flow rate of 1 L min^{-1}. After incubating the bottles at 37 °C and 200 rpm for 24 h, the inoculum was grown under the same conditions, but using, in this case, the production medium.

The study of the influence of operating conditions was carried out in a 2 L stirred tank bioreactor (STBR) BIOSTAT B-Plus (Sartorius AG, Göttingen, Germany) with a working volume of 1 L. In all runs, 0.1 vvm CO_2 was bubbled into the broth while operation was performed at 37 °C and a pH of 6.8 controlled by automatic addition of 5 M NaOH. The production medium was (in grams per liter): 3 K_2HPO_4, 0.43 $MgCl_2.6H_2O$, 0.2 $CaCl_2$, 1 NaCl, 40 glucose, 2.5/5/7.5/10 yeast extract. Experiments were carried out at different stirring speeds: 150, 200, 250, and 300 rpm, as well as at three different initial biomass concentrations: 0.05, 0.075, and 0.1 g L^{-1}.

In the fed-batch-type operation, the fermentation was carried out in the same reactor and with the same culture medium as in the batch type operations, working at 37 °C, 300 rpm, and pH 6.8 (5M NaOH), with an initial biomass concentration of 0.05 g L^{-1} and a yeast extract concentration of 10 g L^{-1}. After the first stage, a concentrated glucose solution was fed at the start of each of the following stages.

2.4. Analytical Methods

A spectrophotometer (Shimadzu UV-vis spectrophotometer UV-1603) was used to measure the biomass concentration, obtaining optical density data at 600 nm.

Glucose, succinic acid, and by-products (formic and acetic acids) were quantified by high-performance liquid chromatography (HPLC) (Agilent Technologies 100 series). The column employed for this analysis was a REZEX ROA-Monosaccharide H+ (8%) column (300 × 7.8 mm, Phenomenex, Torrance, CA, USA) at 80 °C pumping a H_2SO_4 5 mM solution as mobile phase at a flow rate of 0.5 mL min^{-1}. The refraction index detector worked at 55 °C.

3. Results and Discussion

The results for succinic acid titer (C_{SA}), yield with respect to the initial concentration of the carbon source (Y_{SA}), productivity (P_{SA}), and selectivity (S_{SA}) of the batch experiments carried out under different conditions of CO_2 flow, agitation speed, and yeast extract and initial biomass concentrations are shown in Table 1.

Table 1. Summary of succinic acid titers, yields, productivities, and selectivity values under different operational conditions. The first run is the reference one.

Run	Type of Operation	$C_{biomass}$ ($g{\cdot}L^{-1}$)	Agitation (rpm)	CO_2 Flow ($L{\cdot}min^{-1}$)	C_{YE} ($g{\cdot}L^{-1}$)	C_{SA} ($g{\cdot}L^{-1}$)	Y_{SA} ($g{\cdot}g^{-1}$)	P_{SA} ($g{\cdot}L^{-1}{\cdot}h^{-1}$)	S_{SA} ($g{\cdot}g^{-1}$)
1	Batch	0.05	300	0.1	10	27.4	0.68	0.83	0.62
2	Batch	0.075	300	0.1	10	28.5	0.71	0.96	0.64
3	Batch	0.1	300	0.1	10	28.3	0.70	0.76	0.66
4	Batch	0.05	300	0.5	10	27.6	0.69	0.84	0.63
5	Batch	0.05	300	1	10	26.1	0.65	0.81	0.63
6	Batch	0.05	150	0.1	10	23.6	0.59	0.72	0.61
7	Batch	0.05	200	0.1	10	26.4	0.66	0.78	0.62
8	Batch	0.05	250	0.1	10	28.5	0.71	0.84	0.62
9	Batch	0.05	300	0.1	2.5	23.8	0.59	0.48	0.68
10	Batch	0.05	300	0.1	5	26.8	0.66	0.53	0.66
11	Batch	0.05	300	0.1	7.5	28.9	0.72	0.58	0.64
12	Fed-batch	0.05	300	0.1	10	39.7	0.67	0.72	0.69

It should be noted that the first experiment was taken as the reference one, because its operating conditions are those that can be considered standard or intermediate between those more used in batch-type fermentation for the production of succinic acid using *A. succinogenes* as a biocatalyst [10,11,13,23,26,37–42].

In runs 1, 2, and 3, the initial concentration of biomass is a factor whose variation mainly affected the productivity of the process. It was observed that at an initial concentration of 0.075 g L^{-1}, the intermediate value of those studied, the productivity of succinic acid reaches a maximum value of 0.96 g L^{-1} h^{-1}. However, if the cell concentration is further increased to 0.1 g L^{-1}, productivity is reduced by 21%. In addition, a slight increase in selectivity was observed at higher initial amounts of biomass in the reactor.

Until now, an optimal initial concentration of biomass in the production process of succinic acid by *A. succinogenes* has not been determined in the literature, although a couple of studies have been carried out in which an attempt has been made to determine the optimum inoculum size, arriving at certainly different conclusions. On the one hand, Wan et al. [43] observed that an inoculum size of 10% compared to 2% or 5% led to higher yields of succinic acid, which can be justified by relating a greater amount of cell density with reduced latency time. However, Anwar et al. [44] studied the effect of the inoculum size in a succinic acid production process through simultaneous saccharification and fermentation, concluding that increasing the inoculum size from 5% to 15% reduced the final concentration of succinic acid generated by 50%. The latter authors attributed this reduction in yield to the strong competition for nutrients that occurs in the culture broth when cell density is very high.

Considering these observations and the results of this work, it seems that an initial biomass concentration of an intermediate value is required that, on the one hand, is sufficient to avoid long latency times, but, on the other hand, does not lead to too rapid consumption of the carbon and nitrogen source. Other authors have also observed, in studies carried out in different fermentation processes, that excessive initial amounts of biomass also lead to inhibitions by the product and accumulation of metabolites, which considerably impair the performance of the process [45,46].

However, these reflections do not seem to be sufficient to justify the increase in selectivity that occurs at higher initial biomass concentrations. This reduction in the number of by-products seems to be associated with possible variations in metabolism. To clarify this matter, it would be necessary to conduct metabolic analysis by carrying out fermentations with *A. succinogenes* under different inoculum size conditions. This was undertaken by Din et al. [47] with *Saccharomyces cerevisiae*, who observed large changes in glucose metabolism intermediates, amino acids, and metabolites related to the structure of the cell membrane by modifying the inoculum size.

In runs 1, 4, and 5, the CO_2 gas flow rate was increased from 0.1 to 1 L min^{-1}, values that include those that can be found in the literature on succinic acid production [10,11,13,23,26,37–42]. In this study, no significant variations in yield, productivity, or generation of by-products were observed in the range of flows studied. Taking into account the work of Xi et al. [17] and Zou et al. [18], who observed differences in the production of succinic acid working with mixtures of N_2 and CO_2 until reaching saturation of the latter gas, it is concluded that, as in the present work, fermentations with an excess of CO_2 do not favor the deviation of the metabolism towards the generation of succinic acid to the detriment of other metabolites. This means that most of the studies on succinic acid production by *A. succinogenes* published to date were carried out under conditions that involve higher economic costs and do not offer any additional advantage.

The increase in the agitation speed between 150 and 300 rpm, the range of values typically used in the literature [10,11,13,23,37–41,48,49], of runs 1, 6, 7, and 8, shows a considerable improvement in performance and productivity at high stirring values, reaching the best results at a stirring speed of 250 rpm. This operating condition is one of the factors with the greatest impact on the transfer of gases in liquid media, decreasing mixing time and improving mass and heat transfer rates [50,51]. Therefore, it can be deduced that the stirring values of 150 and 200 rpm are insufficient to achieve an adequate transfer of CO_2 in the culture broth. Taking this phenomenon into account, it could be deduced that the higher the stirring speed, the greater the generation of succinic acid; however, excessive shearing forces can lead to cell damage and, as a consequence, to the reduction in the process performance [52]. This seemed to happen in the run performed at 300 rpm, in which the effect of hydrodynamic stress appeared to be reflected.

In runs 1, 9, 10, and 11, the effect of the initial concentration of the nitrogen source in the culture medium was compared. As the YE concentration increased from 2.5 to 10 g L^{-1}, so did the succinic acid productivity. However, the yield of succinic acid with respect to the initial concentration of the carbon source reached its maximum in the fermentations carried out with 7.5 g L^{-1} of YE. However, it should be noted that Jiang et al. [33] achieved their maximum yield of succinic acid at around 20 g L^{-1} of YE, although they do not provide data on productivity or generation of by-products to be able to make a more detailed comparison with the present study. On the other hand, a tendency to improve selectivity was observed as the quantity of the nitrogen source decreased, which agrees with the conclusions of Ventrone et al [53], who observed that high C:N ratios lead to a reduction in by-product formation. In fact, in operations in the absence of a nitrogen source, with resting cells, Escanciano et al. [28] reduced the generation of by-products by 27.5% compared to the equivalent operation with cells in a state of growth, that is, in the presence of a nitrogen source.

The fed-batch run (run 12) reduced the number of by-products generated compared to the reference experiment, increasing the selectivity from 0.62 g g^{-1} (run 1) to 0.69 g g^{-1} (run 12). During their fed-batch production of succinic acid from citrus peel waste, Patsalou et al. [54] observed that the by-products were produced mainly in the first 24 h in a fermentation lasting more than 60 h, while the succinate continued to be generated throughout the entire process. In addition, these authors obtained a marked drop in productivity, as occurred in the present work, which decreased from 0.83 g L^{-1} h^{-1} in the batch operation type to 0.72 g L^{-1} h^{-1} in the fed-batch fermentation. The loss in the production rate of succinic acid in this type of operation is a conclusion shared by more authors; for example, Kanchanasuta et al. [55] also saw a depletion in yield, a trend that was also observed in this work, although in a less pronounced way. Taking into account that succinic acid production appears to be favored in a non-growth steady state [56,57] and that nutrient depletion does not appear to be an obstacle to succinic acid production when there is already a high biomass concentration [28], it seems that the main problem caused by this type of operation is the excessive accumulation of metabolites, which can generate cell damage and strong inhibitions by product [7,13,37,58,59].

3.1. Development of a Simple Kinetic Model

For the development of the kinetic model, a reaction scheme was proposed based on the time course of the biomass, substrate (glucose), and fermentation products (succinic, acetic, and formic acids) of the reference experiment (run 1), whose evolution throughout over time is shown in Figure 1A. It is observed that the biomass grows until reaching its maximum at 10 h of fermentation; however, both succinic acid and by-products continue to increase over time until the carbon source is exhausted around 33 h, indicating that production is not associated with growth. However, the rate of formation of acetic and formic acids slows down after approximately 20 h of fermentation, while the rate of production of succinic acid only suffers a slight reduction in the last hours of the process. In addition, although a greater amount of acetic acid is produced than formic acid, both compounds follow the same growth trend, which is why it was decided to combine both acids, as shown in Figure 1B, with the aim of proposing a model with the metabolite "by-products" (BPs) that allows further reduction in the number of kinetic parameters and the development of a more useful model from the point of view of chemical engineering.

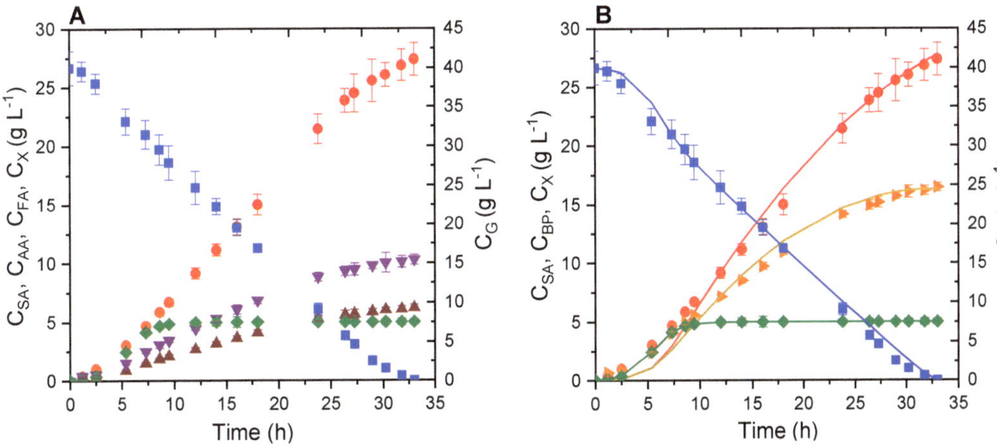

Figure 1. Evolution of the concentration of succinic acid (SA), acetic acid (AA), formic acid (FA), biomass (X), and glucose (G) over time in run 1 (reference). (**A**) representation of AA and FA concentrations separately. (**B**) combining of AA and FA concentrations into by-product (BP) concentration and kinetic model prediction. Data points: SA (●), G (■), AA (▼), FA (▲), X (♦), BP (►), model predictions shown as lines.

Based on these data, a very simple reaction scheme based only on Equations (1)–(3) is proposed. It is an unstructured non-segregated model, that is, the microorganism is considered as a single component, the biomass. This scheme is made up of a first reaction (r_1) for the consumption of glucose (S) for the generation of biomass, a second reaction (r_2) for the generation of succinic acid (P) and by-products (BPs), and a last reaction (r_3) for the independent generation of by-products. Their corresponding rates are shown in Equations (4)–(6), while the consumption and formation rates (R_j) of compounds 'j' are described in Equations (7)–(10). Therefore, the biomass has a growth rate based on the logistic equation, whose kinetic parameters are the specific growth rate (μ) and the maximum biomass concentration (C_{Xm}). The generation of succinic acid and by-products is governed by potential equations independently of the growth of the microorganism in a proportional way to their kinetic constants K_{p1} and K_{p2}. In addition, together with these parameters, the formation and consumption rates are also defined by the macroscopic yields $Y_{S/X}$, $Y_{S/P1}$, $Y_{S/BP}$, and $Y_{S/P2}$.

<u>Reaction network</u>

$$Y_{S/X}S \xrightarrow{r_1} X \quad (1)$$

$$Y_{S/P1}S \xrightarrow{r_2} P + Y_{S/BP}\cdot BP \quad (2)$$

$$Y_{S/P2}S \xrightarrow{r_3} BP \quad (3)$$

<u>Reaction rates</u>

$$r_1 = \mu\cdot C_X\cdot\left(1 - \frac{C_X}{C_{Xm}}\right) \quad (4)$$

$$r_2 = k_{P1}\cdot C_S\cdot C_X \quad (5)$$

$$r_3 = k_{P2}\cdot C_S\cdot C_X \quad (6)$$

<u>Production and consumption rates</u>

$$R_S = \frac{dC_S}{dt} = -Y_{S/X}\cdot r_1 - Y_{S/P1}\cdot r_2 - Y_{S/P2}\cdot r_3 \quad (7)$$

$$R_P = \frac{dC_P}{dt} = r_2 \quad (8)$$

$$R_X = \frac{dC_X}{dt} = r_1 \quad (9)$$

$$R_{BP} = \frac{dC_{BP}}{dt} = Y_{S/BP}\cdot r_2 + r_3 \quad (10)$$

Applying this model, the calculation of its kinetic parameters (Table 2) was performed from the data of the reference experiment (run 1). The simulation of the evolution of the biomass, substrate, product, and by-products is shown in Figure 1B, together with the experimental data, obtaining an excellent fit despite the simplicity of the model and the reduced number of parameters. In addition, Table 3 presents the statistical parameters that reflect the goodness of fit. A value of Fisher's F (F_{95}) of 41,242 was obtained, much higher than its tabulated value at 95% (where 8.55 is the value of F tabulated at that probability) and a percentage of explained variation very close to 100% (99.5%). In addition, very low values were obtained in those parameters that should be as close to zero as possible, with a sum of squared residuals (SSR) of 6.47 and a residual mean squared error (RMSE) of 0.67.

Table 2. Summary of kinetic parameters under different operational conditions.

Run	Type of Operation	$C_{biomass}$ ($g \cdot L^{-1}$)	Agitation (rpm)	CO_2 Flow ($L \cdot min^{-1}$)	C_{YE} ($g \cdot L^{-1}$)	Cxm ± Error ($g_X \cdot L^{-1}$)	Kp_1 ± Error ($L \cdot g^{-1} \cdot h^{-1}$)	Kp_2 ± Error ($L \cdot g^{-1} \cdot h^{-1}$)	μ ± Error (h^{-1})	$Y_{S/P1}$ ± Error ($g \cdot g^{-1}$)	$Y_{S/P2}$ ± Error ($g \cdot g^{-1}$)	$Y_{S/BP}$ ± Error ($g \cdot g^{-1}$)	$Y_{S/X}$ ± Error ($g \cdot g^{-1}$)
1	Batch	0.05	300	0.1	10	5.02 ± 0.02	0.007 ± 0.001	0.019 ± 0.001	0.85 ± 0.04	0.24 ± 0.02	2.27 ± 0.22	1.10 ± 0.09	1.65 ± 0.15
2	Batch	0.075	300	0.1	10	5.81 ± 0.03	0.009 ± 0.001	0.013 ± 0.001	0.85 ± 0.03	0.24 ± 0.02	2.20 ± 0.19	1.08 ± 0.07	1.63 ± 0.16
3	Batch	0.1	300	0.1	10	6.79 ± 0.08	0.004 ± 0.000	0.008 ± 0.001	0.85 ± 0.02	0.24 ± 0.02	2.25 ± 0.19	1.09 ± 0.08	1.63 ± 0.16
4	Batch	0.05	300	0.5	10	5.07 ± 0.03	0.008 ± 0.001	0.018 ± 0.001	0.85 ± 0.01	0.25 ± 0.02	2.27 ± 0.22	1.09 ± 0.09	1.66 ± 0.16
5	Batch	0.05	300	1	10	5.07 ± 0.03	0.008 ± 0.001	0.018 ± 0.001	0.85 ± 0.01	0.25 ± 0.02	2.27 ± 0.22	1.09 ± 0.09	1.66 ± 0.16
6	Batch	0.05	150	0.1	10	5.09 ± 0.09	0.004 ± 0.001	0.019 ± 0.001	0.85 ± 0.04	0.26 ± 0.02	2.30 ± 0.21	1.10 ± 0.08	1.56 ± 0.13
7	Batch	0.05	200	0.1	10	5.05 ± 0.06	0.006 ± 0.001	0.018 ± 0.001	0.85 ± 0.06	0.26 ± 0.02	2.27 ± 0.18	1.04 ± 0.07	1.54 ± 0.11
8	Batch	0.05	250	0.1	10	5.00 ± 0.05	0.008 ± 0.001	0.018 ± 0.001	0.85 ± 0.05	0.24 ± 0.02	2.31 ± 0.18	1.07 ± 0.09	1.60 ± 0.13
9	Batch	0.05	300	0.1	2.5	2.40 ± 0.02	0.008 ± 0.001	0.022 ± 0.001	0.85 ± 0.05	0.24 ± 0.02	1.45 ± 0.10	0.86 ± 0.01	0.98 ± 0.02
10	Batch	0.05	300	0.1	5	5.01 ± 0.03	0.004 ± 0.001	0.011 ± 0.001	0.85 ± 0.03	0.25 ± 0.01	1.76 ± 0.11	0.95 ± 0.02	1.65 ± 0.02
11	Batch	0.05	300	0.1	7.5	5.08 ± 0.03	0.005 ± 0.001	0.014 ± 0.001	0.85 ± 0.04	0.24 ± 0.01	2.06 ± 0.17	1.00 ± 0.04	1.65 ± 0.04
12	Fed-batch cycle 1	0.05	300	0.1	10	5.02 ± 0.02	0.007 ± 0.001	0.019 ± 0.001	0.85 ± 0.04	0.24 ± 0.02	2.27 ± 0.22	1.10 ± 0.09	1.65 ± 0.15
12	Fed-batch cycle 1	0.05	300	0.1	10	5.02 ± 0.02	0.007 ± 0.001	0.019 ± 0.001	0.85 ± 0.04	0.24 ± 0.02	2.27 ± 0.22	0.59 ± 0.02	1.65 ± 0.15
12	Fed-batch cycle 1	0.05	300	0.1	10	5.02 ± 0.02	0.004 ± 0.001	0.005 ± 0.000	0.85 ± 0.04	0.24 ± 0.02	2.27 ± 0.22	0.37 ± 0.01	1.65 ± 0.15

Table 3. Summary of statistical parameters under different operational conditions.

Run	Type of Operation	$C_{biomass}$ ($g{\cdot}L^{-1}$)	Agitation (rpm)	CO_2 Flow ($L{\cdot}min^{-1}$)	C_{YE} ($g{\cdot}L^{-1}$)	F_{95}	RMSE	SSR	VE %
1 REF.	Batch	0.05	300	0.1	10	41,242	0.67	6.47	99.5
2	Batch	0.075	300	0.1	10	13,660	1.09	11.26	98.2
3	Batch	0.1	300	0.1	10	40,640	1.00	10.04	98.7
4	Batch	0.05	300	0.5	10	8457	1.14	12.17	98.5
5	Batch	0.05	300	1	10	8457	1.14	12.17	98.5
6	Batch	0.05	150	0.1	10	19,384	0.92	9.01	99.0
7	Batch	0.05	200	0.1	10	19,684	1.03	9.99	98.6
8	Batch	0.05	250	0.1	10	11,751	1.07	11.70	98.5
9	Batch	0.05	300	0.1	2.5	5037	1.19	14.29	97.3
10	Batch	0.05	300	0.1	5	22,441	1.10	11.57	98.9
11	Batch	0.05	300	0.1	7.5	17,270	1.00	6.32	98.9
12	Fed-batch cycle 1	0.05	300	0.1	10	41,242	0.67	6.47	99.5
12	Fed-batch cycle 1	0.05	300	0.1	10	13,512	1.05	8.97	98.4
12	Fed-batch cycle 1	0.05	300	0.1	10	24,175	0.95	6.09	99.6

After verifying the validity of this model with the reference experiment, it was applied to all the other fermentations carried out under different operating conditions (runs 2–12), with the aims of verifying its robustness, studying the variation in the parameters, and knowing in greater depth the real impact of each of the operating conditions in the succinic acid production process. The estimated kinetic parameters for each of the experiments are shown in Table 2, while the statistical parameters are presented in Table 3.

Until now, the kinetic models of succinic acid production by *A. succinogenes* have been scarce and of the unstructured non-secreted type. In addition, they are usually limited to the study of the rate of biomass formation, leaving aside the evolution of the metabolites present in the broth [12,60,61], although some authors such as Pateraki et al. [13] Lin et al. [37], and Vlysidis et al. [58] have also focused on the carbon source, as well as on the evolution of succinic acid and the generated by-products. Lin et al [37] proposed a biomass growth equation from glucose based on a combination of the Monod equation and the Luong equations for substrate and product inhibition. Vlysidis et al. [58] and Pateraki et al. [13] coincided in using the same combination of equations but used a Haldane–Andrews-type substrate inhibition term. Despite its low value, in all these works Pirt's maintenance coefficient is taken into account when estimating the consumption of the carbon source. In addition, they have in common that they resort to the Luedeking–Piret expression to predict the generation of succinic acid and by-products [7].

Although, on the one hand, they are very complete models that can provide a large amount of information about the fermentation process, on the other hand, the fact that they are precisely made up of long equations with a large number of parameters is impractical when, for example, carrying out an industrial scaling process, designing a control system, or carrying out a techno-economic analysis. In these circumstances, as demonstrated in this work, it is also possible to predict the evolution of the biomass and the same quantity of metabolites with a simpler system of equations based on the previous study of the relationship between the species present in the culture broth.

3.2. Kinetic Study Based on the Initial Biomass Concentration

After applying the kinetic model to runs 1, 2, and 3, a good adjustment to the experimental data was achieved, as shown by the corresponding statistical parameters in Table 3. In these experiments, carried out at increasing initial biomass concentrations, it was observed that most of the estimated kinetic parameters did not suffer variations despite the modification of this operating condition (Table 2). However, three parameters of the model experienced considerable modifications and their variation is represented in Figure 2.

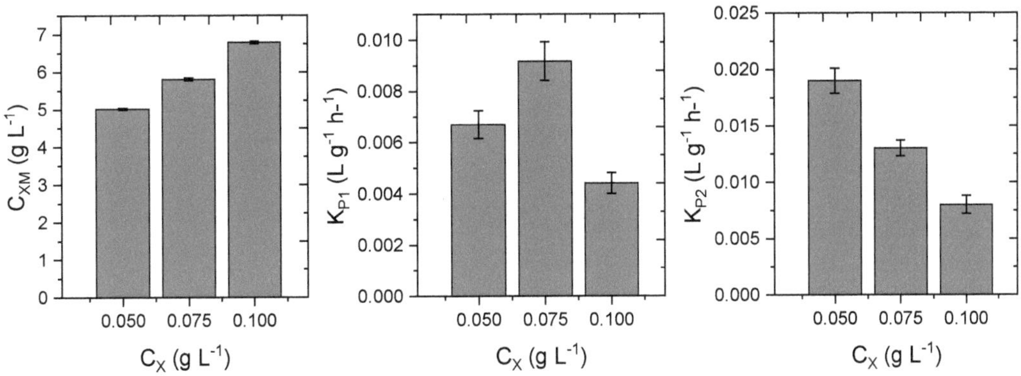

Figure 2. Kinetic parameters that are modified as a function of the initial biomass concentration.

On the one hand, the C_{xm} parameter shows that the increase in the initial biomass also leads to a higher maximum concentration of biomass in the culture broth once the stationary phase of its growth has been reached, with a practically linear trend, as expected. On the other hand, a drop in the kinetic constant of the by-product formation reaction (K_{P2}) is also observed, independent of that of succinic acid as the cell density increases in the broth, which agrees with the upward trend in selectivity previously discussed. Finally, the correlation between succinic acid productivity (Table 1) and the kinetic constant of the succinic acid formation equation (K_{P1}) is observed, reaching its maximum at an initial biomass concentration of 0.075 g L^{-1}.

3.3. Kinetic Study Based on the CO_2 Flow

As discussed in Section 3.1, despite modifying the CO_2 flow between values typically used in the literature [10,11,13,23,37–41,48,49], no variations were observed in the yield, productivity, or selectivity of the process (Table 1). The reason for this is that it seems that the excess of this gas does not propitiate the displacement of the metabolic route towards the formation of succinic acid. This deduction was confirmed with the application of the kinetic model to runs 1, 4, and 5, which enabled a simultaneous estimation of the three experiments without variation in the kinetic parameters (Table 2) and adequate goodness of fit (Table 3). Figure 3 shows the evolution of the experimental data of biomass, glucose, succinic acid, and by-products over time of runs 1, 4, and 5, together with the representation of the prediction of the evolution of their concentrations made on the whole.

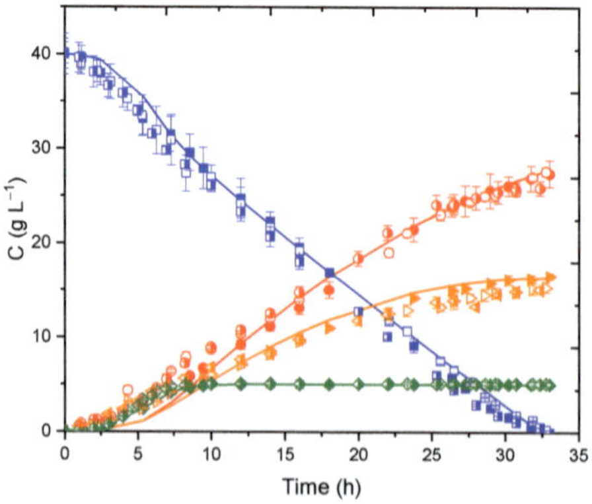

Figure 3. Kinetic model of the evolution of succinic acid (SA), by-products (BPs), biomass (X), and glucose concentrations (G) over time depending on CO_2 flow (0.1, 0.5, 1 L min^{-1}). Data points: SA (•), G (■), X (♦), BP (►), 0.1 L min^{-1} close symbol, 0.5 L min^{-1} open symbol, 1 L min^{-1} half open symbol, model predictions shown as lines.

3.4. Kinetic Study Based on the Stirring Speed

Agitation is an operating condition whose increase favors the transfer of gas in the culture broth and the homogeneity of the compounds, in turn improving the productivity and yield of succinic acid (Table 1); in this case, the maximum reached was 250 rpm. However, above this speed, the cells seem to suffer damage, negatively affecting the development of the process. This behavior is reflected exactly in the parameters of the kinetic model, showing growth in the kinetic constant of the reaction for the formation of succinic acid (K_{P1}) until reaching a maximum at 250 rpm, and then a considerable reduction at 300 rpm, as shown in Figure 4.

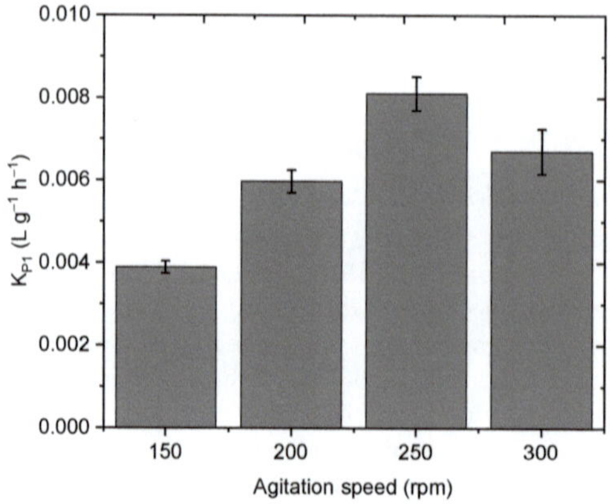

Figure 4. Kinetic parameters that are modified as a function of the stirring speed.

Despite the fact that, in this case, it was possible to observe an increase in productivity and possible cell damage, it should be noted that these conclusions differ from those of other authors such as Bevilaqua et al. [62], who did not observe significant variations in the performance of the process despite increasing the agitation up to 300 rpm, using the same microorganism as a biocatalyst although using hydrochloric hydrolysates of RH as the substrate instead of glucose. Gonzales et al [30] also studied the effect of stirring speed, performing experiments between 100 and 300 rpm. They only observed variations in the biomass concentration and determined that the optimum was to operate at low agitation speeds (100 rpm). These differences in the conclusions clearly open a door to the extension of the study of the influence of this operating condition, whose real impact seems yet to be determined.

3.5. Kinetic Study Based on the Yeast Extract Concentration

As discussed in Section 3.1, the concentration of yeast extract is a variable that has a great impact on the performance and productivity of the process, so that as the nitrogen source increases, higher productivity is achieved. However, the succinic yield peak was not reached at the maximum concentration studied (10 g L^{-1}), but at 7.5 g L^{-1}. Figure 5 shows that the kinetic estimations made in runs 1, 9, 10, and 11 led to the variation in six kinetic parameters. First, in the experiment at 2.5 g L^{-1} of YE (run 9), the values of the maximum biomass concentration (C_{xm}) and the macroscopic yield of biomass production ($Y_{S/X}$) were approximately half of those corresponding to all other fermentations carried out at higher concentrations of the nitrogen source (runs 1, 10, 11). Therefore, taking into account that there was not a great decrease in yield between run 9 and the rest of the experiments, it can be deduced that the amount of succinic acid and by-products generated per gram of biomass is much higher at 2.5 g L^{-1} of YE than in the experiments carried out at higher concentrations of YE, which results in an increase in the values of K_{P1} and K_{P2}.

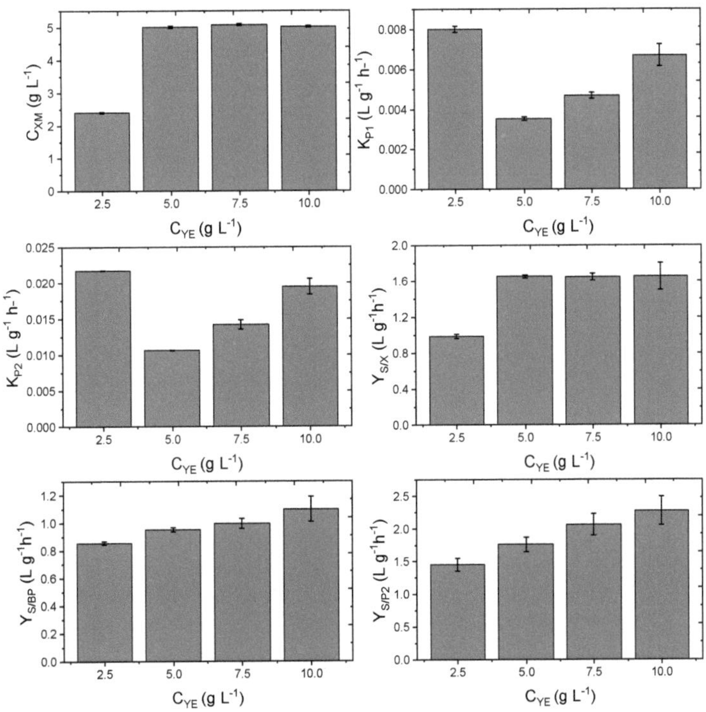

Figure 5. Kinetic parameters that were modified as a function of the yeast extract concentration.

It can also be observed that starting from 5 g L^{-1} of YE, the increase in the nitrogen source leads to the same increasing evolution of K_{P1} and K_{P2} as that of productivity. Despite this, an increase in the yield parameters related to the by-products ($Y_{S/BP}$, $Y_{S/P2}$) was observed as the initial concentration of YE increased, both in the simultaneous generation reaction of succinic acid and by-products (r_2), as well as in the isolated by-product generation reaction (r_3). This is consistent with the data presented in Table 1, that is, with the increase in selectivity with the decrease in the initial amount of YE.

3.6. Kinetic Estimation of the Stages of a Fed-Batch Type Operation

To check the robustness of the model for long fermentation times, a fed-batch fermentation (run 12) was performed. Kinetic estimates were made for each of the three stages, and simulations of the evolution of the biomass, substrate, product, and by-products over time were performed from maximum and minimum values of the kinetic parameters within the confidence interval. Figure 6 shows the experimental concentrations of biomass and metabolites over time, along with the prediction lines of the kinetic model and those simulated from the confidence interval. The yield values of succinic acid with respect to consumed glucose ($Y_{S/Gc}$) are also included, as well as the productivity and selectivity of succinic acid in the three stages. Figure 7 shows the kinetic parameters that underwent modifications throughout the three stages of the process.

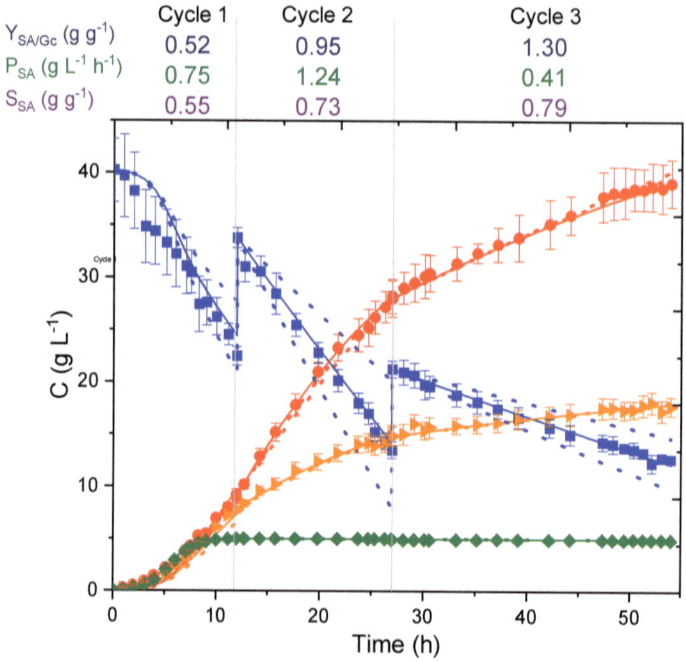

Figure 6. Evolution of the concentration of succinic acid (SA), by-products (BPs), biomass (X) and glucose (G) over time in run 11 (fed-batch). $Y_{SA/Gc}$: yield as a function of glucose consumed, P_{SA}: succinic acid productivity, S_{SA}: selectivity. Data points: SA (•), G (■), X (♦), BP (►), model estimations and simulations shown as lines.

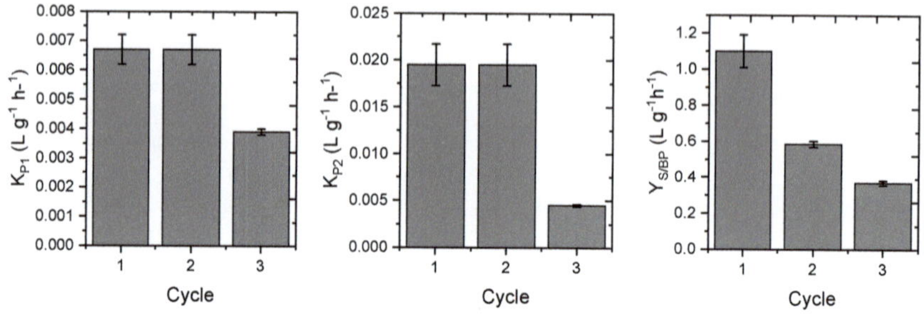

Figure 7. Kinetic parameters that were modified as a function of the fed-batch cycle.

It should be noted that, in the second stage, in line with the increase in selectivity with respect to the previous stage, there was a decrease in the yield parameter towards by-products (YS/BP) of reaction 2. Therefore, due to a greater use of glucose in succinic acid instead of by-products, it is logical that there is an increase in yield and productivity in this intermediate stage, without the need to alter any other kinetic parameter of the model.

Due to the increase in the selectivity of the third stage with respect to the previous stages, the macroscopic performance regarding by-products of reaction 2 ($Y_{S/BP}$) suffers a considerable reduction. However, the drop in succinic acid productivity is only reflected in the decrease in the kinetic constants of reactions 2 and 3 (K_{P1} and K_{P2}), parameters that decrease the reaction rate in a directly proportional manner. The simulation of the first cycle of run 12 could be carried out with the parameters estimated for run 1, since they share the same operating conditions.

4. Conclusions

In this work, an exhaustive study of critical variables in the bioproduction process of succinic acid by *A. succinogenes* was carried out to determine the effect of variables such as the initial concentration of biomass, the agitation speed, the concentration of yeast extract, and the CO_2 flow. A simple but robust kinetic model was developed which, unlike the models currently found in the literature, is capable of predicting the evolution of glucose, succinic acid, by-products, and biomass with few kinetic parameters. Its application to a reference run allowed verification of the goodness of fit, obtaining high values of F_{95} (41,242) and VE (99.5%), and values of RMSE and SSR close to zero. Subsequently, its validity was also demonstrated by estimating the evolution of metabolites in experiments in which the initial biomass concentration, the yeast extract concentration, agitation, and CO_2 flow were modified. Until now, these variables have not been used in the literature for the development of a succinic acid production kinetic model. Finally, the model was applied to a fed-batch type operation, performing simulations based on the confidence intervals of the estimated parameters for each of the stages. In this way, it was possible to develop and validate a model, having significant robustness and simplicity, which is very useful from the point of view of chemical engineering for the scaling of the process, the design of a control system, or the performance of techno-economic analyses.

Author Contributions: Conceptualization, V.E.S. and M.L.; methodology, I.A.E.; software, I.A.E. and M.L.; validation, V.E.S., M.L. and Á.B.; formal analysis, V.E.S. and M.L.; investigation, I.A.E.; resources, V.E.S., M.L. and Á.B.; data curation, I.A.E. and V.E.S.; writing—original draft preparation, I.A.E.; writing—review and editing, V.E.S., M.L. and Á.B.; supervision, V.E.S. and M.L.; project administration, M.L. and Á.B.; funding acquisition, V.E.S., M.L. and Á.B. All authors have read and agreed to the published version of the manuscript.

Funding: This research was supported by the Community of Madrid (Spain) with the research project S2018/EMT-4459 and by the Spanish Science and Innovation Ministry through the research project PID2020-114365RB-C21, funding that is gratefully acknowledged.

Institutional Review Board Statement: Not applicable.

Informed Consent Statement: Not applicable.

Data Availability Statement: All data will be provided to any interested part upon its requesting.

Conflicts of Interest: The authors declare no conflict of interest.

References

1. Kover, A.; Kraljić, D.; Marinaro, R.; Rene, E.R. Processes for the valorization of food and agricultural wastes to value-added products: Recent practices and perspectives. *Syst. Microbiol. Biomanufacturing* **2022**, *2*, 50–66. [CrossRef]
2. Usmani, Z.; Sharma, M.; Awasthi, A.K.; Lukk, T.; Tuohy, M.G.; Gong, L.; Nguyen-Tri, P.; Goddard, A.D.; Bill, R.M.; Nayak, S.; et al. Lignocellulosic biorefineries: The current state of challenges and strategies for efficient commercialization. *Renew. Sustain. Energy Rev.* **2021**, *148*, 111258. [CrossRef]

3. Saxena, R.K.; Saran, S.; Isar, J.; Kaushik, R. Production and Applications of Succinic Acid. In *Current Developments in Biotechnology and Bioengineering: Production, Isolation and Purification of Industrial Products*; Elsevier Inc.: Amsterdam, The Netherlands, 2016; pp. 601–630. ISBN 9780444636621.
4. Pateraki, C.; Patsalou, M.; Vlysidis, A.; Kopsahelis, N.; Webb, C.; Koutinas, A.A.; Koutinas, M. Actinobacillus succinogenes: Advances on succinic acid production and prospects for development of integrated biorefineries. *Biochem. Eng. J.* **2016**, *112*, 285–303. [CrossRef]
5. Shen, N.; Li, S.; Qin, Y.; Jiang, M.; Zhang, H. Optimization of succinic acid production from xylose mother liquor (XML) by Actinobacillus succinogenes using response surface methodology. *Biotechnol. Biotechnol. Equip.* **2022**, *36*, 439–447. [CrossRef]
6. Banger, G.; Kaya, K.; Omwene, P.; Shakoory, S.; Karagündüz, A.; Keskinler, B.; Nikerel, E. Delactosed Whey Permeate as Substrate for Succinic Acid Fermentation by Actinobacillus succinogenes. *Waste Biomass-Valorization* **2021**, *12*, 5481–5489. [CrossRef]
7. Escanciano, I.A.; Wojtusik, M.; Esteban, J.; Ladero, M.; Santos, V.E. Modeling the succinic acid bioprocess: A review. *Fermentation* **2022**, *8*, 368. [CrossRef]
8. Salma, A.; Djelal, H.; Abdallah, R.; Fourcade, F.; Amrane, A. Well Knowledge of the Physiology of Actinobacillus succinogenes to improve succinic acid production. *Appl. Microbiol.* **2021**, *1*, 304–328. [CrossRef]
9. Sillaparassamee, O.; Chinwetkitvanich, S.; Kanchanasuta, S.; Pisutpaisal, N.; Champreda, V. Metabolic flux analysis on succinic acid production from crude glycerol by Actinobacillus succinogenes. *Biomass-Convers. Biorefinery* **2021**, 1–12. [CrossRef]
10. Luthfi, A.A.I.; Jahim, J.M.; Harun, S.; Tan, J.P.; Manaf, S.F.A.; Shah, S.S.M. Kinetics of the bioproduction of succinic acid by Actinobacillus succinogenes from oil palm lignocellulosic hydrolysate in a bioreactor. *Bioresources* **2018**, *13*, 8279–8294. [CrossRef]
11. Salvachúa, D.; Mohagheghi, A.; Smith, H.; Bradfield, M.F.A.; Nicol, W.; Black, B.A.; Biddy, M.J.; Dowe, N.; Beckham, G.T. Succinic acid production on xylose-enriched biorefinery streams by Actinobacillus succinogenes in batch fermentation. *Biotechnol. Biofuels* **2016**, *9*, 28. [CrossRef]
12. Ferone, M.; Raganati, F.; Olivieri, G.; Salatino, P.; Marzocchella, A. Biosuccinic acid from lignocellulosic-based hexoses and pentoses by Actinobacillus succinogenes: Characterization of the conversion process. *Appl. Biochem. Biotechnol.* **2017**, *183*, 1465–1477. [CrossRef] [PubMed]
13. Pateraki, C.; Almqvist, H.; Ladakis, D.; Lidén, G.; Koutinas, A.A.; Vlysidis, A. Modelling succinic acid fermentation using a xylose based substrate. *Biochem. Eng. J.* **2016**, *114*, 26–41. [CrossRef]
14. Omwene, P.I.; Yağcıoğlu, M.; Öcal-Sarihan, Z.B.; Ertan, F.; Keris-Sen, Ü.D.; Karagunduz, A.; Keskinler, B. Batch fermentation of succinic acid from cheese whey by Actinobacillus succinogenes under variant medium composition. *3 Biotech* **2021**, *11*, 389. [CrossRef]
15. Bumyut, A.; Champreda, V.; Singhakant, C.; Kanchanasuta, S. Effects of immobilization of Actinobacillus succinogenes on efficiency of bio-succinic acid production from glycerol. Biomass-Convers. *Biorefinery* **2020**, *12*, 643–654. [CrossRef]
16. Kuglarz, M.; Rom, M. Influence of carbon dioxide and nitrogen source on sustainable production of succinic acid from miscanthus hydrolysates. *Int. J. Environ. Sci. Dev.* **2019**, *10*, 362–367. [CrossRef]
17. Xi, Y.-L.; Chen, K.-Q.; Li, J.; Fang, X.-J.; Zheng, X.-Y.; Sui, S.-S.; Jiang, M.; Wei, P. Optimization of culture conditions in CO_2 fixation for succinic acid production using Actinobacillus succinogenes. *J. Ind. Microbiol. Biotechnol.* **2011**, *38*, 1605–1612. [CrossRef]
18. Zou, W.; Zhu, L.-W.; Li, H.-M.; Tang, Y.-J. Significance of CO2 donor on the production of succinic acid by Actinobacillus succinogenes ATCC 55618. *Microb. Cell Factories* **2011**, *10*, 87. [CrossRef]
19. Herselman, J.; Bradfield, M.F.; Vijayan, U.; Nicol, W. The effect of carbon dioxide availability on succinic acid production with biofilms of Actinobacillus succinogenes. *Biochem. Eng. J.* **2017**, *117*, 218–225. [CrossRef]
20. Amulya, K.; Kopperi, H.; Mohan, S.V. Tunable production of succinic acid at elevated pressures of CO_2 in a high pressure gas fermentation reactor. *Bioresour. Technol.* **2020**, *309*, 123327. [CrossRef]
21. Vigato, F.; Angelidaki, I.; Woodley, J.M.; Alvarado-Morales, M. Dissolved CO_2 profile in bio-succinic acid production from sugars-rich industrial waste. *Biochem. Eng. J.* **2022**, *187*, 108602. [CrossRef]
22. Filippi, K.; Papapostolou, H.; Alexandri, M.; Vlysidis, A.; Myrtsi, E.D.; Ladakis, D.; Pateraki, C.; Haroutounian, S.A.; Koutinas, A. Integrated biorefinery development using winery waste streams for the production of bacterial cellulose, succinic acid and value-added fractions. *Bioresour. Technol.* **2021**, *343*, 125989. [CrossRef] [PubMed]
23. Kim, S.Y.; Park, S.O.; Yeon, J.Y.; Chun, G.-T. Development of a cell-recycled continuous fermentation process for enhanced production of succinic acid by high-yielding mutants of Actinobacillus succinogenes. *Biotechnol. Bioprocess Eng.* **2020**, *26*, 125–136. [CrossRef]
24. Jokodola, E.O.; Narisetty, V.; Castro, E.; Durgapal, S.; Coulon, F.; Sindhu, R.; Binod, P.; Banu, J.R.; Kumar, G.; Kumar, V. Process optimisation for production and recovery of succinic acid using xylose-rich hydrolysates by Actinobacillus succinogenes. *Bioresour. Technol.* **2021**, *344*, 126224. [CrossRef]
25. Ercole, A.; Raganati, F.; Salatino, P.; Marzocchella, A. Continuous succinic acid production by immobilized cells of Actinobacillus succinogenes in a fluidized bed reactor: Entrapment in alginate beads. *Biochem. Eng. J.* **2021**, *169*, 107968. [CrossRef]
26. Mokwatlo, S.C.; Nicol, W. Structure and cell viability analysis of Actinobacillus succinogenes biofilms as biocatalysts for succinic acid production. *Biochem. Eng. J.* **2017**, *128*, 134–140. [CrossRef]
27. Cao, W.; Wang, Y.; Luo, J.; Yin, J.; Xing, J.; Wan, Y. Succinic acid biosynthesis from cane molasses under low pH by Actinobacillus succinogenes immobilized in luffa sponge matrices. *Bioresour. Technol.* **2018**, *268*, 45–51. [CrossRef]

28. Escanciano, I.A.; Ladero, M.; Santos, V.E. On the succinic acid production from xylose by growing and resting cells of Actinobacillus succinogenes: A comparison. *Biomass-Convers. Biorefinery* **2022**, 1–14. [CrossRef]
29. Shen, N.; Zhang, H.; Qin, Y.; Wang, Q.; Zhu, J.; Li, Y.; Jiang, M.-G.; Huang, R. Efficient production of succinic acid from duckweed (Landoltia punctata) hydrolysate by Actinobacillus succinogenes GXAS137. *Bioresour. Technol.* **2018**, *250*, 35–42. [CrossRef]
30. Gonzales, T.A.; de Carvalho Silvello, M.A.; Duarte, E.R.; Santos, L.O.; Alegre, R.M.; Goldbeck, R. Optimization of anaerobic fermentation of Actinobacillus succinogenes for increase the succinic acid production. *Biocatal. Agric. Biotechnol.* **2020**, *27*, 101718. [CrossRef]
31. Tan, J.P.; Jahim, J.M.; Wu, T.Y.; Harun, S.; Mumtaz, T. Use of corn steep liquor as an economical nitrogen source for biosuccinic acid production by Actinobacillus succinogenes. *IOP Conf. Ser. Earth Environ. Sci.* **2016**, *36*, 012058. [CrossRef]
32. Xi, Y.-L.; Chen, K.-Q.; Dai, W.-Y.; Ma, J.-F.; Zhang, M.; Jiang, M.; Wei, P.; Ouyang, P.-K. Succinic acid production by Actinobacillus succinogenes NJ113 using corn steep liquor powder as nitrogen source. *Bioresour. Technol.* **2013**, *136*, 775–779. [CrossRef] [PubMed]
33. Jiang, M.; Chen, K.; Liu, Z.; Wei, P.; Ying, H.; Chang, H. Succinic acid production by Actinobacillus succinogenes using spent brewer's yeast hydrolysate as a nitrogen source. *Appl. Biochem. Biotechnol.* **2009**, *160*, 244–254. [CrossRef]
34. Delgado-Noboa, J.; Bernal, T.; Soler, J.; Peña, J. Kinetic modeling of batch bioethanol production from CCN-51 Cocoa Mucilage. *J. Taiwan Inst. Chem. Eng.* **2021**, *128*, 169–175. [CrossRef]
35. Louasté, B.; Eloutassi, N. Succinic acid production from whey and lactose by Actinobacillus succinogenes 130Z in batch fermentation. *Biotechnol. Rep.* **2020**, *27*, e00481. [CrossRef] [PubMed]
36. Jiang, M.; Dai, W.; Xi, Y.; Wu, M.; Kong, X.; Ma, J.; Zhang, M.; Chen, K.; Wei, P. Succinic acid production from sucrose by Actinobacillus succinogenes NJ113. *Bioresour. Technol.* **2014**, *153*, 327–332. [CrossRef]
37. Lin, S.K.C.; Du, C.; Koutinas, A.; Wang, R.; Webb, C. Substrate and product inhibition kinetics in succinic acid production by Actinobacillus succinogenes. *Biochem. Eng. J.* **2008**, *41*, 128–135. [CrossRef]
38. Guarnieri, M.T.; Chou, Y.-C.; Salvachúa, D.; Mohagheghi, A.; John, P.C.S.; Peterson, D.J.; Bomble, Y.J.; Beckham, G.T. Metabolic engineering of Actinobacillus succinogenes provides insights into succinic acid biosynthesis. *Appl. Environ. Microbiol.* **2017**, *83*, e00996-17. [CrossRef]
39. Lubsungneon, J.; Srisuno, S.; Rodtong, S.; Boontawan, A. Nanofiltration coupled with vapor permeation-assisted esterification as an effective purification step for fermentation-derived succinic acid. *J. Membr. Sci.* **2014**, *459*, 132–142. [CrossRef]
40. Cao, W.; Wang, Y.; Luo, J.; Yin, J.; Xing, J.; Wan, Y. Effectively converting carbon dioxide into succinic acid under mild pressure with Actinobacillus succinogenes by an integrated fermentation and membrane separation process. *Bioresour. Technol.* **2018**, *266*, 26–33. [CrossRef]
41. Rigaki, A.; Webb, C.; Theodoropoulos, C. Double substrate limitation model for the bio-based production of succinic acid from glycerol. *Biochem. Eng. J.* **2019**, *153*, 107391. [CrossRef]
42. Zhang, W.; Tao, Y.; Wu, M.; Xin, F.; Dong, W.; Zhou, J.; Gu, J.; Ma, J.; Jiang, M. Adaptive evolution improves acid tolerance and succinic acid production in Actinobacillus succinogenes. *Process. Biochem.* **2020**, *98*, 76–82. [CrossRef]
43. Wan, C.; Li, Y.; Shahbazi, A.; Xiu, S. Succinic Acid Production from Cheese Whey using Actinobacillus succinogenes 130 Z. *Appl. Biochem. Biotechnol.* **2007**, *145*, 111–119. [CrossRef] [PubMed]
44. Anwar, N.A.K.K.; Hassan, N.; Yusof, N.M.; Idris, A. High-titer bio-succinic acid production from sequential alkalic and metal salt pretreated empty fruit bunch via simultaneous saccharification and fermentation. *Ind. Crop. Prod.* **2021**, *166*, 113478. [CrossRef]
45. Argun, H.; Kargi, F.; Kapdan, I.K.; Oztekin, R. Batch dark fermentation of powdered wheat starch to hydrogen gas: Effects of the initial substrate and biomass concentrations. *Int. J. Hydrogen Energy* **2008**, *33*, 6109–6115. [CrossRef]
46. Eker, S.; Sarp, M. Hydrogen gas production from waste paper by dark fermentation: Effects of initial substrate and biomass concentrations. *Int. J. Hydrogen Energy* **2017**, *42*, 2562–2568. [CrossRef]
47. Ding, M.-Z.; Tian, H.-C.; Cheng, J.-S.; Yuan, Y.-J. Inoculum size-dependent interactive regulation of metabolism and stress response of Saccharomyces cerevisiae revealed by comparative metabolomics. *J. Biotechnol.* **2009**, *144*, 279–286. [CrossRef]
48. Almqvist, H.; Pateraki, C.; Alexandri, M.; Koutinas, A.; Lidén, G. Succinic acid production by Actinobacillus succinogenes from batch fermentation of mixed sugars. *J. Ind. Microbiol. Biotechnol.* **2016**, *43*, 1117–1130. [CrossRef]
49. Corona-González, R.I.; Miramontes-Murillo, R.; Arriola-Guevara, E.; Guatemala-Morales, G.; Toriz, G.; Pelayo-Ortiz, C. Immobilization of Actinobacillus succinogenes by adhesion or entrapment for the production of succinic acid. *Bioresour. Technol.* **2014**, *164*, 113–118. [CrossRef] [PubMed]
50. Garcia-Ochoa, F.; Gomez, E. Bioreactor scale-up and oxygen transfer rate in microbial processes: An overview. *Biotechnol. Adv.* **2009**, *27*, 153–176. [CrossRef]
51. Garcia-Ochoa, F.; Gomez, E.; Santos, V.E.; Merchuk, J.C. Oxygen uptake rate in microbial processes: An overview. *Biochem. Eng. J.* **2010**, *49*, 289–307. [CrossRef]
52. Garcia-Ochoa, F.; Gomez, E.; Santos, V.E. Fluid dynamic conditions and oxygen availability effects on microbial cultures in STBR: An overview. *Biochem. Eng. J.* **2020**, *164*, 107803. [CrossRef]
53. Ventrone, M.; Schiraldi, C.; Squillaci, G.; Morana, A.; Cimini, D. Chestnut shells as waste material for succinic acid production from Actinobacillus succinogenes 130Z. *Fermentation* **2020**, *6*, 105. [CrossRef]
54. Patsalou, M.; Chrysargyris, A.; Tzortzakis, N.; Koutinas, M. A biorefinery for conversion of citrus peel waste into essential oils, pectin, fertilizer and succinic acid via different fermentation strategies. *Waste Manag.* **2020**, *113*, 469–477. [CrossRef]

55. Kanchanasuta, S.; Champreda, V.; Pisutpaisal, N.; Singhakant, C. Optimization of bio-succinic fermentation process from crude glycerol by Actinobacillus succinogenes. *Environ. Eng. Res.* **2021**, *26*, 200121. [CrossRef]
56. Bradfield, M.F.A.; Mohagheghi, A.; Salvachúa, D.; Smith, H.; Black, B.A.; Dowe, N.; Beckham, G.T.; Nicol, W. Continuous Succinic Acid Production by Actinobacillus succinogenes on Xylose-Enriched Hydrolysate. *Biotechnol. Biofuels* **2015**, *8*, 181. [CrossRef] [PubMed]
57. Bradfield, M.F.A.; Nicol, W. Continuous succinic acid production from xylose by Actinobacillus succinogenes. *Bioprocess Biosyst. Eng.* **2016**, *39*, 233–244. [CrossRef] [PubMed]
58. Vlysidis, A.; Binns, M.; Webb, C.; Theodoropoulos, C. Glycerol utilisation for the production of chemicals: Conversion to succinic acid, a combined experimental and computational study. *Biochem. Eng. J.* **2011**, *58*, 1–11. [CrossRef]
59. Song, H.; Jang, S.H.; Park, J.M.; Lee, S.Y. Modeling of batch fermentation kinetics for succinic acid production by Mannheimia succiniciproducens. *Biochem. Eng. J.* **2008**, *40*, 107–115. [CrossRef]
60. Li, Q.; Wang, D.; Wu, Y.; Yang, M.; Li, W.; Xing, J.; Su, Z. Kinetic evaluation of products inhibition to succinic acid producers Escherichia coli NZN111, AFP111, BL21, and Actinobacillus succinogenes 130ZT. *J. Microbiol.* **2010**, *48*, 290–296. [CrossRef]
61. Corona-González, R.I.; Bories, A.; González-Álvarez, V.; Pelayo-Ortiz, C. Kinetic study of succinic acid production by Actinobacillus succinogenes ZT-130. *Process. Biochem.* **2008**, *43*, 1047–1053. [CrossRef]
62. Bevilaqua, D.B.; Montipó, S.; Pedroso, G.B.; Martins, A.F. Sustainable succinic acid production from rice husks. *Sustain. Chem. Pharm.* **2015**, *1*, 9–13. [CrossRef]

fermentation

MDPI

Article

Continuous Bioproduction of Alginate Bacterial under Nitrogen Fixation and Nonfixation Conditions

Pablo Contreras-Abara [1], Tania Castillo [2], Belén Ponce [1], Viviana Urtuvia [1], Carlos Peña [2] and Alvaro Díaz-Barrera [1,*]

[1] Escuela de Ingeniería Bioquímica, Pontificia Universidad Católica de Valparaíso, Valparaíso 2340000, Chile

[2] Departamento de Ingeniería Celular y Biocatálisis, Instituto de Biotecnología, Universidad Nacional Autónoma de México, Ave. Universidad 2001, Col. Chamilpa Cuernavaca, Cuernavaca 62210, Morelos, Mexico

* Correspondence: alvaro.diaz@pucv.cl

Abstract: Alginate is a biomaterial produced by *Azotobacter vinelandii*, a diazotroph that, under nitrogen-fixing conditions, can fix nitrogen under high oxygen levels. In *A. vinelandii*, alginate is synthesized from fructose-6P via synthesis of precursor, polymerization, and modification/exportation. Due to its viscosifying, gelling, and thickening characteristics, alginate is widely used in food, pharmaceutical, and cosmetical industries. This study aimed to develop a continuous bioprocess and a comparative analysis of alginate production under diazotrophic and nondiazotrophic conditions. Continuous cultures were developed at three dilution rates (0.06, 0.08 and 0.10 h^{-1}). In steady state, the respiratory activity, alginate production, alginate molecular weight and the genes encoding alginate polymerase were determined. Under the conditions studied, the specific oxygen uptake rate and respiratory quotient were similar. The diazotrophic conditions improved the conversion of sucrose to alginate and the specific productivity rate, which was 0.24 ± 0.03 g g^{-1} h^{-1}. A higher alginate molecular weight (725 ± 20 kDa) was also achieved under diazotrophic conditions, which can be explained by an increase in the gene expression of genes *alg8* and *alg44* (encoding polymerase). The results of this work show the feasibility of enhancing alginate production (yields and specific productivity rates) and quality (molecular weight) under nitrogen-fixing conditions, opening the possibility of developing a continuous bioprocess to produce alginate with specific characteristics under conditions of diazotrophy.

Keywords: alginate; mean molecular weight; diazotrophy; chemostat

Citation: Contreras-Abara, P.; Castillo, T.; Ponce, B.; Urtuvia, V.; Peña, C.; Díaz-Barrera, A. Continuous Bioproduction of Alginate Bacterial under Nitrogen Fixation and Nonfixation Conditions. *Fermentation* **2023**, *9*, 426. https://doi.org/10.3390/fermentation9050426

Academic Editors: Miguel Ladero and Victoria E. Santos

Received: 11 April 2023
Revised: 26 April 2023
Accepted: 26 April 2023
Published: 28 April 2023

1. Introduction

Alginates are polysaccharides formed by monomers of β-D-mannuronic acid (M) and its epimer α-L-guluronic acid (G). Alginate applicability for practical biomaterials has been proven in applications such as hydrogels for three-dimensional extrusion bioprinting [1], due to their ability to form gels. *Azotobacter vinelandii* is a bacterium able to produce polymers such as poly-3-hydroxybutyrate (P3HB) and alginate during cellular growth [2]. Alginate monomers can be arranged in chains of different molecular weights and distributed in MMMM, GGGG, and MGMG blocks [3].

Alginates are also produced by brown algae, but their production presents several disadvantages, which may limit their industrial use. Algal alginates are complex polymer mixtures with a wide range of molecular weights and compositions of M and G residues. Thus, alginates with a well-defined composition cannot be obtained. Current information about the costs of alginate production is not readily available; it is estimated that commercial alginates used in the food and cosmetic industries can be acquired at prices below 5 USD per kilogram. From a commercial standpoint, the most important characteristics of alginates are their viscosity in solutions and their capacity as gelling agents. The aforementioned alginate properties depend largely on the relative content of the two monomers (G and M), the degree of acetylation, as well as the molecular weight of the polymer [4]. The manipulation of the molecular weight and its distribution can improve the physical properties of resultant gels [5]. One strategy to produce alginates with varied and reproducible physicochemical characteristics is through the manipulation of culture conditions during the bioprocess.

The biosynthesis of alginate in *A. vinelandii* involves three enzymatic stages: synthesis of precursor, polymerization, and modification/secretion [6]. The alginate polymerase complex is composed of glycosyl-transferase/polymerase (Alg8) and copolymerase (Alg44), which are encoded by the genes *alg8* and *alg44* [6]. Díaz-Barrera et al. [7] observed an increase in the relative expression of *alg8* in chemostat cultivations of *A. vinelandii* at $D = 0.1\ h^{-1}$ and a correlation between the mean molecular weight and the gene expression of *alg8*, while in their later work [8] conducted under chemostat cultivations at $D = 0.07\ h^{-1}$, the alginate molecular weight synthesized in those cultivations did not correlate with a high relative expression of *alg8* or *alg44*, which can be indicative of a balanced transcription of alginate polymerization genes.

A. vinelandii is a bacterium able to fix atmospheric nitrogen, a reaction catalyzed by the enzyme nitrogenase, which converts a molecule of N_2 into two molecules of ammonia [9]. If the cellular growth is developed in the absence of a fixed nitrogen source (e.g., without ammonium) the culture is performed under diazotrophic conditions, whereas if the culture medium is supplemented with a fixed nitrogen source, the growth condition is nondiazotrophic. Although the enzyme nitrogenase is highly sensitive to oxygen, *A. vinelandii* has a greater capacity to fix nitrogen because of its ability to fix nitrogen even at high oxygen levels [9]. This capability has been related to a higher respiration rate when this bacterium is grown under diazotrophic conditions and high oxygen levels [10]. The effect of oxygen availability in diazotrophic cultivations on the biosynthesis of alginate has been studied by several authors [11–17]. A modality adequate to evaluate alginate production is the use of continuous cultures, in particular, chemostat cultures, in which the cells can be cultivated at a constant specific growth rate (which can be established by the different dilution rates) which allows for distinguishing between the effect of a specific nutrient such as ammonium or oxygen, as well as for monitoring the specific growth rates [8,11].

Based on genome-scale modeling, Tec-Campos et al. [18] described that under diazotrophic conditions, the carbon and nitrogen fluxes through *A. vinelandii* could decrease as a response to nitrogen limitation, whereas when ammonium is supplemented, the cell growth rate and alginate production rate increase. Although there are studies in the literature about alginate synthesis by *A. vinelandii* under diazotrophic and nondiazotrophic conditions [7,14,19], those works are not comparable due to differences in their operating conditions. For this reason, the present study aimed to develop a continuous bioprocess to produce alginate with different molecular weights using *A. vinelandii* in cultures under diazotrophic and nondiazotrophic conditions. In this sense, a bioprocess to produce alginate under similar conditions of cultivation, such as culture medium composition, agitation rate, and particularly specific growth rate (in chemostat-mode) is implemented, because only the condition diazotrophic or nondiazotrophic was varied in this study.

2. Materials and Methods

2.1. Microorganism, Culture Medium, and Inoculum Preparation

A. vinelandii ATCC 9046 (wild-type strain) was used for this study. Chemostats were developed under diazotrophic (nitrogen-fixing) and nondiazotrophic (using ammonium in the culture medium) conditions. The diazotrophic conditions were evaluated using a culture medium that contained gaseous nitrogen coming from the air incorporated as a nitrogen source alone. The culture medium for diazotrophic conditions contained (in $g\ L^{-1}$): sucrose, 20; $K_2HPO_4{\cdot}3H_2O$, 0.2; $CaSO_4{\cdot}2H_2O$, 0.056; NaCl, 0.2; $MgSO_4{\cdot}7H_2O$, 0.2; $Na_2MoO_4{\cdot}2H_2O$, 0.0029; $FeSO_4{\cdot}7H_2O$, 0.027. Solutions of medium were dissolved and autoclaved at 121 °C for 20 min. To avoid precipitation, NaCl, $MgSO_4{\cdot}7H_2O$, $Na_2MoO_4{\cdot}2H_2O$, $FeSO_4{\cdot}7H_2O$, and $CaSO_4{\cdot}2H_2O$ were separated for sterilization, while the rest of the nutrients were dissolved and autoclaved in the bioreactor. For nondiazotrophic conditions, the culture medium described supplemented with 0.8 g L^{-1} of $(NH_4)_2SO_4$ was used. *A. vinelandii* cells were incubated at 200 rpm and 30 °C in an orbital incubator shaker (Daihan LabTech CO, Namyangju, Kyungki-Do, Republic of Korea) in a 500 mL Erlenmeyer flask with 100 mL of culture medium. After 20 h of cultivation, the bioreactor was inoculated with 10% v/v inoculum.

2.2. Culture Conditions

Chemostat cultures were conducted in a 3 L bioreactor (Applikon, Schiedam, The Netherlands) with a working volume of 1.5 L. The pH was controlled at 7.0 ± 0.2 using a 2 N NaOH solution via a peristaltic pump coupled to an EZ-2-Control unit. The bioreactor was equipped with two Rushton turbines and aerated at 1.5 L min^{-1}. Dissolved oxygen tension (DOT) was measured by polarographic and was not controlled. After 24 h of cellular growth, the fresh culture medium was fed to the bioreactor, and the broth culture was removed from the bioreactor via a continuously operated peristaltic pump (Cole-Parmer, Vernon Hills, IL, USA).

The chemostat was operated at D values of 0.06, 0.08, and 0.10 h^{-1}. After at least 3 residence times, 20 mL samples were taken from the reactor at different times. Steady-state conditions were reached once the optical density reached 540 nm and the sucrose concentration had a variation coefficient below 10%. The results shown are the mean value of two independent chemostat runs, and error bars correspond to the range among the replicates.

2.3. Analytical Methods

Biomass and alginate concentrations were estimated gravimetrically. A 20 mL sample of culture broth was mixed with 2 mL EDTA (Ethylenediaminetetraacetic acid disodium salt) (0.1 M) and 2 mL NaCl (0.1 M) and then centrifuged at 10,000 rpm (Thermo Scientific SL-16R, Waltham, MA, USA) for 10 min. The pellet of biomass was washed three times using distilled water and then dried at 100 °C to a constant weight. For alginate quantification, 10 mL of supernatant was mixed with cold propan-2-ol in a 3:1 vol ratio. The resultant precipitate was filtered through 0.45 µm Millipore filter paper and dried at 70 °C to a constant weight. The sucrose concentration was determined by the dinitrosalicylic acid (DNS) reagent [20] after acid hydrolysis with HCl. The alginate mean molecular weight (MMW) was determined by gel permeation chromatography (GPC) in an HPLC with a differential refractometer detector (Jasco, Mary's Court Easton, MD, USA), according to Díaz-Barrera et al. [19]. The P3HB content was determined by extracting the P3HB from the cell and hydrolyzing it to crotonic acid, which was measured by an HPLC-UV (Jasco, Mary's Court Easton, MD, USA) system [21]. The ammonium concentration in the supernatant was determined by the phenol-hypochlorite method [22], and the phosphate concentration was determined by an automatic analyzer Random Access Y15 (BioSystem, Barcelona, Spain).

2.4. Gene Expression Analysis

The cells were harvested by centrifugation at 10,000 rpm (4 °C) for 4 min (Thermo Scientific SL-16R), and the pellet was washed three to five times. The mRNA was stabilized and protected by adding RNAlater solution (Thermo Fisher Scientific, Waltham, MA, USA) to the biomass pellet and stored at −80 °C for posterior RNA isolation. RNA was isolated using a High Pure RNA Isolation Kit (Roche Life Sciences, Penzberg, Germany) and treated with RNase-free DNase (Roche) according to the fabrication protocol. The RNA was quantified using a BioSpec-nano system (Shimadzu, Kyoto, Japan). cDNA was synthesized using a RevertAid H Minus First Strand cDNA Synthesis Kit (Thermo Fisher Scientific, Waltham, MA, USA) according to the fabrication protocol. Reverse transcription-real-time PCR (RT-qPCR) was carried out with specific primers shown in Table 1.

Table 1. Primers designed for gene expression by qPCR.

Gene	Primers	Gene	Primers
***alg8*-F**	5′-TGTTGAACCAGCTCTGGAAG-3′	***alg8*-R**	5′-CCTACCCGCTGATCCTCTAC-3′
***alg44*-F**	5′-CGACAACTTCACCGAAGGG-3′	***alg44*-R**	5′-CGACAACTTCACCGAAGGG-3′
***gyrA*-F**	5′-ACCTGATCACCGAGGAAGAG-3′	***gyrA*-R**	5′-AGGTGCTCGACGTAATCCTC-3′

For RT-qPCR, 100 ng of total RNA was reverse transcribed. Real-time PCR was performed in the AriaMx Real-Time PCR system (Agilent Technologies, Santa Clara, CA, USA) using Brilliant II SYBR@ Green QPCR Master Mix (Agilent Technologies, Santa Clara, CA, USA). The samples were initially denatured at 95 °C for 5 min. A 40-cycle amplification and quantification protocol was used for RT-qPCR (95 °C for 15 s, 59 °C for 15 s and 72 °C for 15 s). Melting curve analyses confirmed the amplification of a single product for each primer pair. The results were analyzed using the $2^{-\Delta\Delta CT}$ method [23,24].

Relative gene expression values were normalized using *gyrA* as a housekeeping gene [16] and presented as fold changes in transcription levels of culture sample growth in diazotrophic conditions with respect to the transcript levels of cultured sample growth in nondiazotrophic conditions.

2.5. Specific Oxygen Uptake Rate and Respiratory Quotient Determination at the Steady State

The respiratory quotient (RQ) was determined using oxygen transfer rate (OTR) and carbon dioxide transfer rate (CTR) data from the ratio CTR/OTR. Gas analysis was performed by measuring oxygen and carbon dioxide in the exit and inlet gas with a BlueVary gas analyzer (BlueSense, Herten, Germany). The OTR (mmol L^{-1} h^{-1}) and CTR (mmol L^{-1} h^{-1}) were calculated using a steady-state gas phase balance, according to Equations (1) and (2):

$$OTR = \frac{C\,F_G^{in}}{V_R V_M}\left(X_{O_2}^{in} - X_{O_2}^{out}\left(\frac{1 - X_{O_2}^{in} - X_{CO_2}^{in}}{1 - X_{O_2}^{out} - X_{CO_2}^{out}}\right)\right) \quad (1)$$

$$CTR = \frac{C\,F_G^{in}}{V_R V_M}\left(X_{CO_2}^{in}\left(\frac{1 - X_{O_2}^{in} - X_{CO_2}^{in}}{1 - X_{O_2}^{out} - X_{CO_2}^{out}}\right) - X_{CO_2}^{out}\right) \quad (2)$$

where C is unit conversion factor (1000), F_G^{in} is the volumetric inlet air flow under standard conditions (L h^{-1}); V_R is the working volume (L); V_M is the mol volume of the ideal gas under standard conditions (L $mmol^{-1}$); $X_{O_2}^{in}$ is the molar fraction of oxygen in the inlet air (mol mol^{-1}); $X_{O_2}^{out}$ is the molar fraction of oxygen in the outlet fermentation gas of the bioreactor (mol mol^{-1}); $X_{CO_2}^{in}$ is the molar fraction of carbon dioxide in the inlet air (mol mol^{-1}); and $X_{CO_2}^{out}$ is the molar fraction of carbon dioxide in the outlet fermentation gas of the bioreactor (mol mol^{-1}).

During continuous operation in the steady state, $\frac{dC_L}{dt} = 0$ (C_L, dissolved oxygen concentration in the liquid) and the OTR is considered to be equal to the oxygen uptake rate (OUR).

2.6. Carbon Balance

Carbon distribution to the different steady states evaluated was determined from reactor mass balances. For the analysis, the calculation was performed considering that the carbon source is converted mainly into biomass, P3HB, alginate and CO_2 [25].

2.7. Fermentation Parameters

The yields of biomass and alginate based on sucrose ($Y_{x/s}$ and $Y_{p/s}$, respectively), the yield of alginate based on biomass ($Y_{p/x}$), specific oxygen uptake rate (qO_2), and specific production rate (q_p) were calculated at a steady state for each condition, considering the dilution rate (D; h^{-1}), alginate concentration in the steady state (P; g L^{-1}), OTR values (mmol L^{-1} h^{-1}), biomass concentration (X; g L^{-1}) in the steady state, sucrose concentration in the steady state (S; g L^{-1}), and sucrose concentration in the feed medium (Sr; g L^{-1}), as indicated by the following equations:

$$Y_{x/s} = \frac{X}{(S_r - S)} \tag{3}$$

$$Y_{p/s} = \frac{P}{(S_r - S)} \tag{4}$$

$$Y_{p/x} = \frac{P}{X} \tag{5}$$

$$q_p = \frac{DP}{X} \tag{6}$$

$$qO_2 = \frac{OTR}{X} \tag{7}$$

3. Results and Discussion

3.1. Biomass, P3HB, and Alginate Concentration in the Steady State under Diazotrophic and Nondiazotrophic Conditions

Figure 1 shows the biomass, alginate, and P3HB concentrations at steady state in the chemostat cultures performed under diazotrophic and nondiazotrophic conditions. In all the conditions evaluated, the biomass concentration was higher under nondiazotrophic conditions (Figure 1a), which could be related to the availability of ammonium to stimulate the production of biomass. In this context, under nondiazotrophic conditions, the energetic requirements to produce cells are lower than those in cultivations in which nitrogen fixation occurs [26,27]. Under the conditions evaluated, low P3HB production was obtained (Figure 1b). Thus, the highest P3HB concentration (0.35 ± 0.05 g L^{-1}) and polymer content in the cells (34.3 ± 3.0% w w^{-1}) were obtained under diazotrophic conditions at a D of 0.06 h^{-1}. In the steady states conducted at a D of 0.08 and 0.10 h^{-1}, a low P3HB concentration (less than 0.09 g L^{-1}) and P3HB content (less than 8.5 ± 2.0% w w^{-1}) were obtained. Similarly, Díaz-Barrera et al. [8] observed that in diazotrophic chemostat cultures, the highest P3HB content was achieved at 0.07 h^{-1}, and it dropped at 0.09 h^{-1}. These results agree with other studies in which it has been reported that P3HB biosynthesis is mainly enhanced at low specific growth rates [14].

It is known that acetyl-CoA is the acetyl donor for alginate acetylation and is a precursor for P3HB biosynthesis [6]. In this regard, it is possible that under diazotrophic conditions and a low D (in which more P3HB was produced, such as in Figure 1b), a higher proportion of acetyl-CoA could be canalized to synthesize P3HB instead of being used for acetylation of alginate. Other analyses of alginate composition could be carried out to evaluate this possibility.

It is clear from the results that the cultivation conditions were not propitious to produce P3HB. Considering the P3HB content in the cells, the biomass concentration values without including the P3HB content were also highest under nondiazotrophic conditions (Figure 1c).

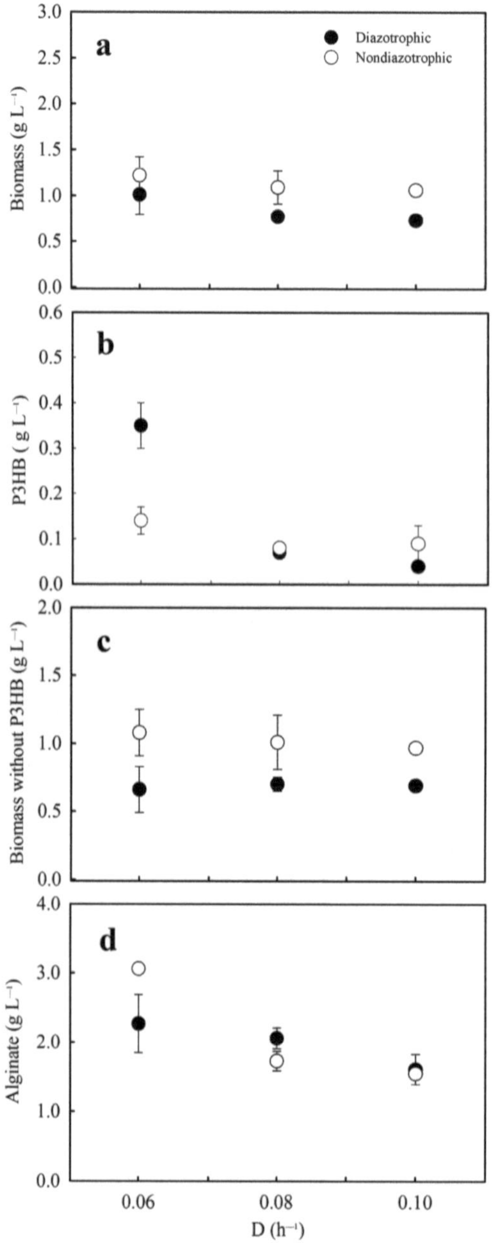

Figure 1. Steady-state biomass concentration (**a**), P3HB concentration (**b**), biomass concentration without P3HB (**c**) and alginate concentration (**d**) under diazotrophic (black circles) and nondiazotrophic (white circles) conditions in *A. vinelandii* continuous cultures.

Regarding the alginate concentration in the steady state, independent of the cultivation conditions, the alginate concentration decreased with increasing D (Figure 1d). The highest alginate concentration (3.06 ± 0.02 g L^{-1}) was obtained at the lowest D assayed under nondiazotrophic conditions (Figure 1d). In this condition, the steady-state carried out to the higher D (0.10 h^{-1}) showed the lowest alginate production, which reached

1.55 ± 0.02 g L^{-1}. Previously, under noncontrolled oxygen conditions (as in this study), a decrease in the alginate concentration by increasing D was observed [8,28]. However, under DOT control (for example, DOT of 1%), by increasing D (from 0.06 to 0.10 h^{-1}), the alginate concentration also increased [19]. From a productive point of view, this evidence is important because to develop a bioprocess without DOT control (and hence develop it in a way that is less expensive), it is necessary to perform cultures with a lower D at the highest residence time, but also at a higher alginate production rate. Chemostat cultures provide nutrient-limiting conditions specific for a single nutrient in a medium with stable levels of the nonlimiting components [29]. In our study, the cultures were not limited by carbon because between 8.1 and 16 g L^{-1} sucrose remained in the bioreactor at a steady state (Table 2). In addition, the ammonium concentration in the steady state was not detected in all the conditions evaluated (data not shown), and phosphate levels under nondiazotrophic conditions were tenfold lower than those reached during diazotrophic conditions (Table 2). In light of this evidence, it is possible that under nondiazotrophic conditions, phosphate could be a nutrient limiting cellular growth.

Table 2. Sucrose and phosphate concentrations in steady state during continuous cultures of *A. vinelandii* conducted under diazotrophic and nondiazotrophic conditions.

D (h^{-1})	Diazotrophy		Nondiazotrophy	
	Residual Sucrose ($g\ L^{-1}$)	**Residual Phosphate ($mg\ L^{-1}$)**	**Residual Sucrose ($g\ L^{-1}$)**	**Residual Phosphate ($mg\ L^{-1}$)**
0.06	14.2 ± 0.6	38.7 ± 11.9	8.9 ± 0.2	3.4 ± 0.6
0.08	14.2 ± 0.5	24.2 ± 0.4	8.1 ± 0.5	2.7 ± 0.9
0.10	16.1 ± 0.6	29.6 ± 5.2	9.0 ± 3.1	0.7 ± 0.2

3.2. qO_2 and RQ at Different Dilution Rates under Diazotrophic and Nondiazotrophic Conditions

In all the steady states evaluated, the DOT was nearly zero (data not shown), which has been previously reported [7]. This condition (DOT~0) is indicative that the steady states were conducted under oxygen limitation. In each culture developed, the qO_2 and RQ were evaluated at the steady state in the chemostat cultures (Figure 2). Comparing diazotrophic and nondiazotrophic conditions reveals that the qO_2 did not show significant differences at the three D evaluated (Figure 2a), which was not expected. Previous evidence has indicated that a higher qO_2 is expected during diazotrophic cultivation as a response to the protection of the nitrogenase complex during nitrogen fixation [9,11].

Nevertheless, recent quantitative mathematical models of nitrogen fixation in *A. vinelandii* have shown that oxygen-scavenging respiration is not a "switch-on/switch-off" mechanism but is performed at different carbon-nitrogen ratios (C/N), even when the cells do not fix nitrogen [10]. Inomura et al. [10] proposed controlling respiration via the C/N ratio, in which excess substrate respiration increases with the C/N ratio until nitrogenase can be derepressed when the C/N ratio is high. It is known that *A. vinelandii* possesses the capacity to fix nitrogen at high oxygen concentrations [30], which depends on different mechanisms to protect nitrogenase from inactivation by oxygen [9]. One mechanism involves respiratory protection, which is based on a high respiration rate to maintain a low oxygen concentration. In light of the evidence, a similar qO_2 under diazotrophic and nondiazotrophic cultivation conditions (Figure 2a) could be explained by the excess respiration mechanism remaining active independent of nitrogen-fixing activity. In this regard, Alleman et al. [31], suggested that the respiratory protection mechanism might be a core principle of metabolism using a genome-scale metabolic model.

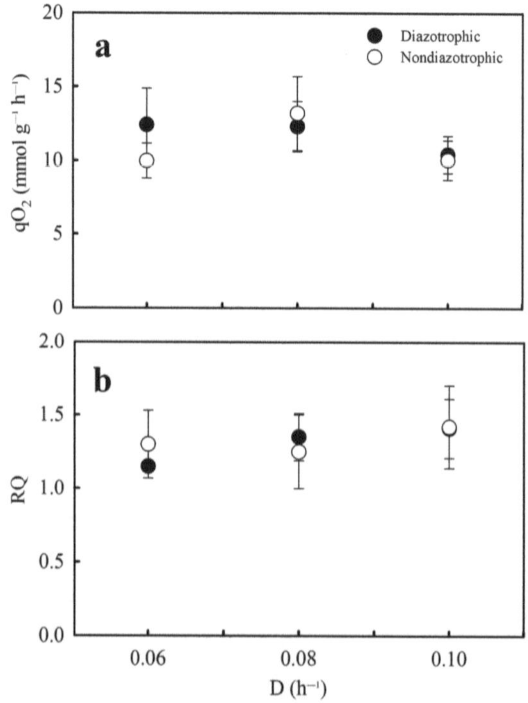

Figure 2. The specific oxygen consumption rate (qO_2) (**a**) and respiratory quotient (RQ) (**b**) under diazotrophic (black circles) and nondiazotrophic (white circles) conditions in *A. vinelandii* continuous cultures.

Similar RQ values over 1.0 were obtained under the conditions studied, reaching between 1.2 ± 0.02 and 1.42 ± 0.28 (Figure 2b). Similar values were reported by Sabra et al. [28] in oxygen-limiting chemostat cultivations under diazotrophic conditions. Those authors [28] calculated a theoretical RQ value of 0.8 associated with higher conversion of the carbon source to alginate, whereas for PHB synthesis, an RQ value of 1.33 was calculated. In concordance, similar RQ values obtained in our experiments can be related to a similar alginate concentration reached in the steady state under diazotrophic and nondiazotrophic conditions (Figure 1d). Our evidence (RQ above 1.0) shows that it is possible to manipulate culture conditions (for example, increasing the agitation rate and hence OTR) to decrease the RQ during the steady state to improve alginate production. Similarly, Díaz-Barrera et al. [17], in batch cultures of *A. vinelandii*, related a decrease in the RQ from 1.3 to 0.8 with an increase in the alginate concentration from 2.1 to 3.3 g L^{-1}.

3.3. Alginate-Specific Production Rate and Yields in Continuous Cultures

Figure 3 shows the alginate-specific production rate and the yields in the steady state under diazotrophic and nondiazotrophic conditions. Under diazotrophic conditions, a higher q_P was obtained, with values that varied between 0.21 and 0.24 g g^{-1} h^{-1} (Figure 3a). It is clear that under diazotrophic conditions, a lower biomass concentration was reached (Figure 1a,c), which can explain the highest specific production rate obtained in the steady state conducted under diazotrophy. Previously, García et al. [32] described an increase in q_P under ammonium limitation in chemostat cultures of *A. vinelandii* compared to cultures with an excess ammonium concentration at steady state.

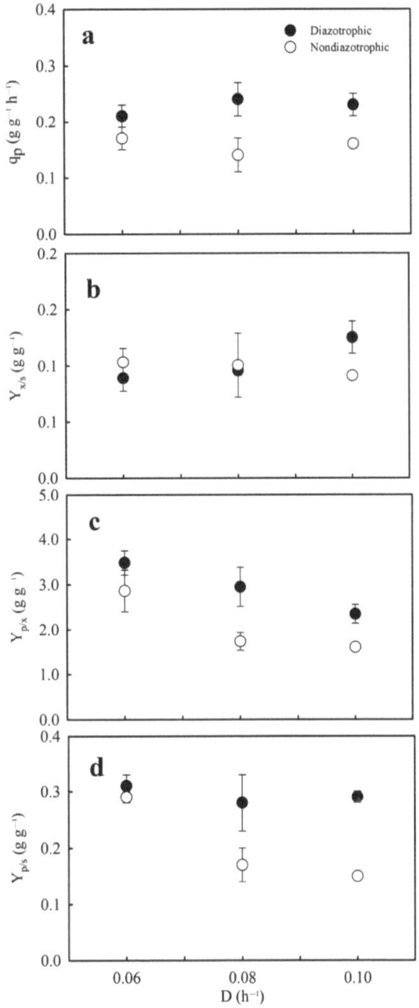

Figure 3. Alginate specific production rate (q_P) (**a**), yield of biomass based on sucrose ($Y_{x/s}$) (**b**), yield of alginate based on biomass ($Y_{p/x}$) (**c**), yield of alginate-based on sucrose ($Y_{p/s}$) (**d**) in continuous cultivations of *A. vinelandii* grown under diazotrophic (black circles) and nondiazotrophic conditions (white circles).

Similar $Y_{x/s}$ values were obtained under both conditions at a D of 0.06 and 0.08 h^{-1} (Figure 3a). Under nondiazotrophic conditions, the $Y_{x/s}$ was similar at all the D values studied, reaching a value of approximately 0.10 g g^{-1} (Figure 3b). However, under diazotrophic conditions, a higher $Y_{x/s}$ (0.13 ± 0.01 g g^{-1}) was observed when the cultures were conducted at a D of 0.10 h^{-1} (Figure 3b). In this case, this increase in the $Y_{x/s}$ can be explained by a higher level of carbon diverted to the biomass (see below Table 3). Under diazotrophic conditions, the $Y_{p/x}$ decreased from 3.5 ± 0.3 to 2.3 ± 0.2 g g^{-1} as D increased from 0.06 h^{-1} to 0.10 h^{-1} (Figure 3c). In cultures conducted at a D of 0.06 and 0.10 h^{-1}, the $Y_{p/x}$ was significantly higher under diazotrophic conditions, which could be associated with the protection of nitrogenase for nitrogen fixation, as previously described by Sabra et al. [11]. In this sense, it is possible that under diazotrophy the quantity of alginate by a unit of cells produced is increased as a mechanism to decrease the oxygen diffusion to the cells. Comparing the alginate yields reveals that the yield of alginate based on biomass obtained in our study is about 25-fold higher than those reported from algae [33]. A possible reason for this difference could be related to the process of extraction of alginate algal, whereas the produced *A. vinelandii* cells are excreted, and depending on the growth conditions, its biosynthesis could be favored over biomass production.

Table 3. The carbon distribution in *A. vinelandii* continuous cultures under diazotrophic and nondiazotrophic conditions.

Condition	**Diazotrophy**			**Nondiazotrophy**		
D (h^{-1})	**0.06**	**0.08**	**0.10**	**0.06**	**0.08**	**0.10**
Biomass (%C-mol)	10.0 ± 1.3	10.7 ± 0.5	14.0 ± 1.6	11.7 ± 1.6	11.2 ± 3.2	10.1 ± 0.1
Alginate (%C-mol)	27.5 ± 1.5	25.3 ± 4.8	26.1 ± 0.7	26.8 ± 0.1	15.2 ± 2.60	13.0 ± 0.2
P3HB (%C-mol)	6.2 ± 0.1	1.2 ± 0.3	1.0 ± 0.1	1.8 ± 0.3	1.0 ± 0.1	1.2 ± 0.4
CO_2 (%C-mol)	58.9 ± 3.1	56.0 ± 3.9	51.5 ± 4.9	63.1 ± 7.3	57.3 ± 17.2	35.7 ± 2.3
C-recovered (%mol)	103	93	93	103	85	60

Similar to the evidence obtained for q_p (Figure 3a), under diazotrophic conditions, $Y_{p/s}$ was not affected by changes in D, reaching a value of approximately 0.30 g g^{-1} (Figure 3d). However, under nondiazotrophic conditions, the $Y_{p/s}$ decreased from 0.29 g g^{-1} to 0.15 g g^{-1} by increasing the D. Comparing the evidence obtained in the conditions evaluated indicates that under diazotrophy, a higher conversion of sucrose to alginate can possibly be obtained at D values of 0.08 and 0.10 h^{-1} (Figure 3d). Based on this, the effect of D on the $Y_{p/s}$ could be related to changes in carbon flux that allow a higher proportion of sucrose to be diverted to the alginate. To the best of our knowledge, the evidence obtained in this study is the first to evaluate alginate production (yields and specific production rates) under diazotrophic and nondiazotrophic conditions in chemostat cultures limited by oxygen.

3.4. Carbon Balance at Different Dilution Rates under Diazotrophic and Nondiazotrophic Conditions

The carbon distribution (percentage of carbon atoms from sucrose converted to alginate, biomass, P3HB and CO_2) at each steady-state condition is shown in Table 3. A higher percentage of carbon diverted to biomass (14.0 ± 1.6%) was obtained at a D of 0.10 h^{-1} under diazotrophic conditions. In all the other conditions evaluated, the carbon diverted to biomass only varied between 10 and 11% (Table 3). In agreement with this evidence, Inomura et al. [10] observed that the carbon flux destined for biomass formation is similar for both diazotrophic and nondiazotrophic conditions using mathematical modeling. Under diazotrophic conditions, the carbon diverted to alginate reached values between 25.3% and 27.5%. Nevertheless, under nondiazotrophic conditions, the carbon diverted to alginate decreased approximately twofold by changing the D from 0.06 to 0.10 h^{-1} (Table 3). Independent of the conditions evaluated, more carbon was diverted to CO_2 (above 51% under diazotrophic conditions), which could be indicative of a carbon flux more active through the TCA cycle. This finding confirms that under the conditions assayed, it is necessary to operate the chemostat under diazotrophic conditions to improve the carbon into alginate.

Under diazotrophic conditions, the carbon balances were between 93–103%, which indicates that the cells utilized sucrose efficiently to produce biomass, alginate, and CO_2. Under nondiazotrophic conditions and D values of 0.08 and 0.10 h^{-1}, the carbon balance was close to 85% and 60%, respectively (Table 3), which is a clear indication that other carbon products were produced. Therefore, a change in metabolism could improve the production of organic acids or amino acids excreted into the culture medium [7,34]. Based on this evidence, further research could be developed to evaluate the effect of nitrogen fixation on the metabolism of carbon in *A. vinelandii* cells.

3.5. Alginate Molecular Weight and Gene Expression of alg8 and alg44 in Continuous Cultures

The algal alginates have several problems concerning their production which may limit their use in many interesting contexts, especially in the pharmaceutical and chemical industries, where polymers with a very well-defined composition are required. In the cultivations of *A. vinelandii*, the G/M ratio, molecular weight, and acetylation degree might be manipulated through the manipulation of culture conditions during the bioprocess. Specifically, the molecular weight is one of the most important chemical characteristics of alginates because this characteristic determines the rheological properties of this polymer [35]. In the present study, independent of the D assayed, higher alginate MMW values were obtained under diazotrophic conditions, particularly for D values of 0.08 and 0.10 h^{-1} (Figure 4a). The alginate MMW values obtained from the steady state under diazotrophic conditions showed a bell-shaped behavior, achieving a maximum value of 725 $\pm$ 20 kDa at D = 0.08 h^{-1} (Figure 4a). Under nondiazotrophic conditions, no significant differences were found in the alginate MMW produced at 0.06 and 0.08 h^{-1}. The higher values of alginate MMW achieved under diazotrophic conditions compared to nondiazotrophic conditions might be related to the protection of the nitrogenase complex, as has been previously proposed [11,26]. Sabra et al. [11] observed that alginate plays an important role in the protection of nitrogenase because it can form a capsule around the cells, decreasing oxygen diffusion to the cell. Those authors concluded that the alginate quality, not quantity, is the most important characteristic for the protection of nitrogenase for the diazotrophic growth of *A. vinelandii* in terms of the MMW.

Considering that at D = 0.08 h^{-1}, a greater difference was found in the alginate MMW synthesized under the conditions studied, the expression of the *alg8* and *alg44* genes was evaluated (Figure 4b). Figure 4b clearly shows that the expression of both *alg44* and *alg8* was higher under diazotrophic conditions, in which the expression of the *alg8* and *alg44* genes was 2.2- and 1.8-fold higher. These data suggest that both *alg44* and *alg8* gene expression can be influenced by nitrogen fixation conditions. Therefore, in cultures of *Pseudomonas aeruginosa*, the expression of the operon of alginate biosynthesis is negatively regulated by the sigma factor RpoN when ammonium is present in the medium [36].

Similar evidence has been reported in batch and continuous cultures of *A. vinelandii* [7,37]. In chemostat cultures, higher *alg8* expression can be related to a higher alginate molecular weight [7], suggesting an explanation at the cellular level for the changes in molecular weight. In agreement, Flores et al. [37] evaluated the transcription of genes involved in alginate polymerization and depolymerization under controlled DOT in batch cultures of *A. vinelandii*. These authors found that in cultures in which a high-molecular-weight alginate (1200 kDa) was synthesized, the transcriptions of *alg8* and *alg44* was considerably greater compared to those in cultures conducted at 5% DOT. The higher expression of *alg8* and *alg44* observed under diazotrophic conditions could explain the highest molecular weight obtained when the cells grew diazotrophically. In this study, we have demonstrated the feasibility of enhancing alginate production and quality (molecular weight) under nitrogen-fixing conditions in a continuous mode of operation. Although at present there is no information on production costs of alginates from *A. vinelandii*, since there is no commercial production, a process like the one described in the present study could be competitive for medical applications. This being said, pharmaceutical grade alginates with defined composition (molecular weight and G/M profiles) can have costs about 480 USD per gram (NovaMatrix, web catalog 2023).

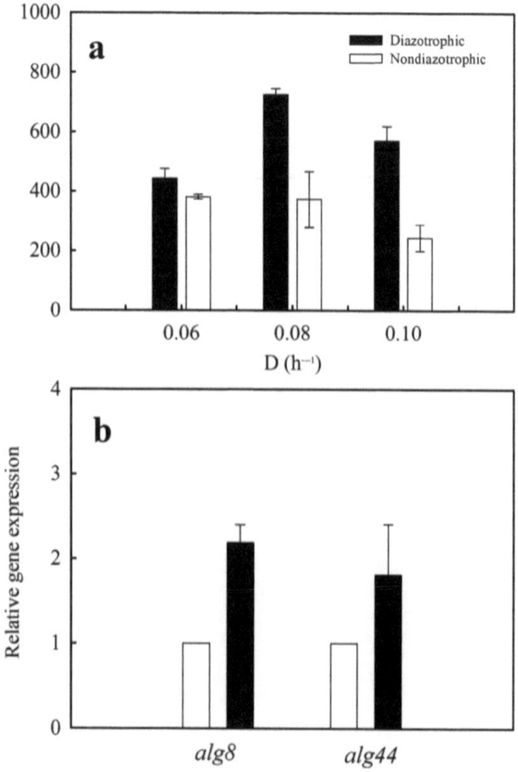

Figure 4. The mean molecular weight (MMW) of alginate a different dilution rate (**a**) and gene expression of *alg8* and *alg44* (**b**) in continuous cultures of *A. vinelandii* operated to D of 0.08 h^{-1}. Cultures under diazotrophic (black) and nondiazotrophic (white) conditions.

4. Conclusions

A comparative analysis of bacterial alginate production under diazotrophic and non-diazotrophic conditions was performed in a continuous bioprocess. A higher conversion of sucrose to alginate and alginate-specific production rates were obtained under diazotrophy (nitrogen-fixing condition). Under diazotrophic conditions the specific productivity was 0.24 ± 0.03 g g^{-1} h^{-1}. A higher alginate molecular weight (725 ± 20 kDa) was produced under diazotrophic conditions. Greater gene expression of *alg8* and *alg44* (encoding polymerases) can explain the higher molecular weight obtained. This study demonstrates that under nitrogen-fixing conditions, alginate production can be enhanced. The ability to maintain constant molecular weight in bacterial alginate production on batch cultures is recognized as a problem. The findings obtained in this study using continuous culture operating in a steady state indicate that this modality can be an appropriate way to control the characteristics of the alginate synthesized, and thus, this modality could be used for a potential industrial production of alginate under diazotrophic conditions.

Author Contributions: P.C.-A.: Investigation, Writing. V.U.: Investigation, Writing. B.P.: Investigation, Writing. T.C. and C.P.: Writing—review & editing. A.D.-B.: Conceptualization, Resources, Funding acquisition, Supervision, Writing—review & editing. All authors have read and agreed to the published version of the manuscript.

Funding: This research was funded by the Agencia Nacional de Investigación y Desarrollo (ANID) Chile: Beca de Doctorado Nacional ANID N°21201148 and Red ANID FOVI210018.

Institutional Review Board Statement: Not applicable.

Informed Consent Statement: Not applicable.

Data Availability Statement: All data will be provided to any interested part upon its requesting.

Conflicts of Interest: The authors declare no conflict of interest.

References

1. Varaprasad, K.; Karthikeyan, C.; Yallapu, M.M.; Sadiku, R. The significance of biomacromolecule alginate for the 3D printing of hydrogels for biomedical applications. *Int. J. Biol. Macromol.* **2022**, *212*, 561–578. [CrossRef] [PubMed]
2. Galindo, E.; Peña, C.; Núñez, C.; Segura, D.; Espín, G. Molecular and bioengineering strategies to improve alginate and polydydroxyalkanoate production by *Azotobacter vinelandii*. *Microb. Cell. Fact.* **2007**, *16*, 7. [CrossRef] [PubMed]
3. Ertesvåg, H. Alginate-modifying enzymes: Biological roles and biotechnological uses. *Front. Microbiol.* **2015**, *6*, 523. [CrossRef] [PubMed]
4. Geddie, J.L.; Sutherland, I.W. The effect of acetylation on cation binding by algal and bacterial alginates. *Biotechnol. Appl. Biochem.* **1994**, *20*, 117–129.
5. LeRoux, M.A.; Guilak, F.; Setton, L.A. Compressive and shear properties of alginate gel: Effects of sodium ions and alginate concentration. *J. Biomed. Mater. Res.* **1999**, *47*, 46–53. [CrossRef]
6. Urtuvia, V.; Maturana, N.; Acevedo, F.; Peña, C.; Díaz-Barrera, A. Bacterial alginate production: An overview of its biosynthesis and potential industrial production. *World J. Microbiol. Biotechnol.* **2017**, *33*, 198. [CrossRef]
7. Díaz-Barrera, A.; Soto, E.; Altamirano, C. Alginate production and *alg8* gene expression by *Azotobacter vinelandii* in continuous cultures. *J. Ind. Microbiol. Biotechnol.* **2012**, *39*, 613–621. [CrossRef]
8. Díaz-Barrera, A.; Martínez, F.; Guevara-Pezoa, F.; Acevedo, F. Evaluation of gene expression and alginate production in response to oxygen transfer in continuous culture of *Azotobacter vinelandii*. *PLoS ONE* **2014**, *9*, e105993. [CrossRef]
9. Oelze, J. Respiratory protection of nitrogenase in *Azotobacter* species: Is a widely held hypothesis unequivocally supported by experimental evidence? *FEMS Microbiol. Rev.* **2000**, *24*, 321–333. [CrossRef]
10. Inomura, K.; Bragg, J.; Riemann, L.; Follows, M.J. A quantitative model of nitrogen fixation in the presence of ammonium. *PLoS ONE* **2018**, *13*, e0208282. [CrossRef]
11. Sabra, W.; Zeng, A.P.; Lunsdorf, H.; Deckwer, W.D. Effect of oxygen on formation and structure of *Azotobacter vinelandii* alginate and its role in protecting nitrogenase. *Appl. Environ. Microbiol.* **2000**, *66*, 4037–4044. [CrossRef] [PubMed]
12. Díaz-Barrera, A.; Silva, P.; Berrios, J.; Acevedo, F. Manipulating the molecular weight of alginate produced by *Azotobacter vinelandii* in continuous cultures. *Bioresour. Technol.* **2010**, *101*, 9405–9408. [CrossRef] [PubMed]
13. Díaz-Barrera, A.; Aguirre, A.; Berrios, J.; Acevedo, F. Continuous cultures for alginate production by *Azotobacter vinelandii* growing at different oxygen uptake rates. *Process Biochem.* **2011**, *46*, 1879–1883. [CrossRef]
14. Castillo, T.; Galindo, E.; Peña, C. The acetylation degree of alginates in *Azotobacter vinelandii* ATCC9046 is determined by dissolved oxygen and specific growth rate: Studies in glucose-limited chemostat cultivations. *J. Ind. Microbiol.* **2013**, *40*, 715–723. [CrossRef]
15. Jiménez, L.; Castillo, T.; Flores, C.; Segura, D.; Galindo, E.; Peña, C. Analysis of respiratory activity and carbon usage of a mutant of *Azotobacter vinelandii* impaired in poly-β-hydroxybutyrate synthesis. *J. Ind. Microbiol. Biotechnol.* **2016**, *43*, 1167–1174. [CrossRef]
16. Ponce, B.; Urtuvia, V.; Maturana, N.; Peña, C.; Díaz-Barrera, A. Increases in alginate production and transcription levels of alginate lyase (*alyA1*) by control of the oxygen transfer rate in *Azotobacter vinelandii* cultures under diazotrophic conditions. *Electron. J. Biotechnol.* **2021**, *52*, 35–44. [CrossRef]
17. Díaz-Barrera, A.; Sánchez-Rosales, F.; Padilla-Córdova, C.; Andler, R.; Peña, C. Molecular weight and guluronic/mannuronic ratio of alginate produced by *Azotobacter vinelandii* at two bioreactor scales under diazotrophic conditions. *Bioprocess Biosyst. Eng.* **2021**, *44*, 1275–1287. [CrossRef] [PubMed]
18. Tec-Campos, D.; Zuñiga, C.; Passi, A.; Del Toro, J.; Tibocha-Bonilla, J.; Zepeda, A.; Betenbaugh, M.; Zengler, K. Modeling of nitrogen fixation and polymer production in the heterotrophic diazotroph *Azotobacter vinelandii* DJ. *Metabolic Chem. Eng. Commun.* **2020**, *11*, e00132. [CrossRef]
19. Díaz-Barrera, A.; Maturana, N.; Pacheco-Leyva, I.; Martínez, I.; Altamirano, C. Different responses in the expression of alginases, alginate polymerase and acetylation genes during alginate production by *Azotobacter vinelandii* under oxygen-controlled conditions. *J. Ind. Microbiol. Biotechnol.* **2017**, *44*, 1041–1051. [CrossRef]
20. Miller, G. Use of dinitrosalicylic acid reagent for determination of reducing sugar. *Anal. Chem.* **1969**, *31*, 426–428. [CrossRef]
21. Díaz-Barrera, A.; Urtuvia, V.; Padilla-Córdova, C.; Peña, C. Poly (3-hydroxybutyrate) accumulation by *Azotobacter vinelandii* under different oxygen transfer strategies. *J. Ind. Microbiol. Biotechnol.* **2019**, *46*, 13–19. [CrossRef] [PubMed]
22. Kaplan, A. The determination of urea, ammonia, and urease. *Methods Biochem. Anal.* **1969**, *17*, 311–324. [CrossRef] [PubMed]
23. Livak, K.J.; Schmittgen, T.D. Analysis of relative gene expression data using real-time quantitative PCR and the $2^{-\Delta\Delta CT}$ method. *Methods* **2001**, *25*, 402–408. [CrossRef]
24. Schmittgen, T.D.; Livak, K.J. Analyzing real-time PCR data by the comparative C_T method. *Nat. Protoc.* **2008**, *3*, 1101–1108. [CrossRef]
25. Díaz-Barrera, A.; Andler, R.; Martínez, I.; Peña, C. Poly-3-hydroxybutyrate production by *Azotobacter vinelandii* strains in batch cultures at different oxygen transfer rates. *J. Chem. Technol. Biotechnol.* **2016**, *91*, 1063–1071. [CrossRef]

26. Inomura, K.; Bragg, J.; Follows, M.J. A quantitative analysis of the direct and indirect costs of nitrogen fixation: A model based on *Azotobacter vinelandii. ISME J.* **2017**, *11*, 166–175. [CrossRef]
27. Wu, C.; Herold, R.A.; Knoshaug, E.P.; Wang, B.; Xiong, W.; Laurens, L. Fluxomic analysis reveals central carbon metabolism adaptation for diazotroph *Azotobacter vinelandii* ammonium excretion. *Sci. Rep.* **2019**, *9*, 13209. [CrossRef]
28. Sabra, W.; Zeng, A.P.; Sabry, S.; Omar, S.; Deckwer, W.D. Effect of phosphate and oxygen concentrations on alginate production and stoichiometry of metabolism of *Azotobacter vinelandii* under microaerobic conditions. *Appl. Microbiol. Biotechnol.* **1999**, *52*, 773–780. [CrossRef]
29. Hoskisson, P.A.; Hobbs, G. Continuous culture-making a comeback? *Microbiology* **2005**, *151*, 3153–3159. [CrossRef]
30. Post, E.; Kleiner, D.; Oelze, J. Whole cell respiration and nitrogenase activities in *Azotobacter vinelandii* growing in oxygen controlled continuous culture. *Arch. Microbiol.* **1983**, *134*, 68–72. [CrossRef]
31. Alleman, A.B.; Mus, F.; Peters, J.W. Metabolic model of the nitrogen-fixing obligate aerobe *Azotobacter vinelandii* predicts its adaptation to oxygen concentration and metal availability. *Mbio* **2021**, *12*, e02593-21. [CrossRef] [PubMed]
32. García, A.; Ferrer, P.; Albiol, J.; Castillo, T.; Segura, D.; Peña, C. Metabolic flux analysis and the NAD(P)H/NAD(P)$^+$ ratios in chemostat cultures of *Azotobacter vinelandii. Microb. Cell Factories* **2018**, *17*, 10. [CrossRef] [PubMed]
33. Mazumder, A.; Holdt, S.L.; De Francisci, D.; Alvarado-Morales, M.; Mishra, H.N.; Angelidaki, I. Extraction of alginate from *Sargassum muticum*: Process optimization and study of its functional activities. *J. Appl. Phycol.* **2016**, *28*, 3625–3634. [CrossRef]
34. Kuhla, J.; Dingler, C.; Oelze, J. Production of extracellular nitrogen-containing components by *Azotobacter vinelandii* fixing dinitrogen in oxygen-controlled continuous culture. *Arch. Microbiol.* **1985**, *141*, 297–302. [CrossRef]
35. Liparoti, S.; Speranza, V.; Marra, F. Alginate hydrogel: The influence of the hardening on the rheological behaviour. *J. Mech. Behav. Biomed. Mater.* **2021**, *116*, 104341. [CrossRef]
36. Damron, F.H.; Owings, J.P.; Okkotsu, Y.; Varga, J.J.; Schurr, J.R.; Goldberg, J.B.; Schurr, M.J.; Yu, H.D. Analysis of the *Pseudomonas aeruginosa* regulon controlled by the sensor kinase KinB and sigma factor RpoN. *J. Bacteriol.* **2012**, *194*, 1317–1330. [CrossRef]
37. Flores, C.; Moreno, S.; Espín, G.; Peña, C.; Galindo, E. Expression of alginases and alginate polymerase genes in response to oxygen, and their relationship with the alginate molecular weight in *Azotobacter vinelandii. Enzym. Microb. Technol.* **2013**, *1053*, 85–91. [CrossRef]

fermentation

MDPI

Article

The *TgRas1* Gene Affects the Lactose Metabolism of *Trichoderma guizhouense* NJAU4742

Jiaxi Miao [1,2,†], Chen Chen [1,2,†], Yajing Gu [1,2], Han Zhu [1,2], Haiyang Guo [1,2], Dongyang Liu [1,2,*] and Qirong Shen [1,2]

[1] Jiangsu Provincial Key Lab of Solid Organic Waste Utilization, Jiangsu Collaborative Innovation Center of Solid Organic Wastes, Educational Ministry Engineering Center of Resource-Saving Fertilizers, Nanjing 210095, China

[2] College of Resources and Environmental Sciences, Nanjing Agricultural University, Nanjing 210095, China

* Correspondence: liudongyang@njau.edu.cn; Tel.: +86-25-84396853; Fax: +86-25-84395212

† These authors contributed equally to this work.

Abstract: *Trichoderma* is one of the fungi commonly used in fermentation engineering. The hydrolytic enzymes secreted by *Trichoderma* have great economic value. *Trichoderma guizhouense NJAU4742* is a branch of *Trichoderma harzianum,* which also has application potential. Lactose can induce fungi to secrete cellulase. Unfortunately, neither the lactose-inducing effect nor the mechanism of lactose metabolism in the study of *Trichoderma guizhouense NJAU4742* is clear. Our study showed that carbon sources such as glucose, galactose, and sucrose could not induce cellulase secretion from *Trichoderma guizhouense NJAU4742*. Lactose induced the filter paper activity of the cellulase secreted by *Trichoderma* to reach 4.13 ± 0.11 $U{\cdot}mL^{-1}$. The ratio of 0.4% lactose–0.6% straw is the best way to induce cellulase and is better than adding only straw or lactose. TgRas family genes respond differently to different carbon sources at the gene level, and these proteins may be involved in different carbon source metabolisms. The results of transcriptional responses under different growth conditions showed that *TgRas1* occupies a dominant position among TgRas family genes. The growth of the Δ*TgRas1* mutant on the plate was inhibited, and the hyphae were dense, thick, and swollen. Under the condition of lactose, the biomass of Δ*TgRas1* was severely inhibited in liquid fermentation, and its biomass decreased by 91.43% compared with WT. The liquid fermentation of Δ*TgRas1* under other carbon source conditions was not affected.

Keywords: *Trichoderma guizhouense* NJAU4742; *TgRas1*; cellulase; fermentation; lactose metabolism

Citation: Miao, J.; Chen, C.; Gu, Y.; Zhu, H.; Guo, H.; Liu, D.; Shen, Q. The *TgRas1* Gene Affects the Lactose Metabolism of *Trichoderma guizhouense* NJAU4742. *Fermentation* **2023**, *9*, 440. https://doi.org/10.3390/fermentation9050440

Academic Editors: Miguel Ladero and Victoria E. Santos

Received: 31 March 2023
Revised: 29 April 2023
Accepted: 30 April 2023
Published: 4 May 2023

1. Introduction

Trichoderma is a fungus with high utilization value and is often used in fermentation engineering [1]. The secondary metabolites of *Trichoderma* also have the effect of inhibiting the growth of pathogenic microorganisms and stimulating plant growth [2–4]. *Trichoderma harzianum* and *Trichoderma reesei* can secrete a variety of lignocellulose-decomposing enzymes, which are widely used in the production of biological products such as cellulase [5]. *Trichoderma guizhouense* NJAU4742, which evolutionarily belongs to one of the clades of *Trichoderma harzianum,* was isolated from soil samples in Guizhou Province, China [6]. *Trichoderma guizhouense* NJAU4742 has been elucidated to promote root growth and development and activate plant innate immune response [7]. In addition, *Trichoderma guizhouense* NJAU4742 retains the straw-degrading characteristics of *Trichoderma*, which also has application value. The expression and secretion mechanisms of cellulases have been studied for decades [8]. Studies have shown that the expression of cellulase genes is not only dependent on the induction of cellulose, but can also be induced by other carbon sources [9]. Lactose is an inexpensive inducer that has attracted widespread attention because of its low cost.

Lactose is composed of D-galactose and D-glucose linked by β-1,4 glycosidic linkages. Previous studies have shown that lactose is an inexpensive carbon source. Not only can it be used for fermentation to produce cellulase [10], but it is also the only cheap and soluble substrate currently available for *Trichoderma reesei* fermentation engineering [11,12]. Lactose as a carbon source has also been shown to enhance the ability of *Vibrio cellulolyticus C-1* to produce cellulase [13]. Although the cheap properties of lactose can be used in microbial fermentation technology, not all microorganisms can effectively utilize lactose, which is one of the reasons that limits the development of the fermentation industry.

Ras family genes are ubiquitous in mammalian and eukaryotic cells. Ras genes play a key role in fungal carbon metabolism. A large number of studies have shown that Ras protein is involved in the process of fungal growth and development. There are two types of Ras proteins in *Candida albicans*; knocking out *Ras1* can inhibit the hyphal production and the overexpression of Ras can enhance the hyphal formation [14]. There are two Ras genes in corn smut (*U. maydis*). The *Ras1* mutants have been reported to be lethal, and Δ*ras2* exhibits a reduced ability to form hyphae with a round shape [15]. It has been reported that Ras genes in *Aspergillus* and *Aspergillus fumigatus* regulate growth and development more extensively. The *RasA* gene of both filamentous fungi is involved in the process of mitosis, as well as processes that regulate the asexual developmental cycle [16]. The *Ras1* and *Ras2* in *Cryptococcus neoformans* not only participate in the regulation of high temperature growth process, but also participate in mycelium formation and type II transformation, which affects the pathogenicity [17,18]. Ras regulates cellular carbon sensing by regulating its affinity for GTP and GDP. Therefore, Ras may also play an important role in fungal lactose metabolism. This study aimed to investigate the feasibility of lactose-induced cellulase in *Trichoderma guizhouense* NJAU4742 and the effect of *TgRas1* gene on lactose metabolism.

2. Methods and Materials

2.1. Determination of Enzymatic Activity of T. guizhouense NJAU4742 Solid Fermentation

Determination of endoglucanase activity: Take 480 μL 0.5% carboxymethyl cellulose solution (CMC-Na) in a 2 mL centrifuge tube, add 500 μL 50 mM acetic acid buffer, and finally add 20 μL crude enzyme solution. After mixing evenly, react in a 50 °C water bath for 10 min, then add 1 mL of DNS reagent in a boiling water bath for 10 min and cool down. Its absorbance value is measured at a wavelength of OD_{520} nm. The amount of enzyme required to produce 1 μmol of reducing sugar per 1 min is defined as 1 unit of enzyme activity (U).

Determination of exoglucanase activity: Take 10 μL 5 mM pNPC and add it to the 96-well plate, add 40 μL crude enzyme solution and 50 μL 50 mM acetic acid buffer, and use the inactivated crude enzyme solution as a blank control. The 96-well plate is reacted in a 50 °C water bath for 10 min, and 100 μL 1 M of Na_2CO_3 solution is added to terminate the reaction. Its absorbance value is measured at OD_{402} nm. The content of p-nitrophenol is calculated based on the marking. The amount of enzyme required to produce 1 μmol of p-nitrophenol per minute for 1 min is defined as one unit of enzyme activity (U).

Determination of xylanase activity: Take 480 μL of xylanase in a 2 mL centrifuge tube, add 500 μL of 50 mM acetic acid buffer, and finally add 20 μL of crude enzyme solution. After mixing evenly, react in a 50 °C water bath for 20 min, then add 1 mL of DNS reagent in a boiling water bath for 10 min and cool down. Its absorbance value is measured at OD_{520} nm. The amount of enzyme required to produce 1 μmol of reducing sugar per 1 min is defined as 1 unit of enzyme activity (U).

Determination of enzyme activity of filter paper: Take 50 μL of crude enzyme solution, add 500 μL of 50 mM acetic acid buffer and 450 μL of deionized water, and put two small discs of filter paper (Whatman No. 1 filter paper; use a hole puncher to obtain a small disc of filter paper; approximately 10 mg) and add it to a 2 mL centrifuge tube. After the reaction in a water bath at 50 °C for 20 min, add 1 mL of DNS reagent and cool in a boiling water bath for 10 min. Its absorbance value is measured at OD_{520} nm. The amount of enzyme required to produce 1 μmol of reducing sugar per 1 min is defined as 1 unit of enzyme activity (U).

The crude enzyme solution inactivated in a 100 °C water bath for 10 min is used as the comparison solution, and the remaining steps are the same. Enzyme activity refers to the results calculated after deducting the reducing sugar in the comparison solution. Drawing of glucose standard curve: Add different volumes of glucose standard solution, deionized water, and DNS reagents to each centrifuge tube and mix well, then place each centrifuge tube in a boiling water bath at the same time and react for 10 min. After cooling, the absorbance value is measured at OD_{520} nm, with glucose content as the abscissa and OD value as the ordinate, and the standard curve of glucose is drawn. The enzyme inactivated in a 100 °C water bath for 10 min is used as the control solution, and the remaining steps are the same. The production of 1 μmol of reducing sugar per minute is defined as 1 enzyme activity (unit U).

Drawing of the standard curve of p-nitrophenol: Different volumes of standard solution of p-nitrophenol (1 mg·mL^{-1}), deionized water and Na_2CO_3 solution are drawn in a 96-well plate, and the absorbance is measured at OD_{402} nm after mixing. The content of p-nitrophenol is used as the abscissa, and the OD value is used as the ordinate to make a standard curve of p-nitrophenol.

2.2. Identification of TgRas Gene in the Genome of T. guizhouense NJAU4742

The TgRas protein sequence was retrieved from the *Trichoderma guizhouense* NJAU4742 database using the accession number search keywords obtained from the *Saccharomyces cerevisiae* database, while the Ras gene family hidden Markov model domain (PF00071) was downloaded from the Pfam database. The NCBI BLASTP homepage (http://www.ncbi.nlm.nih.gov (accessed on 30 March 2023)) was used to perform homologous alignment with the *T. guizhouense* NJAU4742 genome database, and at the same time use HMMER software was used to perform domain alignment and identify all potential Ras genes. All TgRas gene family members of *T. guizhouense* NJAU4742 were subsequently determined using the online tools CDD (https://www.ncbi.nlm.nih.gov/cdd/ (accessed on 30 March 2023)) and SMART (http://smart.embl.de/ (accessed on 30 March 2023)). The software DNAMAN (https://www.lynnon.com/ (accessed on 30 March 2023)) was used to compare the primary structure of the TgRas family proteins and the ExPASy online tool was used to calculate the chemical characteristics of the TgRas protein in *T. guizhouense* NJAU4742 (http://web.expasy.org/protparam/ (accessed on 30 March 2023)), such as theoretical isoelectric point (pI), amino acid number, and molecular weight (Da).

2.3. Phylogenetic Analysis of TgRas Gene, Conserved Motif Analysis, Gene Structure Analysis

MUSCLE software (https://www.ebi.ac.uk/Tools/msa/muscle/ (accessed on 30 March 2023)) was used to perform multiple sequence alignment of the amino acid sequence of TgRas protein, MEGA v7 was used to select the maximum likelihood method for phylogenetic analysis and construct the phylogenetic tree, where the bootstrap value was set to 1000, and finally the online interactive tool iTOL was used for data visualization (https://itol.embl.de/ (accessed on 30 March 2023)). The online tool MEME (http://memesuite.org/ (accessed on 30 March 2023)) was used to identify the conserved protein structure of the TgRas family protein, with the maximum motif value set to 10, and the software TBtools was then used to analyze the motif structure of the TgRas family protein to visualize. The exon/intron and CDs structures of the TgRas gene were predicted using the online tool GSDS (http://gsds.cbi.pku.edu.cn/ (accessed on 30 March 2023)).

2.3.1. Cultivation, Determination of Biomass, and Preparation of Intracellular Material

T. guizhouense NJAU4742 was selected for this study, and it was cultured on PDA medium for conidia production and incubated at 28 °C under static cultivation conditions for seven days. The conidia were then harvested by washing the plate with 10 mL of sterile double-distilled water (ddH_2O) followed by removal of mycelia by filtration through four layers of gauze. The conidia were resuspended, and the concentration was adjusted to 1×10^7 conidia·mL^{-1}. Mandel's salt solution without organic components (1.4 g·L^{-1} $(NH_4)_2SO_4$, 2.0 g·L^{-1} KH_2PO_4, 0.3 g·L^{-1} $CaCl_2$, 0.3 g·L^{-1} $MgSO_4$, 5 mg·L^{-1} $FeSO_4 \cdot 7H_2O$, 20 mg L^{-1} $CoCl_2$, 1.6 mg L^{-1} $MnSO_4$, and 1.4 mg L^{-1} $ZnSO_4$) was supplemented with 1% (w/v) lactose for liquid fermentation. The biomass in submerged cultures was determined by filtering the cultures onto preweighed filter paper (Whatman No. 1), and the harvested biomass was dried at 70 °C for 24 h before being quantified. After being washed once with sterile water, the fresh mycelia were disrupted using glass beads (0.1 mm) and liquid nitrogen and then centrifuged at 10,000× *g* for 10 min to remove cell debris. The supernatant was frozen in liquid nitrogen and stored at –20 °C for subsequent experiments.

2.3.2. Construction of Functional Fractions for Gene Deletion Complementation

A fragment of approximately 1 kb upstream and downstream of the target gene from the genome of NJAU4742 (NCBI accession number: GQ337429) and a hygromycin fragment from the vector of PcDNA13.1 (Invitrogen, Waltham, MA, USA) were amplified with primers. After purification with the AxyPrep DNA Pure Cycle Kit (Axygen, Hangchow, China), the PCR products of these three genes were ligated together by the HiFi enzyme (KAPA Biosystems, KK2102, Wilmington. MA, USA). After agarose gel electrophoresis analysis, the fusion-completed fragments were recovered by an AxyPrep DNA gel purification and extraction kit (Axygen, Hangchow, China). The gene deletion cassette was obtained with the primer's gene-up-F and gene-down-R (Table S1).

Seven-day-old spores were harvested from PDA solid medium with 10 mL of freshly sterilized water. The spore suspension was filtered through four layers of sterile medical gauze and then collected in a 1.5 mL sterile centrifuge tube for subsequent experiments. Fifty microliters of the spore solution were spread on the cellophane-covered plates and then incubated at 28 °C for 16 h. After incubation, the cellophane was transferred into a new Petri dish, and then 5 mL of solution A (0.1 M KH_2PO_4, 1.2 M sorbitol, pH 5.6, filter sterilized 0.15 g lyase (Sigma-Aldrich, St. Louis, MO, USA)) was used to wash the germinated spores on cellophane to make protoplast suspensions. The protoplast was extracted as follows: the harvested suspension was incubated at 28 °C for 100 min and then centrifuged at 100 rpm, after which the cells were gently disrupted with autoclaved tweezer tips; the suspension was filtered with four layers of sterile medical gauze in a 50 mL sterile centrifuge tube, and then the protoplast suspension was obtained by centrifugation at 2000 rpm at 4 °C for 10 min; the supernatant was discarded, and the pelleted protoplasts were resuspended with 5 mL of solution B (1 M sorbitol, 50 mM $CaCl_2$, 10 mM Tris-HCl, pH 7.50) and then centrifuged at 2000 rpm for 10 min at 4 °C. The supernatant was then discarded, and the pelleted protoplasts were resuspended with 200 µL of solution B and kept at 4 °C for the subsequent experiment. The functional fragments were transferred into the protoplasts of *Trichoderma guizhouense* NJAU4742 by the polyethylene glycol (PEG)-mediated method according to Miao et al. [19], with some modifications. The mixture containing 200 µL protoplasts, 10 µL ligated fragments (concentration should be greater than 250 ng·µL^{-1}), and 50 µL PEG (1 M PEG 6000, 50 mM $CaCl_2$, pH 7.50) was placed on the ice bath for 20 min, and then an additional 2 mL of PEG (room temperature) was added and mixed gently. After 5 min of incubation at room temperature, 3 mL of solution B was added to the system and mixed gently, and then 0.5 mL of the mixture was spread on PDA medium containing 1 M sucrose and incubated at 28 °C for 16 h. The germinated protoplasts were overlaid with PDA medium containing 200 ng·mL^{-1} hygromycin B and incubated at 28 °C for 72 h. Some single colonies were selected and then transferred into new PDA medium containing 200 ng·mL^{-1} hygromycin B and incubated at 28 °C for 48 h. The mutants were directly screened for positive deletion mutants by PCR with primer validation (Table S1) and purified into single spores. The construction method of the gene complementary strain is similar to the previous gene-knockout method.

3. Results

3.1. *The Combination of Straw and Lactose Can Effectively Promote the Secretion of Cellulase by T. guizhouense NJAU4742*

Firstly, the differences in cellulase secretion in *Trichoderma guizhouense* NJAU4742 under different carbon source conditions were compared. The enzymatic activities of various cellulase enzymes in the liquid fermentation process were measured to show the induction effect of carbon source. The whole process is based on MM medium, to which 1% glucose, galactose, sucrose, lactose, and straw carbon sources are added, respectively. After culturing for 4 days, the extracellular enzyme solution was taken to measure the enzyme activity. It can be seen from Figure 1 that glucose, galactose, and sucrose cannot induce the secretion of cellulase. Both rice straw and lactose could induce *Trichoderma guizhouense* NJAU4742 to secrete cellulase. Under lactose conditions, the filter paper enzyme activity of the cellulase secreted by *Trichoderma guizhouense* NJAU4742 reached 4.13 ± 0.11 U·mL^{-1}, which was slightly lower than that induced by straw. In addition, under the condition of lactose, the activity of endoglucanase reached 3.66 ± 0.17 U·mL^{-1}, the activity of exoglucanase reached 3.54 ± 0.22 U·mL^{-1}, and the activity of xylanase activity reached 3.60 ± 0.25 U·mL^{-1}.

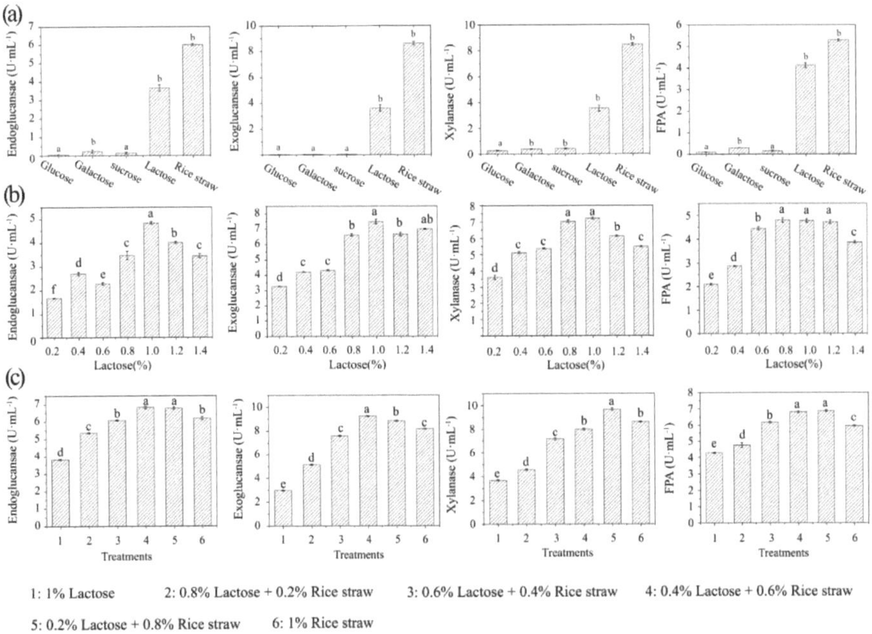

Figure 1. Enzyme activity of cellulase measured under different conditions. (**a**) Cellulase activity measured under different carbon sources; (**b**) cellulase activity measured under different concentrations of lactose; (**c**) cellulase activity measured under different lactose and straw ratios. The "abcde" shows significant differences separately. Different letters represent significant differences from each other.

In addition, we also obtained the best lactose-inducing concentration of *Trichoderma guizhouense* NJAU4742 to secrete cellulase. We added 0.2–1.4% lactose to MM medium. The enzyme activity data showed that both endoglucanase and exoglucanase had the maximum enzymatic activity under the condition induced by 1% lactose, which were 4.84 ± 0.05 U·mL^{-1} and 7.22 ± 0.06 U·mL^{-1}, respectively. At this time, xylanase also had a high activity (7.48 ± 0.15 U·mL^{-1}). It is worth noting that cellulase is often a mixture of various enzymes. We further measured the enzyme activity of filter paper to comprehensively measure the effect of cellulase. The results showed that when the lactose concentration was 0.8%, 1%, and 1.2%, the filter paper enzyme activities were 4.80 ± 0.09 U·mL^{-1}, 5.37 ± 0.08 U·mL^{-1}, and 4.72 ± 0.08 U·mL^{-1}, respectively. Therefore, the addition of lactose can induce *Trichoderma guizhouense* NJAU4742 to secrete cellulase. When the amount of lactose added was approximately 1%, the induction effect was most obvious. However, too much lactose may affect cellulase induction.

Both straw and lactose induction data showed a high activity of cellulase. We then adjusted the ratio of lactose to straw to try to obtain the best solution. The total content of lactose and straw was still 1% in the MM medium, and the content of individual components fluctuated from 0% to 1%. The results showed that the endoglucanase activity of adding 0.4% lactose–0.6% rice straw was not significantly different from that of 0.2% lactose–0.8% rice straw, which were 6.83 ± 0.07 U·mL^{-1} and 6.79 ± 0.06 U·mL^{-1}, respectively. However, the enzyme activities induced by 1% straw increased by 9.22% and 8.69%, respectively. For exoglucanase, only the combination of 0.4% lactose–0.6% rice straw had the maximum induction effect, and the enzyme activity reached 9.68 ± 0.11 U·mL^{-1}. Compared with adding 1% rice straw, the enzyme activity increased by 10.95%. Adding an appropriate proportion of lactose can also improve the activity of xylanase. When 0.2% lactose–0.8% straw was added, the enzyme activity reached 9.26 ± 0.06 U·mL^{-1}, an increase of 11.56% (compared with 1% straw). Among many combinations, the combination of 0.4% lactose and 0.6% rice straw had the largest filter paper enzyme activity, reaching 7.30 ± 0.05 U·mL^{-1}. Compared with the treatment of 1% rice straw, the activity increased by 18.52%. In summary, the induction effect of lactose and straw mixed was significantly better than that of single straw or lactose. When proportioned, it can most effectively increase the enzymatic activity of hydrolytic enzymes such as cellulase.

3.2. *Functional Identification of TgRas Family Genes in T. guizhouense NJAU4742*

Previous studies have shown that the Ras gene family is involved in the growth and development of organisms, including the metabolism of various carbon sources. The TgRas family gene sequence was downloaded from the *T. guizhouense* NJAU4742 database. TgRas family protein domains were further analyzed by searching candidate TgRas family proteins in HMMER and NCBI databases. Twenty-two candidate TgRas genes were identified in the genome of *Trichoderma guizhouense* NJAU4742. They were sorted according to homology. Table 1 shows information such as chromosome location, amino acid (length), molecular weight, and isoelectric point (pI). The results showed that the length of TgRas family genes ranged from 765 (*TgRas9*) to 1260 (*TgRas14*) bases. TgRas family proteins range in length from 194 (*TgRas14*) to 336 (*TgRas22*) amino acids. The PI values were between 4.79 (*TgRas11*) and 9.61 (*TgRas22*), and the protein molecular weights were between 20.761 kDa (*TgRas4*) and 36.096 kDa (*TgRas22*) (Table 1).

Next, the structures of 22 TgRas genes were analyzed. The blue squares represent the upstream and downstream segments of the gene. The yellow squares represent exons of genes. Introns are indicated by black lines. The results showed that the exon numbers of 22 genes varied widely, ranging from one (*TgRas6*) to eight (*TgRas3*). The difference in the number of exons in the TgRas family may occur during the evolution of the TgRas gene, which may also be one of the reasons for the different functions of the TgRas family (Figure 2a). In addition, we identified 10 conserved motifs associated with TgRas. The identified motifs range in length from 11 to 50 amino acids. The color of each square in the figure represents a conserved motif. As can be seen in Figure 2b, the motif combinations of different TgRas proteins are not the same. The conserved motifs and distribution order of *TgRas1* and *TgRas2* are consistent, which means that the similarity is high. *TgRas4*, *TgRas14*, and *TgRas18* have specific Motif 6. It is worth noting that some TgRas proteins also have Motif 8, and all of them are located at the amino acid terminal. Motif 8 is a CAAX motif sequence consisting of a cysteine residue followed by two aliphatic residues and a C-terminal X residue. X can be C, S, M, Q, or A. CAAX motifs are premodified targets for protein transport and have important biological functions in organisms. The results showed that the end of *TgRas1–4* has a CAAX structure (orange Motif8), which is closely related to the membrane localization function. Figure 2c shows the analysis of the amino acid composition of each functional domain. An amino acid with a larger font size (such as C in Motif 8) means that among the 22 TgRas proteins, more TgRas proteins have amino acid C (such as C in Motif 8) at this position. For Ras family proteins, this motif appears to contribute to membrane localization and signal transduction.

Table 1. General information of TgRas family in *Trichoderma guizhouense* NJAU4742.

Gene ID	Name	Location	Gene Length	Amino Acid Quantity	pI	Molecular Weight (kDa)
A1A109149.1	*TgRas1*	TGA1_S20	991	214	4.85	24,205.32
A1A102886.1	*TgRas2*	TGA1_S07	931	237	9.2	26,098.67
A1A109579.1	*TgRas3*	TGA1_S20	1181	217	5.73	24,447.57
A1A102432.1	*TgRas4*	TGA1_S06	822	186	6.43	20,761.59
A1A109628.1	*TgRas5*	TGA1_S20	998	206	5.56	22,977.12
A1A111449.1	*TgRas6*	TGA1_S23	795	264	5.27	28,640.29
A1A103922.1	*TgRas7*	TGA1_S08	1111	202	5.47	22,448.3
A1A104429.1	*TgRas8*	TGA1_S10	922	212	5.61	23,192.99
A1A108119.1	*TgRas9*	TGA1_S18	765	254	7.65	27,544.88
A1A108307.1	*TgRas10*	TGA1_S18	1053	208	7.7	23,332.35
A1A110534.1	*TgRas11*	TGA1_S22	1010	206	4.79	23,136.02
A1A106211.1	*TgRas12*	TGA1_S15	832	229	5.14	24,249.11
A1A103198.1	*TgRas13*	TGA1_S07	926	220	8.74	23,990.25
A1A106127.1	*TgRas14*	TGA1_S15	1260	194	5.83	21,546.93
A1A109539.1	*TgRas15*	TGA1_S20	994	204	8.2	22,445.79
A1A100568.1	*TgRas16*	TGA1_S02	1015	216	6.91	24,354.95
A1A105281.1	*TgRas17*	TGA1_S13	1012	249	8.52	27,671.72
A1A101316.1	*TgRas18*	TGA1_S03	1311	195	6.21	21,515.72
A1A105388.1	*TgRas19*	TGA1_S13	943	200	5.44	21,999.18
A1A110131.1	*TgRas20*	TGA1_S20	1086	307	8.46	33,673.86
A1A106470.1	*TgRas21*	TGA1_S15	979	207	4.68	23,029.93
A1A110255.1	*TgRas22*	TGA1_S21	1221	336	9.61	36,096.6

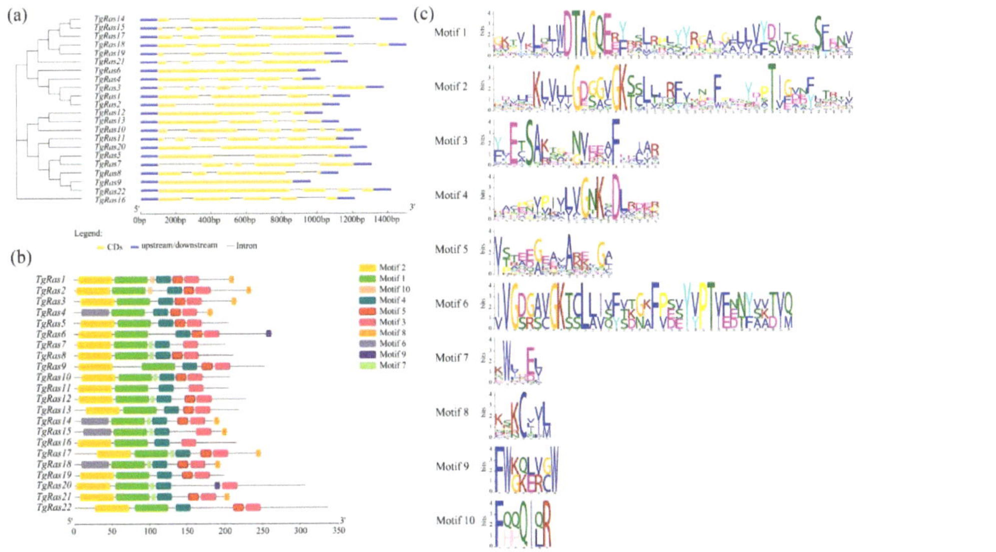

Figure 2. Identification and analysis of TgRas family genes and TgRas family proteins. (**a**) Gene structure of TgRas genes; (**b**) distribution of conserved motif of TgRas proteins; (**c**) motif analysis of TgRas proteins.

We analyzed the expression profile of TgRas family genes when glucose or rice straw was used as a carbon source. The heatmap shows the quantitative results for 22 TgRas genes. The results showed that the type of carbon source did not affect the high-level expression of *TgRas1*, *TgRas2*, *TgRas3*, *TgRas9*, *TgRas21*, and *TgRas22*. Only under the condition of glucose as a carbon source did *TgRas5*, *TgRas7*, *TgRas8*, *TgRas10*, *TgRas11*, *TgRas13*, *TgRas16*, *TgRas17*, *TgRas18*, and *TgRas19* show higher expression levels. *TgRas14* and *TgRas15* only showed higher expression levels on a rice straw medium. In summary, TgRas family genes respond to different types of carbon sources at the gene level, and these proteins may be involved in different carbon source metabolisms (Figure 3).

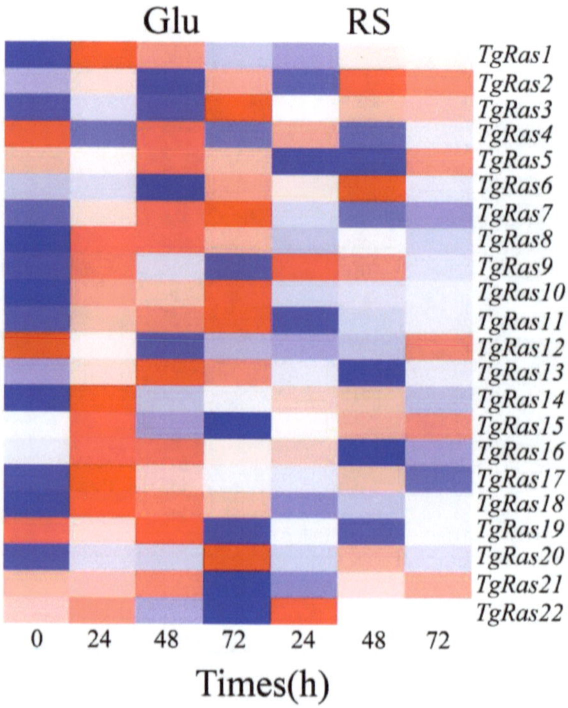

Figure 3. Expression profile of TgRas family genes for different carbon sources and times. Darker red indicates higher gene expression and darker blue indicates lower gene expression.

3.3. The T. guizhouense NJAU4742 Mutant Has a Reduced Ability to Utilize Lactose, Resulting in Abnormal Growth

Based on the TgRas family analysis results of *T. guizhouense NJAU4742*, we selected *TgRas1*, *TgRas2*, *TgRas3*, and *TgRas4* genes for follow-up research. The methods included construction of mutants and validation. These four mutants were named Δ*TgRas1*, Δ*TgRas2*, Δ*TgRas3*, and Δ*TgRas4*. We compared the transcriptional responses of TgRas family genes in wild-type strains under different temperatures and carbon sources.

The results showed that the expression of *TgRas1* had the highest expression level regardless of the carbon source. The results induced by lactose and sucrose were 54.13% and 62.93% higher than those induced by glucose, respectively. The expression of *TgRas4* was low, while the expression of *TgRas2* and *TgRas3* was very low (Figure 4a). Therefore, *TgRas1* may be the key gene of carbon source metabolism in *T. guizhouense NJAU4742*. The wild type was cultured under liquid conditions with ammonium sulfate, ammonium nitrate, and sodium nitrate as nitrogen sources, respectively. The expression of *TgRas1* was still the highest, and there was no significant difference between different nitrogen sources. It is worth noting that *TgRas4* showed an increased expression level, which was close to that of *TgRas1*, when nitrate was used as a nitrogen source (Figure 4b). This means that *TgRas4* may be involved in the regulation of nitrogen sources. For the results of liquid culture at different temperatures, the expression level of *TgRas1* was still high, and the change of temperature seemed to have no effect on TgRas family genes (Figure 4c). In addition, under the growth conditions of pH 9 and pH 3, the expression of *TgRas1* gene was significantly lower than that of pH 6, which decreased by 44.71% and 36.33%, respectively (Figure 4d). Interestingly, the expression of *TgRas1* tended to increase under the light-only culture condition, but that may also be due to experimental error (Figure 4e). In short, the results suggest that in *T. guizhouense NJAU4742*, *TgRas1* may play a leading role and perform a major function in the organism.

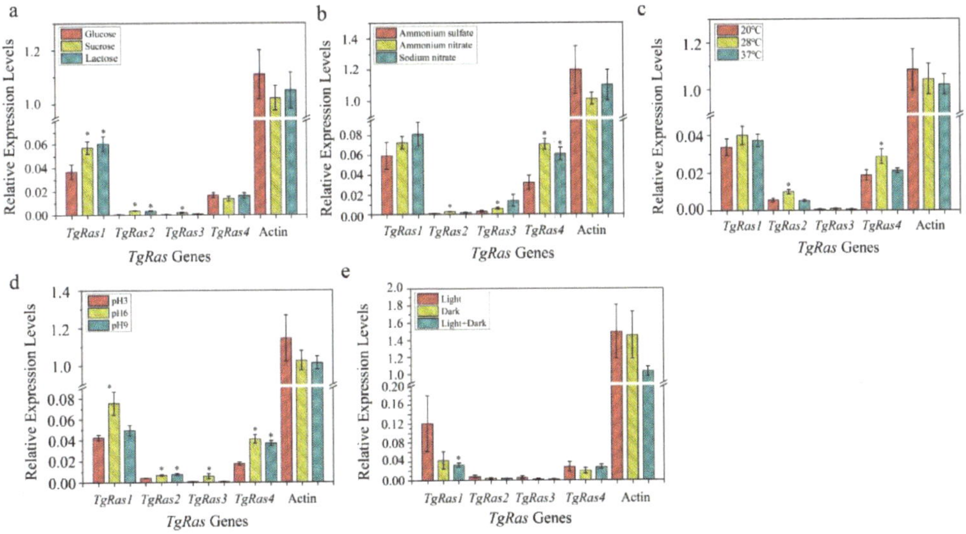

Figure 4. Transcriptional responses of TgRas family genes in wild-type strains under different carbon sources, nitrogen sources, temperatures, pH levels, and light conditions. (**a**) Transcriptional responses of TgRas family genes in wild-type strains under different carbon sources; (**b**) transcriptional responses of TgRas family genes in wild-type strains under different nitrogen sources; (**c**) transcriptional responses of TgRas family genes in wild-type strains under different temperatures; (**d**) transcriptional responses of TgRas family genes in wild-type strains under different pH levels; (**e**) transcriptional responses of TgRas family genes in wild-type strains under different light conditions. The "*" shows significant differences separately.

Furthermore, we evaluated the effect of TgRas family genes on the growth of *T. guizhouense* NJAU4742. Wild types (WT) Δ*TgRas1*, Δ*TgRas2*, Δ*TgRas3*, and Δ*TgRas4* were inoculated on PDA plates for one week, and the daily growth was recorded, as shown in Figure 5a. There was no significant difference in growth between Δ*TgRas3* and WT during 7 days of continuous culture. The phenotype of Δ*TgRas1* had the greatest influence. The hyphae of Δ*TgRas1* could not fill the entire plate, and the colony edge grew irregularly and the colony size was small. The growth of Δ*TgRas2* was slow and the colony edge grew irregularly, indicating that the growth polarity of this strain may be limited. Δ*TgRas4* was less affected but had slower growth and greater spore production. Irregular colony edges were simultaneously observed in Δ*TgRas1*, Δ*TgRas2*, and Δ*TgRas4*, suggesting that the main function of TgRas family genes may be to affect polar growth (Figure 5b).

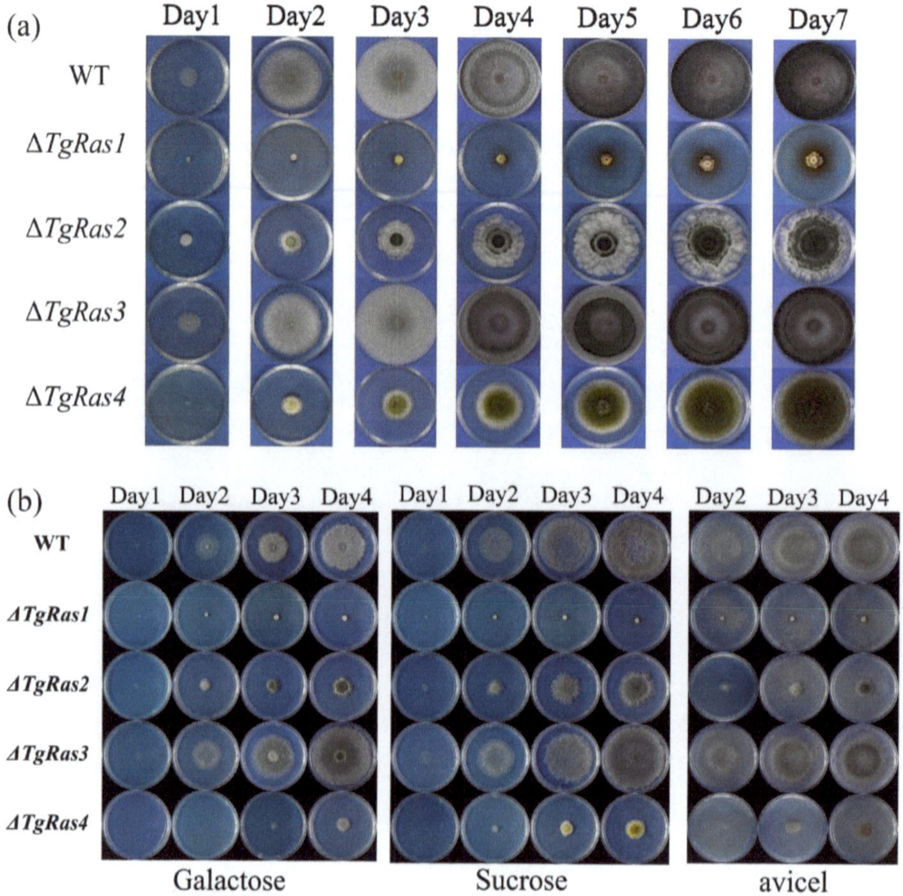

Figure 5. Growth of single-knockout strains of the TgRas family gene on the plate. (**a**) Growth of TgRas family single-knockout strains on PDA; (**b**) growth of TgRas family single-knockout strains on galactose, sucrose, and microcrystalline cellulose.

Further, we observed the hyphal growth and spore germination of each mutant strain on the plate. We observed the fungal border phenotype at 24 h and 48 h time points. Δ*TgRas2*, Δ*TgRas3*, and Δ*TgRas4* all showed denser hyphae numbers with more mycelial branches. Δ*TgRas1* exhibited a fully dysplastic fungal hyphal morphology from the very beginning, with swollen and smooth edges (Figure 6a). The spore germination observed under the microscope showed that as the germination time passed (12–18 h), the hyphae of Δ*TgRas2,* Δ*TgRas3,* and Δ*TgRas4* were similar to those of WT, and all of them could obviously produce hyphae. However, the germination process of Δ*TgRas1* spores was affected, and the spores were large, and no obvious mycelial growth could be seen (Figure 6b). We measured the length of the hyphae to represent the growth rate of the fungus. The results showed that the radial growth rate of Δ*TgRas1* was the slowest, and there were almost no elongated hyphae within 72 h. Δ*TgRas2* and Δ*TgRas4* had slower growth rates, but Δ*TgRas3* had little difference in growth rate from WT (Figure 6c,d). Colony edge growth and spore germination also confirmed that the deletion of the TgRas gene had an effect on the polar growth of hyphae, and *TgRas1* played a major role in the TgRas gene.

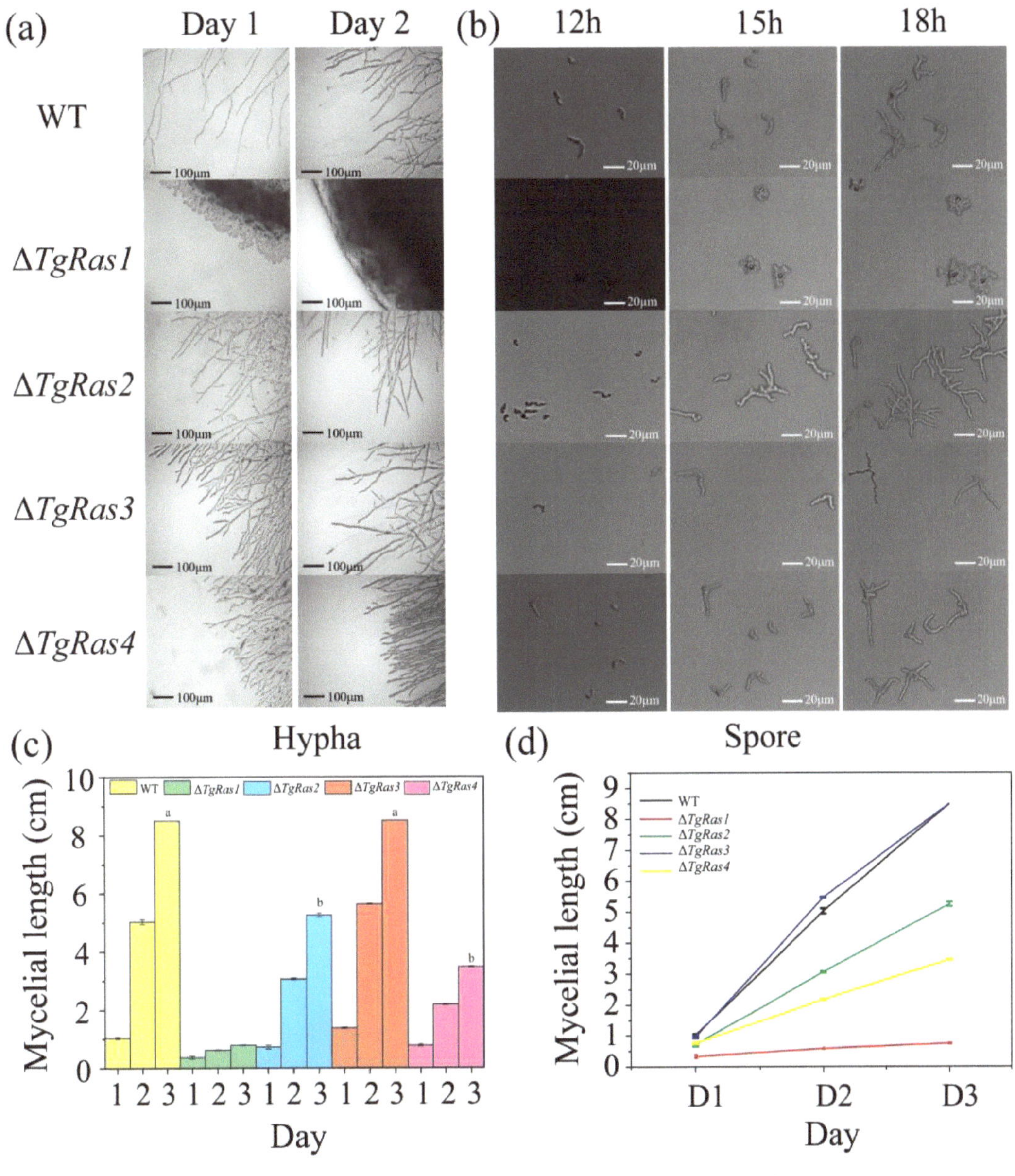

Figure 6. Mycelial growth and spore germination of TgRas family single-knockout strains growing on PDA. (**a**) Mycelial growth at the colony edge of the TgRas family single-knockout strain; (**b**) spore germination of the TgRas family single-knockout strain; (**c**) mycelium length of single-knockout strains of TgRas family genes. The "ab" shows significant differences separately. Different letters represent significant differences from each other.; (**d**) radial growth of mycelium of single-knockout strains of the TgRas family genes.

3.4. Liquid Fermentation of TgRas1 Mutants from other Carbon Sources Was Not Affected

The above experimental results have shown that *TgRas1* is the main functional gene. Therefore, we observed the growth of wild-type and *TgRas1* mutants (Δ*TgRas1*) in a Biolog-FF medium and investigated the response of Δ*TgRas1* to different carbon sources. Each legend and number in Figure 7 represent an individual carbon source. The unmarked numbers indicate that neither wild type nor Δ*TgRas1* can grow on this carbon source. The carbon source corresponding to each number is listed in detail in the appendix. The thickened light green line in Figure 7 represents the growth of glucose, the thickened black line represents the growth of lactose, and the thickened dark blue line represents the growth of arabinose. The wild-type strain grew rapidly under the condition of glucose, but after 96 h, the growth of the wild-type strain under the lactose condition gradually increased (Figure 7a). However, during the growth process of Δ*TgRas1*, the growth of Δ*TgRas1* under lactose conditions was much weaker than that under glucose conditions (Figure 7b), which formed a significant contrast with WT. The results showed that Δ*TgRas1* could not utilize lactose well, but glucose utilization was not affected (Figure 7).

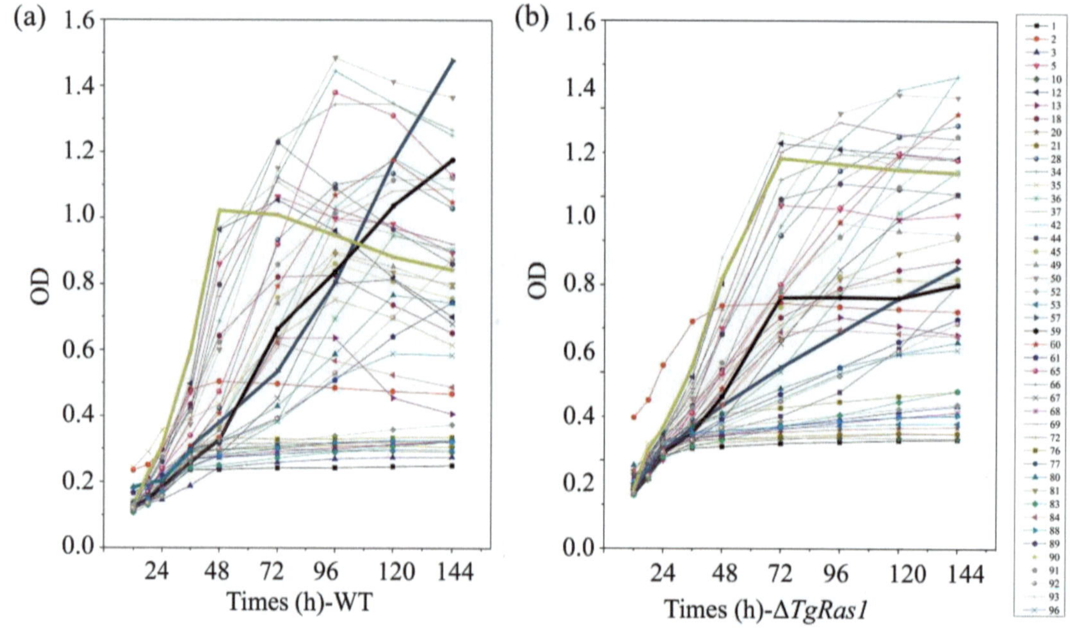

Figure 7. Growth of WT and Δ*TgRas1* on Biolog-FF plates. (**a**) WT; (**b**) Δ*TgRas1*.

According to the growth and development of Δ*TgRas1* on Biolog-FF plates, glucose and lactose were selected as carbon sources for liquid fermentation. The dry weight of the mycelia was measured after 4 days of culture. The results showed that the growth of WT and Δ*TgRas1* was basically the same under glucose liquid culture conditions. After standing still for 10 min, the liquid was still cloudy, which was due to the floating of hyphae scattered throughout the medium. Under the condition of lactose liquid culture, the growth status of WT was not inhibited, and the medium was always turbid. Interestingly, the Δ*TgRas1* liquid medium became clear after standing for 10 min. Δ*TgRas1* sank to the bottom of the medium, and a small amount of granular Δ*TgRas1* could be clearly observed. This indicated that the hyphae of Δ*TgRas1* might be stunted in growth and more likely to sink to the bottom. Under the glucose carbon source condition, the biomass of Δ*TgRas1* increased by 16.36% compared with WT, which is an interesting phenomenon. However, when lactose was used as the sole carbon source, the biomass of Δ*TgRas1* was severely suppressed, decreasing by 91.43% compared with WT (Figure 8a,b).

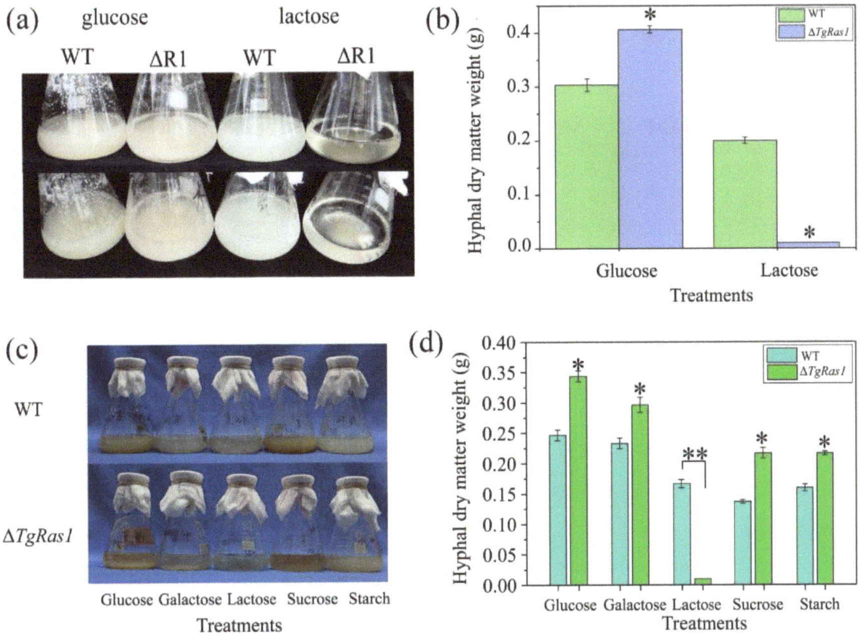

Figure 8. Growth of Δ*TgRas1* strain under different carbon source conditions. (**a**,**b**) Growth of Δ*TgRas1* strain under different carbon source conditions selected according to the results of Biolog-FF plate; (**c**,**d**) growth of Δ*TgRas1* strain under different common carbon source conditions. The "*" shows significant differences separately. Different letters represent significant differences from each other. The "**" represents a very significant difference.

In the process of utilizing carbon sources, fungi may secrete relevant hydrolytic enzymes to decompose carbohydrates. In order to avoid the influence of the type of polysaccharides on the bacteria, we additionally selected common carbohydrates (glucose, galactose, lactose, sucrose, soluble starch) for verification. This cultivation was carried out under liquid conditions. WT was turbid in all media, and mycelia were evenly distributed in the media. For carbon source media that do not affect the growth of Δ*TgRas1*, Δ*TgRas1* will naturally settle to the bottom of the bottle. The growth of Δ*TgRas1* in lactose medium was still inhibited. We further measured the dry weight of mycelia. Δ*TgRas1* exhibited increased biomass under glucose, galactose, sucrose, and starch conditions, which increased by 39.52%, 24.17%, 60.71%, and 36.33%, respectively, compared with WT. Similarly to the previous experimental results, under lactose conditions, the growth of Δ*TgRas1* was significantly inhibited, and its biomass was reduced by 94.12% compared with WT. The reason for the biomass growth of Δ*TgRas1* under other carbon source conditions is unclear. However, it can be clearly shown that only lactose can severely inhibit the growth of Δ*TgRas1*, and *TgRas1* may be involved in the lactose metabolic pathway (Figure 8c,d).

3.5. Growth in Liquid Fermentation and Solid Fermentation with Straw as the Sole Carbon Source

For fungi, liquid culture conditions allow easy access to nutrients for growth. In the above liquid fermentation results, the growth of Δ*TgRas1* under lactose-free conditions was better than that of the wild type, so we further compared the performance of Δ*TgRas1* in liquid fermentation and solid fermentation. The only carbon source in the medium was straw. The results showed that WT could grow normally on straw under solid fermentation, and mycelium extended to the whole surface and crevices of straw, while Δ*TgRas1* only grew on the surface of straw without mycelium growth. Δ*TgRas1* clustered tightly on the straw surface, similarly to the growth form on agar plates (Figure 9a). In liquid culture, the biomass of Δ*TgRas1* was slightly higher than that of WT. However, the growth of Δ*TgRas1* in solid culture was inhibited, and the biomass was 17.91% lower than that of WT (Figure 9b).

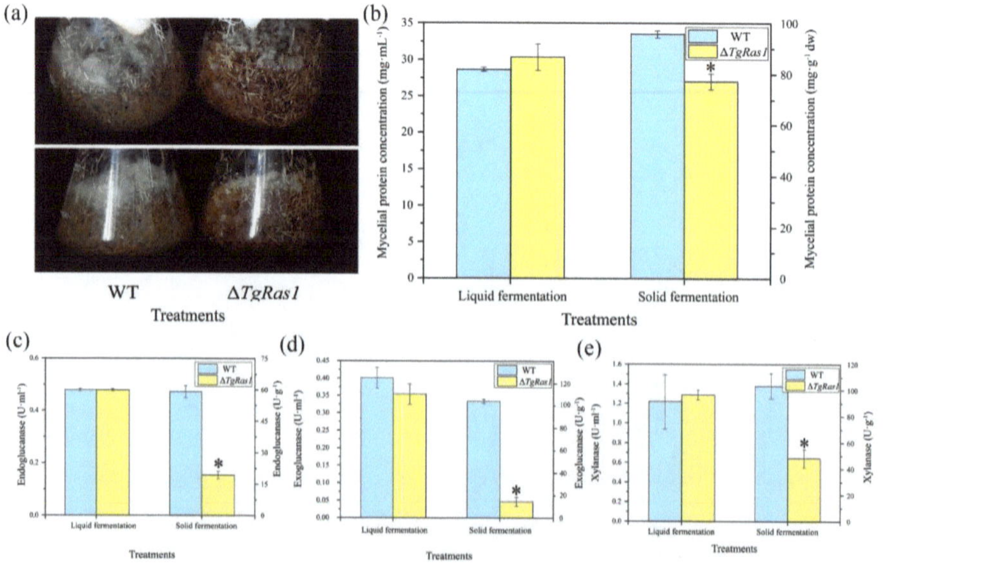

Figure 9. Δ*TgRas1* cultured under liquid and solid conditions with rice straw as the sole carbon source, respectively. (**a**) Growth morphology of bacteria during solid fermentation; (**b**) biomass of Δ*TgRas1* cultured in liquid and solid conditions with rice straw as the sole carbon source; (**c**) en-doglucanase of Δ*TgRas1* cultured under liquid and solid conditions with rice straw as the sole carbon source; (**d**) exoglucanase of Δ*TgRas1* cultured under liquid and solid conditions with rice straw as the sole carbon source; (**e**) xylanase of Δ*TgRas1* cultured under liquid and solid conditions with rice straw as the sole carbon source. The "*" shows significant differences separately. Different letters represent significant differences from each other.

Next, the activities of cellulose-related enzymes were measured. Under liquid fermentation, there was no significant difference in the endoglucanase activities of WT and Δ*TgRas1*. However, under solid fermentation, the enzymatic activity of Δ*TgRas1* was 70.83% lower than that of WT (Figure 9c). Likewise, there was no significant difference in exoglucanase activity between WT and Δ*TgRas1* in liquid fermentation. In solid fermentation, the enzymatic activity of Δ*TgRas1* was 85.45% lower than that of WT (Figure 9d). In the liquid fermentation condition, the xylanase activity of WT and Δ*TgRas1* did not show a significant difference. Under solid fermentation conditions, the enzyme activity of Δ*TgRas1* decreased by 54.29% compared with WT (Figure 9e).

4. Discussion

Most microbial cellulases are inducible hydrolases that are secreted by microorganisms growing on cellulosic substrates [20,21]. Different microorganisms can utilize agricultural residues to produce cellulases and xylanases [22,23]. Straw and bran from cereals, corncobs, and bagasse are the most popular substrates in different regions [24,25]. Enzyme yields can be increased by using a mixture of different substrates instead of a single substrate [26,27]. Lactose is a low-cost additive used to induce fungal secretion of cellulase and xylanase [28]. Lactose has been reported to regulate the secretion of hydrolases from *Trichoderma reesei*, including xyn1, xyn2, cbh1, cbh2, and egl1-encoded xylanase I, xylanase II, cellulosic biohydrolase II, and endoglucan carbohydrase I [29]. Despite their positive impact on cellulase production, the addition of such additives in inappropriate amounts can lead to inhibitory effects. In summary, the secretion of cellulase depends on the medium and culture conditions, and the types of inducers will have different effects.

In almost all eukaryotes, Ras proteins play a key role in sensing and responding to environmental signals [30,31], and the Ras gene is highly conserved in evolution [32]. In particular, the genome of *Saccharomyces cerevisiae* contains highly homologous genes to the mammalian Ras gene, which contains two Ras genes (*Ras1* and *Ras2*), one Rap (Rsr1/Bud1), and one Rheb (Rhb1) direct homologue. It is significantly homologous to the mammalian Ras gene, with *Ras1* and *Ras2* containing 309 and 322 amino acids, respectively, and the first 180 amino acids of these two proteins are nearly 90% homologous [33–35].

There have been few reports of obtaining mutants of deleted genes in filamentous fungi, suggesting that Ras activity in some species is essential [36]. In the reported studies, Ras1 deletion mutants were obtained in *Aspergillus fumigatus*, but there were no successful cases in other filamentous fungi. Ras mutants from other families are easier to obtain [37,38]. In this study, we successfully obtained *TgRas1*-deleted mutants. It was found that the deletion of the *TgRas1* gene affected the polar growth of hyphae and the production and germination of conidia. The marked growth inhibition caused by the Δ*TgRas1* mutant may be due to the critical role of *TgRas1* in maintaining the survival of filamentous fungi. Two proteins in *Saccharomyces cerevisiae*, *Ras1* and *Ras2*, are thought to be involved in a range of fungal physiological processes in an antagonistic or cooperative manner, such as asexual development, polarization morphogenesis, and pathogenesis [39]. Fungal hyphae are usually able to maintain their growth direction for a considerable period of time, and from time to time they establish new growth axes, such as lateral branching. Fungal hyphae typically grow in polarity, and once polarity is established, they spread at an alarming rate.

In fact, conidial germination and mycelial growth of Δ*TgRas*1 were affected to varying degrees on agar solid plates, especially the polar growth of mycelium. Similarly, *RasA1* mutants in yeast severely delay polarity establishment, which is characterized by the hyperswelling of conidia [40]. Previously, the Δ*AoRas2* mutant of *Arthrobotrys oligospora* exhibited hyphal swelling, a phenotype similar to that of Δ*TgRas1* [41]. The abnormal spore characteristic of Δ*TgRas1* is manifested by the swelling of nascent hyphae during germination. These results may imply that the function of *TgRas1* is more focused on the growth and survival of *Trichoderma guizhouense* NJAU4742. Mature hyphae were also found in the Δ*TgRas1* mutant, but the hyphae growth rate was significantly inhibited. The growth speed of the other three TgRas mutants also showed obvious differences.

We observed that Δ*TgRas1* hyphae showed edge swelling during the growth process, and the growth direction of Δ*TgRas1* hyphae was affected. Ras mutants in filamentous fungi such as *Neurospora crassa* and *Schizophyllum* exhibit defects in twisted mycelial growth. In *Neurospora crassa*, there are ropy-like mutants. By destabilizing the apical vesicle supply center, it leads to a deviating growth trajectory [42].

Growth inhibition of the Δ*TgRas1* strain further revealed inappropriate development of conidia structure. TgRas in this study also appears to be involved in a pathway that regulates conidia production. The knockout of the *TgRas1* gene affected the conidial germination and hyphal development of *Trichoderma* and changed the colony morphology. This is similar to previous reports of the effects of Ras on other fungi. For example, disruption of *Ras2* had no effect on the germination pattern of germ tubes in *Fusarium graminearum* but slowed conidia germination [43]. *RasB* knockout mutants in *A. fumigatus* exhibit delayed initiation of germination, but subsequent processes are not observed, in contrast to the slowed germination in the dominant-negative form of *RasA*. Germination was not affected in dominant-active forms of *RasA* or *RasB* [44]. However, in *N. crassa*, the germination rate of Δ*Ras2* did not change significantly [45]. In this study, spore germination was significantly delayed in all four TgRas single-knockout mutants. In Δ*TgRas1* with the most obvious phenotype, not only the germination of spores was delayed but the production of conidia was also greatly delayed. WT will produce green conidia in about 4 days, but the conidia of Δ*TgRas1* will be delayed to about 10 to 14 days. Obviously, the germination of spores and the generation of conidia in *Trichoderma guizhouense* NJAU4742 are positively regulated by TgRas family genes. In addition to being involved in the positive regulation of sporulation, *TgRas1* was also found to alter mycelial branching in *Trichoderma guizhouense* NJAU4742. Two or more germ tubes were observed in Δ*TgRas1*. Fungal germination usually causes germ tubes to sprout from the apical rather than lateral sides of the conidia [46]. However, in our study, multiple germ tubes aberrantly sprouted laterally from the conidia and were dysplastic in Δ*TgRas1*, which shows disordered branching and slow growth. This phenomenon may be due to the instability of the polar axis of cell development. This phenomenon is very similar to the hyperbranched or serrated hyphae observed in mutants of *RasB* of *A. fumigatus*, *Ras2* of *Neurospora crassa*, and Ras of *A. gossypii* [47].

In this study, Δ*TgRas1* was cultured on solid plates with different carbon sources, and a reduction in radial growth, serrated colony edges, and aerial hyphae were observed. This occurs regardless of the type of carbon source. Although the Δ*TgRas1* mutant had reduced radial growth on plates, its growth rate in liquid culture was indistinguishable from its wild type. This situation is similar to that seen in *A. fumigatus* Δ*RasB* mutants. Interestingly, liquid-fermented Δ*TgRas1* showed severe inhibition of mycelial cell growth under a lactose carbon source, but the growth of Δ*TgRas1* under other carbon source liquid fermentation conditions returned to normal levels, and polar growth inhibition was also observed. The above results indicate that *TgRas1* may be related to the lactose metabolism pathway in *Trichoderma guizhouense* NJAU4742. In short, the TgRas family genes are critical to the growth of fungi, and most of the proteins in this family have important effects on cell wall integrity, growth site selection, polarity establishment, and maintenance. At the same time, the results of this study complement the evidence that TgRas family genes play an indispensable role in fungal survival and lactose metabolism.

Supplementary Materials: The following supporting information can be downloaded at: https://www.mdpi.com/article/10.3390/fermentation9050440/s1, Table S1: Primers used to obtain mutant strains.

Author Contributions: Conceptualization, J.M. and D.L.; methodology, J.M. and D.L.; software, C.C. and H.Z.; validation, C.C., Y.G. and J.M.; resources, Q.S.; data curation, Y.G.; writing—original draft preparation, H.G.; writing—review and editing, D.L.; supervision, Q.S.; project administration, D.L. All authors have read and agreed to the published version of the manuscript.

Funding: This research received no external funding.

Institutional Review Board Statement: Not applicable.

Informed Consent Statement: Not applicable.

Data Availability Statement: Not applicable.

Acknowledgments: This research was financially supported by the National Natural Science Foundation of China (31972513 and 32172680) and the Fundamental Research Funds for the Central Universities (KYZ201716). All authors agreed to publish. There is no conflict of interest between the authors.

Conflicts of Interest: The authors declare no conflict of interest.

References

1. Brotman, Y.; Kapuganti, J.G.; Viterbo, A. Trichoderma. *Curr. Biol.* **2010**, *20*, R390–R391. [CrossRef] [PubMed]
2. Contreras-Cornejo, H.A.; Lopez-Bucio, J.S.; Mendez-Bravo, A.; Macias-Rodriguez, L.; Ramos-Vega, M.; Guevara-Garcia, A.A.; Lopez-Bucio, J. Mitogen-Activated Protein Kinase 6 and Ethylene and Auxin Signaling Pathways Are Involved in Arabidopsis Root-System Architecture Alterations by Trichoderma atroviride. *Mol. Plant-Microbe Interact.* **2015**, *28*, 701–710. [CrossRef]
3. Contreras-Cornejo, H.A.; Macias-Rodriguez, L.; Vergara, A.G.; Lopez-Bucio, J. Trichoderma Modulates Stomatal Aperture and Leaf Transpiration Through an Abscisic Acid-Dependent Mechanism in Arabidopsis. *J. Plant Growth Regul.* **2015**, *34*, 425–432. [CrossRef]
4. Zin, N.A.; Badaluddin, N.A. Biological functions of Trichoderma spp. for agriculture applications. *Ann. Agric. Sci.* **2020**, *65*, 168–178. [CrossRef]

5. Peterson, R.; Nevalainen, H. Trichoderma reesei RUT-C30–thirty years of strain improvement. *Microbiology* **2012**, *158*, 58–68. [CrossRef]
6. Li, Q.R.; Tan, P.; Jiang, Y.L.; Hyde, K.D.; McKenzie, E.H.C.; Bahkali, A.H.; Kang, J.C.; Wang, Y. A novel Trichoderma species isolated from soil in Guizhou, T. guizhouense. *Mycol. Prog.* **2013**, *12*, 167–172. [CrossRef]
7. Xu, Y.; Zhang, J.; Shao, J.H.; Feng, H.C.; Zhang, R.F.; Shen, Q.R. Extracellular proteins of Trichoderma guizhouense elicit an immune response in maize (*Zea mays*) plants. *Plant Soil* **2020**, *449*, 133–149. [CrossRef]
8. Saloheimo, M.; Wang, H.; Valkonen, M.; Vasara, T.; Huuskonen, A.; Riikonen, M.; Pakula, T.; Ward, M.; Penttila, M. Characterization of secretory genes ypt1/yptA and nsf1/nsfA from two filamentous fungi: Induction of secretory pathway genes of Trichoderma reesei under secretion stress conditions. *Appl. Environ. Microbiol.* **2004**, *70*, 459–467. [CrossRef]
9. Sternberg, D.; Mandels, G. Induction of cellulolytic enzymes in Trichoderma reesei by sophorose. *J. Bacteriol.* **1979**, *139*, 761–769. [CrossRef]
10. Ivanova, C.; Bååth, J.A.; Seiboth, B.; Kubicek, C.P. Systems analysis of lactose metabolism in Trichoderma reesei identifies a lactose permease that is essential for cellulase induction. *PLoS ONE* **2013**, *8*, e62631. [CrossRef]
11. Porciuncula, J.D.; Furukawa, T.; Shida, Y.; Mori, K.; Kuhara, S.; Morikawa, Y.; Ogasawara, W. Identification of Major Facilitator Transporters Involved in Cellulase Production during Lactose Culture of Trichoderma reesei PC-3-7. *Biosci. Biotechnol. Biochem.* **2013**, *77*, 1014–1022. [CrossRef]
12. Seiboth, B.; Hofmann, G.; Kubicek, C. Lactose metabolism and cellulase production in Hypocrea jecorina: The gal7 gene, encoding galactose-1-phosphate uridylyltransferase, is essential for growth on galactose but not for cellulase induction. *Mol. Genet. Genom.* **2002**, *267*, 124–132. [CrossRef]
13. Karaffa, L.; Fekete, E.; Gamauf, C.; Szentirmai, A.; Kubicek, C.P.; Seiboth, B. d-Galactose induces cellulase gene expression in Hypocrea jecorina at low growth rates. *Microbiology* **2006**, *152*, 1507–1514. [CrossRef]
14. Yaar, L.; Mevarech, M.; Koltint, Y. A Candida albicans RAS-related gene (CaRSRl) is involved in budding, cell morphogenesis and hypha development. *Microbiology* **1997**, *143*, 3033–3044. [CrossRef]
15. Weeks, G.; Spiegelman, G.B. Roles played by Ras subfamily proteins in the cell and developmental biology of microorganisms. *Cell Signal.* **2003**, *15*, 901–909. [CrossRef] [PubMed]
16. Martin-Vicente, A.; Souza, A.C.O.; Al Abdallah, Q.; Ge, W.; Fortwendel, J.R. SH3-class Ras guanine nucleotide exchange factors are essential for Aspergillus fumigatus invasive growth. *Cell Microbiol.* **2019**, *21*, e13013. [CrossRef]
17. Fortwendel, J.R.; Seitz, A.E.; Askew, D.S.; Rhodes, J.C. Aspergillus Fumigatus rasA: A Non-Essential Gene that Regulates Germination, Mitosis and Hyphal Morphology. Ph.D. Thesis, University of Cincinnati, Cincinnati, Ohio, 1999.
18. Zhu, Z.; Ma, G.; Yang, M.; Tan, C.; Yang, G.; Wang, S.; Li, N.; Ge, F.; Wang, S. Ras subfamily GTPases regulate development, aflatoxin biosynthesis and pathogenicity in the fungus Aspergillus flavus. *Environ. Microbiol.* **2021**, *23*, 5334–5348. [CrossRef]
19. Miao, Y.; Xia, Y.; Kong, Y.; Zhu, H.; Mei, H.; Li, P.; Feng, H.; Xun, W.; Xu, Z.; Zhang, N.; et al. Overcoming diverse homologous recombinations and single chimeric guide RNA competitive inhibition enhances Cas9-based cyclical multiple genes coediting in filamentous fungi. *Environ. Microbiol.* **2021**, *23*, 2937–2954. [CrossRef]
20. Kubicek, C.P.; Messner, R.; Gruber, F.; Mach, R.L.; Kubicek-Pranz, E.M. The Trichoderma cellulase regulatory puzzle: From the interior life of a secretory fungus. *Enzym. Microb. Technol.* **1993**, *15*, 90–99. [CrossRef] [PubMed]
21. Mach, R.; Zeilinger, S. Regulation of gene expression in industrial fungi: Trichoderma. *Appl. Microbiol. Biotechnol.* **2003**, *60*, 515–522. [CrossRef]
22. Techapun, C.; Poosaran, N.; Watanabe, M.; Sasaki, K. Thermostable and alkaline-tolerant microbial cellulase-free xylanases produced from agricultural wastes and the properties required for use in pulp bleaching bioprocesses: A review. *Process Biochem.* **2003**, *38*, 1327–1340. [CrossRef]
23. Delabona, P.d.S.; Pirota, R.D.P.B.; Codima, C.A.; Tremacoldi, C.R.; Rodrigues, A.; Farinas, C.S. Using Amazon forest fungi and agricultural residues as a strategy to produce cellulolytic enzymes. *Biomass Bioenergy* **2012**, *37*, 243–250. [CrossRef]
24. Mussatto, S.I.; Ballesteros, L.F.; Martins, S.; Teixeira, J.A. Use of agro-industrial wastes in solid-state fermentation processes. *Ind. Waste* **2012**, *274*. [CrossRef]
25. Santos, F.A.; Carvalho-Gonçalves, L.C.T.d.; Cardoso-Simões, A.L.d.C.; Santos, S.F.d.M. Evaluation of the production of cellulases by Penicillium sp. FSDE15 using corncob and wheat bran as substrates. *Technol. Rep.* **2021**, *14*, 100648. [CrossRef]
26. Adsul, M.G.; Bastawde, K.B.; Varma, A.J.; Gokhale, D.V. Strain improvement of Penicillium janthinellum NCIM 1171 for increased cellulase production. *Bioresour. Technol.* **2007**, *98*, 1467–1473. [CrossRef]
27. Steudler, S.; Werner, A.; Walther, T. It is the mix that matters: Substrate-specific enzyme production from filamentous fungi and bacteria through solid-state fermentation. *Solid State Ferment.* **2019**, *169*, 51–81.
28. Singhania, R.R.; Sukumaran, R.K.; Patel, A.K.; Larroche, C.; Pandey, A. Advancement and comparative profiles in the production technologies using solid-state and submerged fermentation for microbial cellulases. *Enzym. Microb. Technol.* **2010**, *46*, 541–549. [CrossRef]
29. Stricker, A.R.; Steiger, M.G.; Mach, R.L. Xyr1 receives the lactose induction signal and regulates lactose metabolism in Hypocrea jecorina. *FEBS Lett.* **2007**, *581*, 3915–3920. [CrossRef]
30. Fortwendel, J.R.; Juvvadi, P.R.; Rogg, L.E.; Asfaw, Y.G.; Burns, K.A.; Randell, S.H.; Steinbach, W.J. Plasma Membrane Localization Is Required for RasA-Mediated Polarized Morphogenesis and Virulence of Aspergillus fumigatus. *Eukaryot. Cell* **2012**, *11*, 966–977. [CrossRef]

31. Arkowitz, R.A.; Bassilana, M. Regulation of hyphal morphogenesis by Ras and Rho small. GTPases. *Fungal Biol. Rev.* **2015**, *29*, 7–19. [CrossRef]
32. Yang, L.; Li, X.; Xie, M.; Bai, N.; Yang, J.; Jiang, K.; Zhang, K.-Q.; Yang, J. Pleiotropic roles of Ras GTPases in the nematode-trapping fungus Arthrobotrys oligospora identified through multi-omics analyses. *iScience* **2021**, *24*, 102820. [CrossRef] [PubMed]
33. Casas-Flores, S.; Rios-Momberg, M.; Rosales-Saavedra, T.; Martinez-Hernandez, P.; Olmedo-Monfil, V.; Herrera-Estrella, A. Cross talk between a fungal blue-light perception system and the cyclic AMP signaling pathway. *Eukaryot. Cell* **2006**, *5*, 499–506. [CrossRef] [PubMed]
34. Tamanoi, F. Ras Signaling in Yeast. *Genes Cancer* **2011**, *2*, 210–215. [CrossRef]
35. Knapp, G.S.; McDonough, K.A. Cyclic AMP Signaling in Mycobacteria. *Microbiol. Spectr.* **2014**, *2*, 2. [CrossRef]
36. Nogueira, K.M.V.; Costa, M.d.N.; de Paula, R.G.; Mendonça-Natividade, F.C.; Ricci-Azevedo, R.; Silva, R.N. Evidence of cAMP involvement in cellobiohydrolase expression and secretion by Trichoderma reesei in presence of the inducer sophorose. *BMC Microbiol.* **2015**, *15*, 195. [CrossRef]
37. Som, T.; Kolaparthi, V.S. Developmental decisions in Aspergillus nidulans are modulated by Ras activity. *Mol. Cell Biol.* **1994**, *14*, 5333–5348. [CrossRef]
38. Boyce, K.J.; Hynes, M.J.; Andrianopoulos, A. The Ras and Rho GTPases genetically interact to co-ordinately regulate cell polarity during development in Penicillium marneffei. *Mol. Microbiol.* **2005**, *55*, 1487–1501. [CrossRef] [PubMed]
39. Lengeler Klaus, B.; Davidson Robert, C.; D'Souza, C.; Harashima, T.; Shen, W.-C.; Wang, P.; Pan, X.; Waugh, M.; Heitman, J. Signal Transduction Cascades Regulating Fungal Development and Virulence. *Microbiol. Mol. Biol. Rev.* **2000**, *64*, 746–785. [CrossRef]
40. Brown, N.A.; Ries, L.N.A.; Goldman, G.H. How nutritional status signalling coordinates metabolism and lignocellulolytic enzyme secretion. *Fungal Genet. Biol.* **2014**, *72*, 48–63. [CrossRef]
41. Schuster, A.; Tisch, D.; Seidl-Seiboth, V.; Kubicek, C.P.; Schmoll, M. Roles of Protein Kinase A and Adenylate Cyclase in Light-Modulated Cellulase Regulation in Trichoderma reesei. *Appl. Environ. Microbiol.* **2012**, *78*, 2168–2178. [CrossRef] [PubMed]
42. Riquelme, M.; Roberson, R.W.; McDaniel, D.P.; Bartnicki-García, S. The effects of ropy-1 mutation on cytoplasmic organization and intracellular motility in mature hyphae of Neurospora crassa. *Fungal Genet. Biol.* **2002**, *37*, 171–179. [CrossRef]
43. Bluhm, B.H.; Zhao, X.; Flaherty, J.E.; Xu, J.R.; Dunkle, L.D. RAS2 Regulates Growth and Pathogenesis in Fusarium graminearum. *Mol. Plant-Microbe Interactions®* **2007**, *20*, 627–636. [CrossRef] [PubMed]
44. Fortwendel, J.R.; Panepinto, J.C.; Seitz, A.E.; Askew, D.S.; Rhodes, J.C. Aspergillus fumigatus rasA and rasB regulate the timing and morphology of asexual development. *Fungal Genet. Biol.* **2004**, *41*, 129–139. [CrossRef] [PubMed]
45. Kana-uchi, A.; Yamashiro, C.T.; Tanabe, S.; Murayama, T. A ras homologue of Neurospora crassa regulates morphology. *Mol. Gen. Genet. MGG* **1997**, *254*, 427–432. [CrossRef]
46. D'Enfert, C. Fungal Spore Germination: Insights from the Molecular Genetics ofAspergillus nidulansandNeurospora crassa. *Fungal Genet. Biol.* **1997**, *21*, 163–172. [CrossRef]
47. Bauer, Y.; Knechtle, P.; Wendland, J.; Helfer, H.; Philippsen, P. A Ras-like GTPase is involved in hyphal growth guidance in the filamentous fungus Ashbya gossypii. *Mol. Biol. Cell* **2004**, *15*, 4622–4632. [CrossRef]

fermentation

MDPI

Article

Multi-Response Optimization of Thermochemical Pretreatment of Soybean Hulls for 2G-Bioethanol Production

Martín Gil Rolón, Rodrigo J. Leonardi, Bruna C. Bolzico, Lisandro G. Seluy, Maria T. Benzzo and Raúl N. Comelli *

Grupo de Procesos Biológicos en Ingeniería Ambiental (GPBIA), Facultad de Ingeniería y Ciencias Hídricas (FICH), Universidad Nacional del Litoral (UNL), Consejo Nacional de Investigaciones Científicas y Técnicas (CONICET), Ciudad Universitaria CC 242 Paraje El Pozo, Santa Fe 3000, Argentina

* Correspondence: rcomelli@fich.unl.edu.ar

Abstract: Soybean is a major crop used in the production of human food. The soybean hull (SH) is also known as the seed coat and it constitutes about 5–8% of the total seed on a dry weight basis, depending on the variety and the seed size. Dilute sulfuric acid was employed for the thermochemical pretreatment of SH prior to enzymatic saccharification and alcoholic fermentation. Empirical modeling of response surface, severity factor and multi-response desirability function methodology, were used to perform the process optimization. Temperature, acid concentration and reaction time were defined as operational variables, while furfural, 5-hydroxymethylfurfural and solubilized hemicellulose and cellulose were defined as response variables. Mathematical models satisfactorily described the process and optimal conditions were found at 121 °C; 2.5% w/v H_2SO_4 and 60 min. More than 80% and 90% of hemicelluloses and celluloses, respectively, were able to solubilize at this point. The fermentation performance of an industrial *Saccharomyces cerevisiae* strain was also evaluated. The glucose available in the hydrolysates was completely consumed in less than 12 h, with an average ethanol yield of 0.45 $g_{ethanol}/g_{glucose}$. Thus, the thermochemical conditioning of SH with dilute sulfuric acid is a suitable operation for 2G-bioethanol production.

Keywords: soybean hull; pretreatment; bioethanol; multi-response optimization; severity factor

Citation: Gil Rolón, M.; Leonardi, R.J.; Bolzico, B.C.; Seluy, L.G.; Benzzo, M.T.; Comelli, R.N. Multi-Response Optimization of Thermochemical Pretreatment of Soybean Hulls for 2G-Bioethanol Production. *Fermentation* **2023**, *9*, 454. https://doi.org/10.3390/fermentation9050454

Academic Editors: Miguel Ladero and Victoria E. Santos

Received: 31 March 2023
Revised: 5 May 2023
Accepted: 5 May 2023
Published: 10 May 2023

1. Introduction

The transportation sector is an important industrial player in the worldwide economy, which deals with the movement of people and products. This is one of the main energy-claimants: it consumes more than a quarter of the energy produced in the world and also emits high levels of CO_2 [1,2]. It is necessary to transform the transport system towards a more sustainable model, which allows the reduction of greenhouse gasses in the atmosphere. One of the main strategies considered to achieve this goal is the replacement of fossil fuels with biofuels.

Bioethanol is one of the most important renewable fuels. It is added to gasoline to reduce the negative environmental impact generated by the worldwide use of fossil fuels. Several energy crops including sugarcane, corn and jatropha are used as raw materials for bioethanol production [1]. The high worldwide bioethanol demand exerts enormous pressure on primary production capacity. Thus, it is imperative to identify new renewable sources for the production of this 'green' fuel. Several alternatives have received increased focus, such as lignocellulosic biomass [2] and agro-industrial wastewaters [3]. Lignocellulosic biorefineries can be applied to the production of second generation (2G) biofuels, including bioethanol, as well as a number of high value products [4–6].

Soybean (*Glycine max* L.) is one of the most important crops for the production of human food. The United States, Brazil, Argentina, China and India are the top-five producers, whereas Argentina is the world's top exporter of flour and oils. The processing of soybean grains results in two main products: soybean meal and soybean oil (also used for biodiesel production). Thus, soybean processing generates different by-products, such as hulls, molasses, and okara. Soybean hulls (SHs) are the main residue of soybean processing, representing about 5–8% of the whole grain dry matter, and containing about 85% of complex carbohydrates [7]. Because the SH has low weight, large storage areas are necessary to manage this waste. This complicates and makes SH handling more expensive. Thus, the bioconversion of this waste in added-value compounds emerges as a promising alternative to valorize a waste with a circular economy approach [8].

Because SHs are essentially lignocellulosic residues, the polymers cellulose, hemicellulose and lignin are the core components of them. *Cellulose* is a polysaccharide consisting of a linear chain of D-glucose units, whereas *hemicellulose* contains pentoses, hexoses and/or uronic acids. Unlike cellulose, hemicellulose composition varies depending on cell tissue and plant species. The low lignin content (6–8%) of the SH is noteworthy, which makes it a golden waste [9,10].

In general, a process to produce bioethanol from lignocellulosic biomass (LB) comprises four sequential stages: (1) LB pretreatment; (2) saccharification, i.e., enzymatic hydrolysis of cellulose and hemicellulose; (3) fermentation and (4) ethanol distillation and then rectification to achieve fuel specifications. The major drawbacks for the use of LB are: (i) the recalcitrance to degradation of the lignocellulosic matrix, which is comprised of covalent and hydrogen bonds; and (ii) the poor utilization by *Saccharomyces cerevisiae* of D-xylose and L-arabinose, which are the main pentoses present in hemicellulose polymers. This work is focused on the first step of the process: *the pretreatment operations*.

Pretreatment is a crucial step because it breaks down the complex chemical structures of the LB, making cellulose more accessible to subsequent enzymatic hydrolysis. This stage has a significant influence on the overall process economy since it represents up to 35% of the total cost for bioethanol production [11]. The development of efficient, robust, easy-to-upscale and environmentally friendly pretreatment techniques is crucial in order to achieve competitive biorefineries based on LB. The selection of a pretreatment method depends on the chemical composition and physical nature of the material. For this reason, an ideal pretreatment must: consider the sugar loss from pretreated fractions; reduce the production of inhibitory compounds for enzymatic hydrolysis and fermentation; minimize enzyme loading for efficient hydrolysis; decrease energy consumption; and permit the delignification and recovery of other promising compounds. Currently, the thermochemical conditioning processes are the preferred ones for large scale application [12]. Dilute acid hydrolysis was extensively investigated as a pretreatment alternative for LB conversion, with a focus on the use of carbohydrates for bioethanol production [13–18]. This method aims to break and solubilize the hemicelluloses and lignin, allowing cellulose to be more accessible to enzymatic attack.

In this work, the thermochemical pretreatment of SH with dilute sulfuric acid was evaluated. The hydrolysis was performed in batch-type reactors and a multi-response optimization methodology based on a desirability function was also applied. The aim of the factorial-type experimental design was to identify the relationship between the main reaction conditions (temperature, acid concentration and reaction time) and the hemicellulose rupture and solubilization, as well as furans production. The effect on cellulose accessibility to subsequent enzymatic hydrolysis was also considered. Finally, several fermentation tests were performed with an industrial strain of *Saccharomyces cerevisiae* to evaluate the potential inhibitory compounds released.

2. Materials and Methods

2.1. Characterization of SH and Hydrolysates

The SHs were provided by a local soybean processing company (Vicentin S.A.I.C., https://www.vicentin.com.ar, accessed on 13 April 2023). The material was ground and sieved to obtain particles with a size in the range between 1 to 2 mm. The ash and moisture content of SHs were determined following the LAP-NREL/TP-510-42622 procedure reported by the National Renewable Energy Laboratory (NREL) [19].

The carbohydrate content (cellulose and hemicellulose) was measured according to the LAP-NREL/TP-510-42618 protocol [20] with some modifications: a two-step acid hydrolysis was performed using sulfuric acid. Firstly, 3.00 ± 0.01 mL of 72% *v*/*v* H_2SO_4 was added to 300.0 ± 10.0 mg of ground SHs and incubated at 30 °C for 1 h. Then, distilled water was added to reach a 4% *v*/*v* H_2SO_4 solution and it was autoclaved at 121 °C for 1 h. A sample of the liquid phase of the hydrolysate was stored at −20° C for further determinations of soluble hydrolysis products, including pentoses, glucose, furans and acid-soluble lignin. The pH of the remaining hydrolysate was adjusted to 5.0 ± 0.2 using NaOH beads and a commercial enzyme mixture with cellulase activity (Novozyme, Cellic CTec2®) was added following the supplier's recommendations for the dosage (mass of enzymes: mass of solids ratio). The reaction was kept at 50 °C for 72 h. Glucose was measured using an enzymatic kit (SB Lab., Santa Fe, Argentina), whereas the remaining solid fraction in the hydrolysate was employed for acid-insoluble lignin determination. HMF and furfural were quantified by HPLC-UV (Thermo UltiMate 3000) using a C18 ODS Hypersil column, water:acetonitrile (80:20) as mobile phase and detection at 260 nm. The identity of the main monosaccharides released from SHs were evaluated by HPLC-RI using a Hypersil™ APS-2 (250 × 4.6 mm, 5 µm) column and water:acetonitrile (20:80) as mobile phase.

Hydrolysate samples were centrifuged for 10 min at 10,000 rpm and filtered through a 0.45 µm nylon membrane. C_5 sugars were estimated by a colorimetric method using hydrochloric acid, acetic acid and phloroglucinol [21]. A modified Klasson methodology was followed to determine the lignin content [22]. This procedure was selected because it considered the influence of the furans (HMF and furfural) in the medium. Ethanol was quantified by gas chromatography-flame ionization detection (GC–FID, SHIMADZU GC Solution). A 30 m Agilent J&W DB-1 column was used. A sample volume of 1 µL was injected and butanol was used as an internal standard.

2.2. Experimental Design

A Central Composite Design (CCD) that included the time, temperature and sulfuric acid concentration as independent variables (factors) was performed to determine the optimal relationship of these factors. The axial distance (α value) was chosen to be 1.287. For predicting the optimal point, a second-order polynomial function was fitted to correlate the relationship between the independent variables and the response. The values for each factor are shown in Table 1. The experiment was conducted in three blocks, with each condition being studied in triplicate, and the order of the experiments was completely randomized. In total, 48 experimental runs were performed, including 2 central points per block, resulting in 36 degrees of freedom for error estimation. Factor combinations for pretreatment are detailed in Table 2. The reaction time is defined as the period of time in which the temperature is maintained constant, considering the zero time (t = 0) to be the instant where the desired temperature is reached.

Table 1. Factors and higher and lower values defined.

Factors	Units	Lower Value	Higher Value
Time	min	30	180
Temperature	°C	80	125
H_2SO_4 concentration	% *w*/*v*	0.5	2.5

Table 2. Experimental design.

Treatment	Time (min)	Temperature (°C)	H_2SO_4 (% *w*/*v*)
1	105	103	2.79
2	105	103	0.21
3	9	103	1.50
4	30	80	0.50
5	202	103	1.50
6	180	80	2.50
7 [1]	105	103	1.50
8	105	131	1.50
9	180	125	2.50
10	180	125	0.50
11	180	80	0.50
12	105	74	1.50
13	30	80	2.50
14	30	125	0.50
15	30	125	2.50

[1] The central point of each block was performed by duplicate, so this condition was evaluated in sextuplicate.

All analysis of statistical experimental designs and results was performed using Statgraphics Centurion XV v15.2.06 (StatPoint Inc., Suite, VA, USA), R Studio (R Core Team, Vienna, Austria), MATLAB (MathWorks Inc., Natick, MA, USA). The statistical analysis was performed in the form of analysis of variance (ANOVA). The significance of the regression coefficients and the associated probabilities $p(t)$ were determined by Student's t-test. The variance explained by the model is given by the multiple determination coefficients, R^2, and the parameter values should be close to 1.0 for a good statistical model, while a value above 0.75 indicates the suitability of the model.

The percentage of hemicelluloses solubilization (%HC) was defined as follows:

$$\%HC = \frac{C_{pentoses}}{C_{solids} * X_{Hemicelluloses}} * 100 \quad (1)$$

where, $C_{pentoses}$ is the concentration of pentoses in the liquid phase of the hydrolysate (g/L), C_{solids} is the concentration of SHs in the reactor (g/L) and $X_{hemicelulloses}$ is the fraction of hemicellulose in dry SHs.

Acid hydrolysis was carried out in borosilicate glass tubes with screw caps (HACH, Loveland, CO, USA) and using a dry block with heating system for temperature control. The reaction volume was 10 mL, with 150 g_{SH}/L.

2.3. *Enzymatic Saccharification*

After thermochemical pretreatment, the pH of the hydrolysates was adjusted to 5.0 ± 0.2 by adding NaOH beads. Chloramphenicol at final concentration of 2.5 ppm was also incorporated to avoid bacterial proliferation. Then, a cellulase commercial enzyme (Novozyme, Cellic CTec2®) was added as aforementioned. Reactors were kept under orbital shaking at 50 °C for 72 h. Samples of the liquid fraction were taken at regular intervals and the glucose concentration was determined. The cellulose conversion percentage was calculated as follows:

$$\%CEL = \frac{C_{glucose}}{C_{solids} * X_{celullose}} * 100 \quad (2)$$

where, $C_{glucose}$ is the glucose concentration (g/L), C_{solids} is the concentration of solids (g/L) and $X_{cellulose}$ is the fraction of cellulose in the dry material.

2.4. Fermentation

Fermentations were performed in triplicate using a 500-mL glass reactor (300-mL working volume) operated in batch mode under anaerobic conditions and at a constant temperature of 30 °C, as described previously [23]. The pH was initially adjusted to 4.50 ± 0.10, and orbital shaking (150 rpm) was maintained throughout the experiments to avoid biomass precipitation. The initial concentration of yeast in each experiment was 1.00 ± 0.10 g/L. The samples were collected immediately after inoculation (t = 0) and every 2 h until the end of the experiments. During the fermentation experiments, samples (1 mL) were taken in duplicate and immediately centrifuged for 5 min at 1200× *g*. The initial supernatants were transferred to sterile 1.5 mL tubes and stored at −20 °C until the corresponding determinations.

The industrial strain *Saccharomyces cerevisiae* var. Ethanol red (Fermentis Ltd., Lille, France) was selected for fermentation experiments. The performance of this strain on control media and sugar-sweetened high-strength wastewaters from the soft drinks industry was previously studied [24]. Yeast strain was maintained at 4 °C on YPG solid medium (yeast extract 5 g/L, peptone 3 g/L, D-glucose 20 g/L and agar 15 g/L). Pre-culture was obtained by propagation of colonies at 30 °C overnight in a shaken Erlenmeyer flask with liquid YPG. The cells had been previously grown at 30 °C for 12–18 h and were then harvested by centrifugation for 5 min at 4500 rpm, washed twice using distilled water and finally resuspended in a small volume of water before the reactors' inoculation. A biomass concentrated suspension was used to inoculate the fermentation reactors to reach an initial concentration of 1 g/L. The biomass concentration was indirectly determined by turbidity measurements at 600 nm using a VIS spectrophotometer (DR/2010, HACH, Loveland, CO, USA). These measurements were correlated to biomass concentration using a calibration curve built according to the standard technique for determination of Volatile Suspended Solids (VSS) [25]. To build the calibration curve, the yeasts were grown on YPD medium at 30 °C for 12–18 h and were then harvested by centrifugation for 5 min at 1200× *g*, then washed five times using distilled water. Several dilutions of distilled water were made by triplicate. An aliquot of each diluted sample was used for measure of turbidity (at 600 nm) using distilled water as blank. Another aliquot of the well-mixed sample (50-mL) was filtered in vacuum through a weighted standard Whatman GFC glass fiber filter (47 mm diameter and 1.2 mm nominal pore size, Biopore, Buenos Aires, Argentina) and the residue retained on the filter was dried to a constant weight at 103–105 °C. The increase in weight of the filter represents the total suspended solids (TSS). The next step was the combustion of the filter at 500 °C for 15 min and the weight lost after combustion represents the weight of Volatile Suspended Solids (VSS) in the sample [23].

SH hydrolysate was fermented in a batch mode at a constant temperature of 30 °C and stirring at 150 rpm. Previous yeast adaptation to the SH hydrolysate was not required. Neither nitrogen source nor additional nutrients (salts or vitamins) were added to the fermentation media. For comparative purposes, the yeast fermentative performance was also evaluated in a synthetic medium containing D-glucose 60 g/L and commercial yeast extract 5 g/L under the same experimental conditions. Duplicate fermentation experiments were conducted for each medium. Finally, the performance of yeasts was evaluated using several parameters defined in a previous work [26].

3. Results and Discussion

3.1. Characterization of Soybean Hull

Variation in the chemical composition of SHs usually depends on the cultivation characteristics, influence of environmental factors (soil and climate), type of grain and genetic factors, among others [27]. The composition of the employed SHs, determined on a dry basis, consisted of 45.0 ± 1.21% cellulose, 15.1 ± 0.92% hemicelluloses, 4.0 ± 1.03%

lignins, and 2.0 ± 0.13% ashes (Table 3). These results are in agreement with reported ones by different authors [23–28] and support the idea that SHs contain relatively high amounts of carbohydrates (cellulose and hemicellulose) and low lignin content. Thus, the SH is an interesting platform for the development of biorefinery strategies [8], mainly those based on bioprocesses. Because of its potential to obtain fermentable sugars with low generation of lignin degradation compounds, which can be toxic to microbial metabolism, the SH has a high potential as a feedstock for the production of added-value compounds applying biological processes.

Table 3. Soybean hull compositions.

Cellulose (%)	Hemicellulose (%)	Lignin (%)	Ash (%)	Reference
51.2	15.9	1.5–4.2	ND [1]	[28] [2]
36.4	12.5	18.2	4.2	[29]
39.7	25.5	9.1	NI	[9]
39.0	24.0	12.0	NI	[30]
52.3	18.5	3.7	2.5	[31]
40.6	33.8	7.8	4.2	[32]
45.0	15.1	4.0	2.0	This work

[1] ND: Not determined. [2] Estimates by Van Soest method.

3.2. *SH loading Selection for Pretreatment*

The solid loading is an operational variable that commonly has a significant impact on the efficiency of thermochemical pretreatment and enzymatic hydrolysis processes [33–35]. High solid loadings could increase the availability of fermentable sugar for the subsequent fermentation stage, and contribute to decrease the cost of the recovery of ethanol by distillation. However, a high solid concentration is also associated to challenges such as (i) slurry's high viscosity, (ii) limitation in the effectiveness of both the pretreatment and the enzymatic hydrolysis, (iii) improperly enzyme distribution, (iv) sub-optimal solubilization of carbohydrates, and (v) ineffective mixing of slurry and heat/mass transfer. These topics could lead to an increase in consumption of energy due to additional operations needed [33,36,37].

Since the initial solid content can have an impact on carbohydrate recovery, the selection of an appropriate loading is a crucial factor to consider before commencing the process. Then, a thermochemical hydrolysis *screening* was performed using 50, 100, 150 and 200 g_{SH}/L. At the highest loading, the solids completely absorbed the liquid, resulting in a mixture without a free aqueous phase (not shown). Because it was difficult to mix and manage the solids at this loading, it was discarded for future experiments. In the remaining reactors, the proportion of pentoses in the liquid phase was evaluated after chemical hydrolysis. Since the measurement was statistically similar for all three conditions (50, 100 and 150 g_{SH}/L), the loading with the highest initial solid content was selected for further experiments. Thus, the study of the impact of the pretreatment conditions on carbohydrate recovery was performed at 150 g_{SH} /L (dry basis). The best initial loading is out of scope of this work and this topic may be addressed in future works.

3.3. *Relation between Pretreatment and the Hemicellulose Solubilization*

The acid-catalyzed hydrolysis leads to breaking of intermolecular and intramolecular bonds, releasing oligosaccharides and soluble monomers, primarily pentoses, from the hemicellulose. A series of reactions on both the hemicellulose and their hydrolysis products also take place at high temperature and acidic conditions. These reactions include dehydration, oxidation, isomerization, decarboxylation and re-hydration, among others. Thus, in order to study the effect of pretreatment conditions on hemicellulose, the release soluble pentoses were measured, as well as the presence of furfural, the main product of the degradation of C_5 sugars at high acidic and temperature conditions. Xylose and arabinose are the main pentoses present in the hemicellulose structure. Partial hydrolysis of hemicellulose to pentose oligomers like arabinoxylan-oligosaccharides could also occur. Both monosaccharides and oligomers are soluble products of the hydrolysis of hemicellulose. Thus, an encompassing variable named generically "pentoses" can be taken as a direct indicator of the effectiveness of chemical hydrolysis; when more pentoses are present in the liquid phase, more hemicellulose has been solubilized.

The conditions evaluated for thermo-chemical hydrolysis of the SHs were listed in Table 2. The solubilization of hemicellulose (%HC) varied between 1.6% and 87.4%, whereas three different conditions reached mean effectiveness above 80%: (i) 81.17 ± 1.99% for treatment 15 (2.50% *w*/*v* H_2SO_4; 125 °C; 30 min); (ii) 83.42 ± 3.78% for treatment 8 (1.50% *w*/*v* H_2SO_4; 131 °C; 105 min); and (iii) 84.34 ± 0.60% for treatment 9 (2.50% *w*/*v* H_2SO_4; 125 °C; 180 min). Analysis of variance of these treatments indicated that they were not statistically different. (p-value > 0.35; alpha = 0.05). These results are slightly higher than others informed by similar studies that use dilute sulfuric acid to pretreat SHs. A solubilization close to 76% was reported at 1.4% *v*/*v* H_2SO_4, 125 °C and 60 min of hydrolysis [38]. A level of 67% of solubilization was observed at 2.2% *w*/*w* H_2SO_4/135 °C/40 min), whereas more severe conditions (1.7% *w*/*w* H_2SO_4; 153 °C; 60 min) lead to a 77% of hemicellulose solubilization [9]. The highest values in literature fall into the range of 90–96% of pentoses and it was obtained after steam explosion pretreatment with dilute sulfuric acid (2% *w*/*w* H_2SO_4; 140 °C; 30 min) [29].

In order to search the optimal pretreatment operative conditions, a linear regression analysis method was applied. The statistical approach of experimental design allowed data acquisition in proper quantity and quality for the built and evaluation of multiple-regression mathematical models. A quadratic polynomial type model, with interactions of three factors (Time, Temperature and Sulfuric acid concentration) was initially built and evaluated to predict the degree of hemicellulose solubilization (Equation (3)).

$$Y = \beta_0 + \beta_1 T + \beta_2 t + \beta_3 A + \beta_{11} T^2 + \beta_{22} t^2 + \beta_{33} A^2 + \beta_{12} Tt + \beta_{13} TA + \beta_{23} tA, \quad (3)$$

where, Y is the percentage of solubilized hemicellulose (%); βi and βij are regression coefficients, whereas A, T and t are the Sulfuric acid concentration (% *w*/*v*); the temperature (°C) and the reaction time (min), respectively.

Statistical evaluation of the polynomial model was performed through t test, ANOVA and AIC (Akaike Information Criterion), choosing a significance level of 95%. Analysis showed that the interaction terms, those involving the time factor and the square of the temperature, were not significant enough to explain response variations (β_2, β_{11}, β_{22}, β_{12}, β_{13}, β_{23} = 0). These terms were eliminated one by one and the resulting models were re-evaluated after each elimination round. The regression analysis of the data showed that a second order linear model (Equation (4)) was a suitable model to describe the hemicellulose solubilization as a function of temperature (T) and sulfuric acid concentration (A), with adjusted R^2 value of 0.97.

$$\%HC = -81.67 + 0.76\ T + 53.87\ A - 10.07\ A^2,\ R^2 = 0.97 \quad (4)$$

The surface response for different temperatures and acid concentrations is shown in Figure 1.

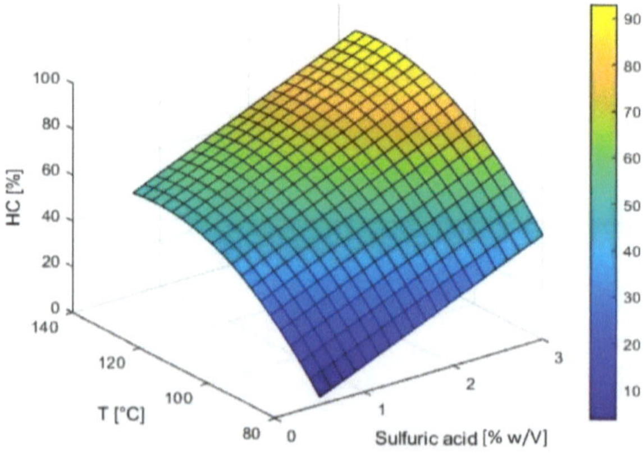

Figure 1. Response surface showing %HC after acid pretreatment.

As the model shows, in the upper limits of the space studied, the predicted and observed values correspond to 80–90 of %HC. Considering that the goal of the treatment with diluted acids is the solubilization of hemicellulose, the results were close to maximum effectiveness conditions. Sulfuric acid concentration and Temperature were the most important factors that had a significant impact on solubilization of the hemicellulose. The interaction between the temperature and the sulfuric acid concentration was particularly interesting (Figure 2). These findings may be mainly due to the range of conditions tested. A wider range of conditions (100–170 °C and 1.4–3.2% w/w H_2SO_4) lead to obtaining response surface models for the release of xylose and arabinose, in which, a significant interaction between temperature and sulfuric acid concentration can be observed [9]. In the range of conditions evaluated in this work, the surfaces resemble planes, presenting curvatures in higher values, and being influenced primarily by the temperature.

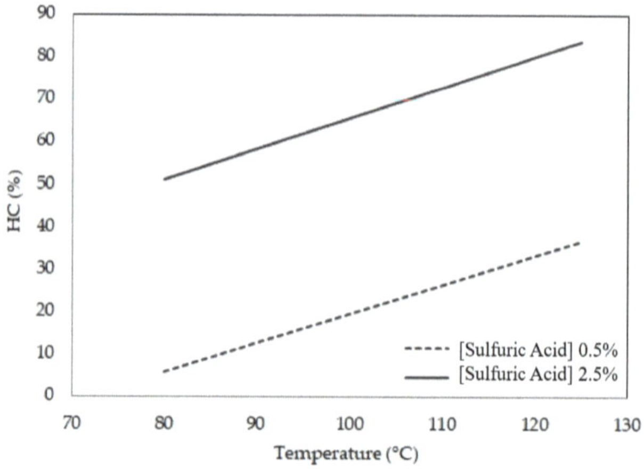

Figure 2. Effects graph for Temperature and Sulfuric acid concentration over %HC. Parallel lines show that no interaction exists.

Figure 3 summarizes the effects of the factors studied: bars correspond to the average values of %HC reached at the corner points of the factorial design. It can be seen that at low and high temperature values, 80 °C and 125 °C, respectively, the %HC increased significantly as the acid concentration increased and the reaction time was maintained constant. On the other hand, non-significant changes can be observed by increasing reaction time and leaving the other variables fixed. This suggests that a significant fraction of the hemicellulose was solubilized relatively quickly and/or during the transient state (heating stage). Similar observations were reported for the hydrolysis of *Pinus taeda* with dilute sulfuric acid at pH = 1.65 and 150 °C; a significant amount of hemicellulose was also solubilized during the heating of the reactor [39].

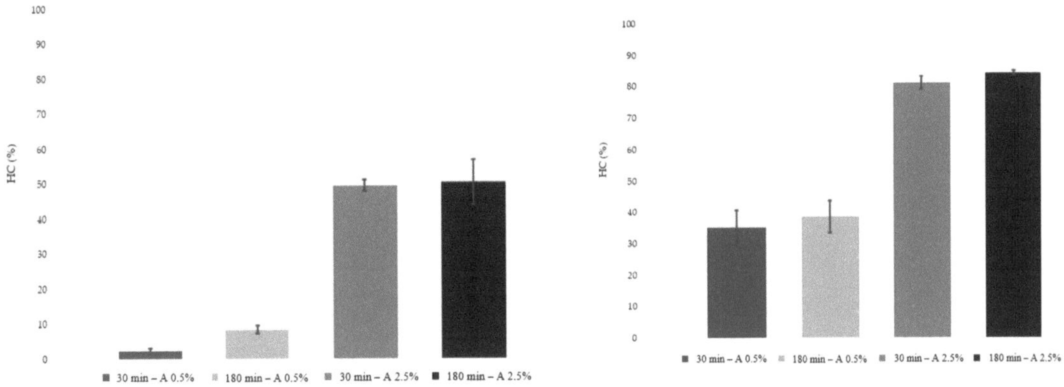

(a) (b)

Figure 3. Bar graph for %HC at T = 80 °C (**a**) and T = 125 °C (**b**). No significant differences were observed when reaction time increased from 30 min to 180 min.

3.4. Furans Formation

There are two main reasons why monitoring the concentrations of HMF and furfural is relevant: *(i)* because they are the products of degradation of hexoses and pentoses, respectively, their presence implies the loss of fermentable sugars, and *(ii)* depending on their concentration levels, these compounds can act as an inhibitor to the metabolism of a wide range of fermenting microorganisms [40–44].

Figure 4 shows the production of furans in all the evaluated treatments. It can be seen that the more severe the treatment, the more furans are produced. As expected, the generation of furfural is always higher than HMF, suggesting that pretreatment mainly produces the hydrolysis of hemicellulose, i.e., the source of pentoses. The maximal production of furfural and HMF were recorded in treatment 9 (2.50% w/v H_2SO_4; 125 °C; 180 min). In these conditions, an average of 392.6 $mg_{Furfural}/L$ and 75.4 mg_{HMF}/L were reached. Additionally, treatment 8 (1.50% w/v H_2SO_4; 131 °C; 105 min) accumulated an average of 229.5 $mg_{Furfural}/L$ and 63.5 mg_{HMF}/L, suggesting an interaction between the factors. ANOVA showed that the productions of furfural and HMF in treatment 9 are significantly higher than treatment 8 (p-value $< 0.4 \times 10^{-3}$ and p-value $< 0.8 \times 10^{-3}$). An average of 60.7 $mg_{Furfural}/L$ and 38.2 mg_{HMF}/L were produced in treatment 15 (2.50% w/v H_2SO_4, 125 °C, 30 min), which ranked third. However, the furfural concentrations in treatment 15 were significantly lower than treatments 8 and 9, but were not significantly superior to any other, whereas the HMF concentrations were significantly lower with respect to treatments 8 and 9, but significantly superior to the rest (p-valor $< 0.2 \times 10^{-4}$).

The furan concentrations presented in this section were in agreement with those reported in the literature. In particular, values of 190 and 280 $mg_{Furfural}/L$ were reported in SH hydrolysates treated 60 min with sulfuric acid (1.4% v/v) at 120 °C and 125 °C, respectively [38].

By applying multiple regression analysis, a mathematical relationship between the furans production and the pretreatment conditions was obtained. Both models are exponential-type and show effects of the three factors and their interactions, satisfactorily representing the data dispersion in the range of study (Equations (5) and (6)).

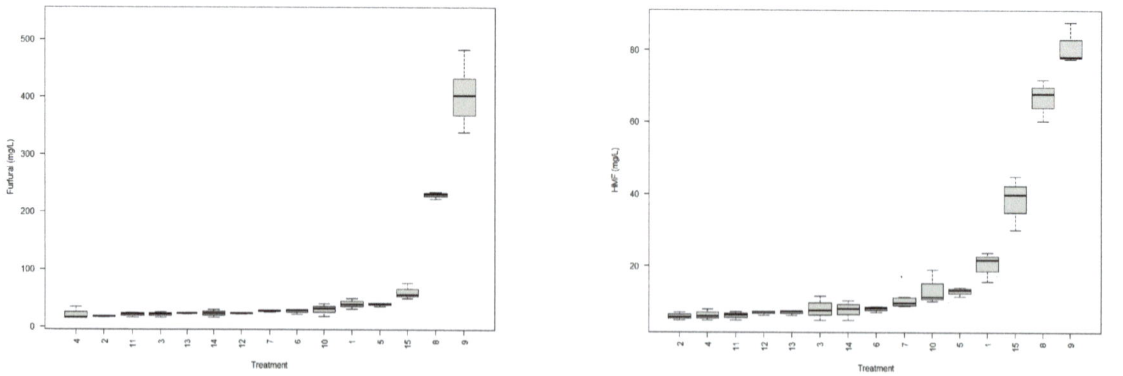

(a) **(b)**

Figure 4. Boxplot that shows the furfural (**a**) and HMF (**b**) production by treatment. The treatment sequence is running from low to high concentrations.

$$\mathrm{Fur}\left(\tfrac{\mathrm{mg}}{\mathrm{L}}\right) = \exp\left(4.49 - 1.70\ \mathrm{A} - 1.43 * 10^{-2}\ \mathrm{T} - 1.55 * 10^{-2}\ \mathrm{t} + 1.78 * 10^{-2}\ \mathrm{AT} + 2.94 * 10^{-3}\ \mathrm{At} + 1.43 * 10^{-4}\ \mathrm{Tt}\right),\quad R^2 = 0.80 \tag{5}$$

$$\mathrm{HMF}\left(\tfrac{\mathrm{mg}}{\mathrm{L}}\right) = \exp\left(2.17 - 0.50 * 10^{-2}\ \mathrm{T} - 1.42\ \mathrm{A} - 0.60 * 10^{-2}\ \mathrm{t} + 0.75 * 10^{-4}\ \mathrm{Tt} + 0.93 * 10^{-3}\ \mathrm{At} + 0.02\ \mathrm{AT}\right),\quad R^2 = 0.86 \tag{6}$$

The surfaces obtained with both models are shown in Figure 5 for a reaction time of 180 min, in which, the exponential effect on the responses occurs towards at the upper values of temperature and acid concentration. The temporal evolution of furfural and HMF concentrations at constant temperature or acid sulfuric concentration, with the other operative conditions free in the studied range are also shown.

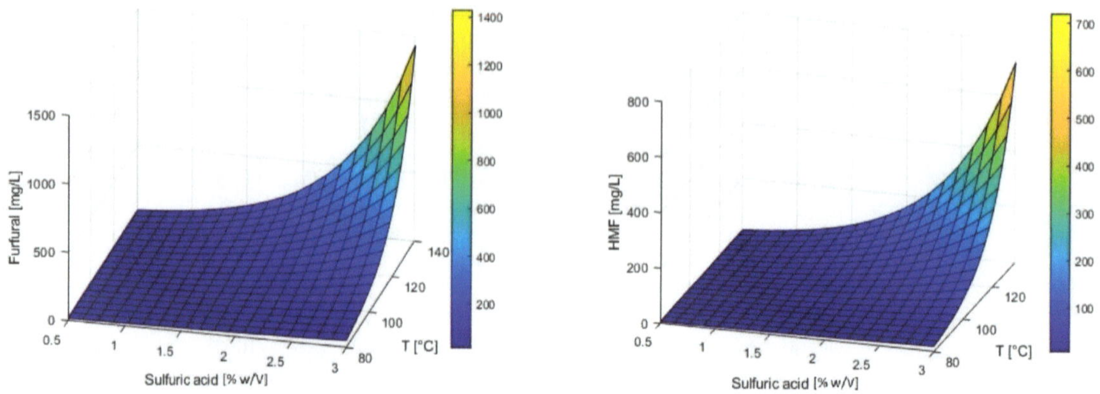

(a) **(d)**

Figure 5. *Cont.*

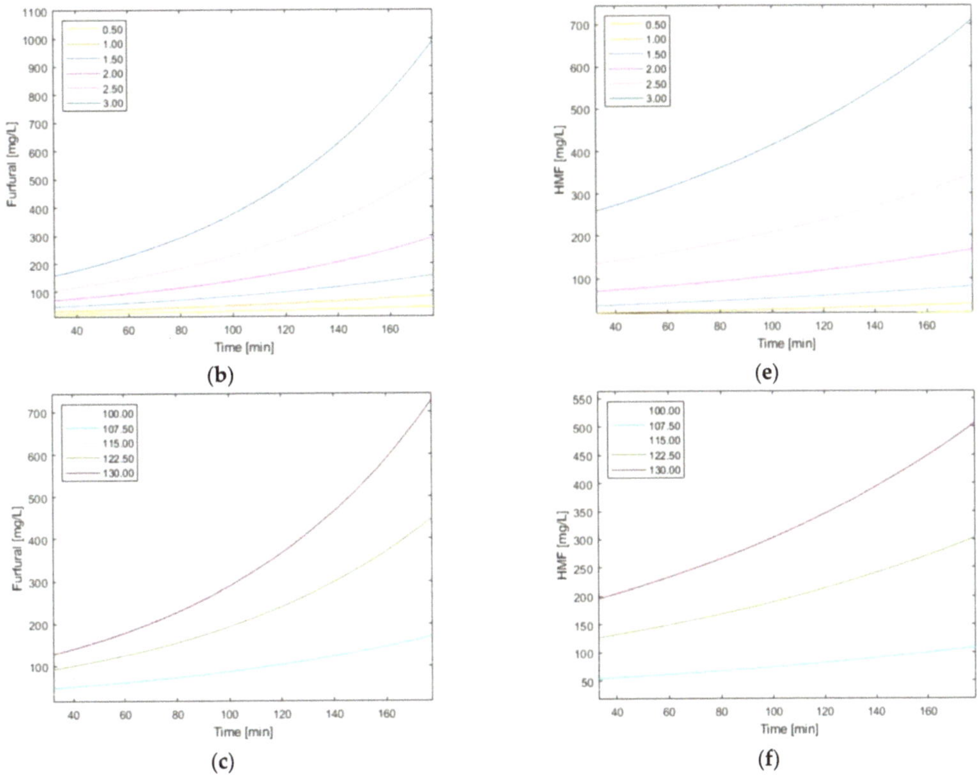

Figure 5. Relations between operational variables (Temperature, Sulfuric acid concentration and Time) over furans production. Response surfaces obtained for furfural (**a**) and HMF (**d**) production for a reaction time of 180.0 min. Temporal evolution of furans production predicted for different sulfuric acid concentrations when T = 135.0 °C, (**b**,**e**); and for different temperatures when sulfuric acid concentration was 3.0% w/v, (**c**,**f**), respectively.

3.5. Enzymatic Saccharification

The glucose concentration in the liquid phase was monitored in all reactors by sampling every 2 h, for a duration of 72 h. In order to analyze the effects of pretreatment on cellulose conversion, glucose concentration was analyzed 12 h after the starting of the enzymatic saccharification. A mathematical relationship with high goodness of fit (R^2 = 0.98) for %CEL (Equation (7)), as a function of *temperature, sulfuric acid concentration* and *time* was encountered. The surface response for different temperatures and acid concentrations is shown in Figure 6.

$$\%\text{CEL} = -26.29\ \text{A} + 0.12\ \text{T} + 0.07\ \text{t} + 0.40\ \text{AT},\ \text{R}^2 = 0.98 \quad (7)$$

Maximum glucose concentration after 12 h of enzymatic hydrolysis was detected in treatment 9 (2.50% w/v H_2SO_4; 125 °C; 180 min), with an average glucose release of 86.90 ± 4.11%; followed by treatments 8 (1.50% w/v H_2SO_4; 131 °C; 105 min) and 15 (2.50% w/v H_2SO_4; 125 °C; 30 min), with 73.85 ± 6.71% and 69.53 ± 4.11%, respectively. ANOVA shows that treatment 9 is the only one significantly different (T9 > T8 ≈ T15). These amounts of glucose released are similar to those reported by Corredor et al. (2008) [30] and Yoo et al. (2011) [45], which report values around 70% of hexose release after 24 h of enzymatic treatment and a previous pretreatment with dilute sulfuric acid (2.50% w/w) at 140 °C for 30 min.

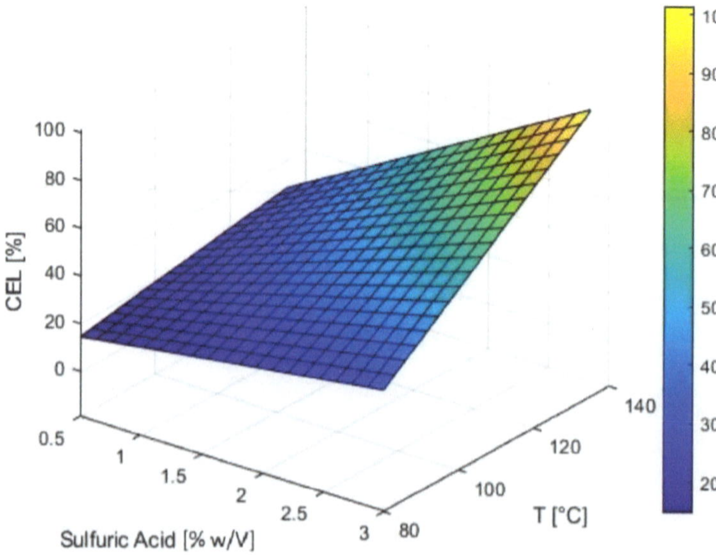

Figure 6. Response surface showing %CEL at 12 h of enzymatic saccharification after acid pretreatment during 30 min using different combinations of temperature and acid concentration.

It is noteworthy that a direct and positive relationship between solubilized hemicellulose during thermochemical pretreatment and the amount of glucose released after 12 h of enzymatic hydrolysis was observed (Figure 7). The high correlation (R^2 = 0.90) between these two variables suggests that the interactions between cellulose and hemicellulose would represent the main bottleneck for the use of lignocellulosic material in general, and SHs in particular.

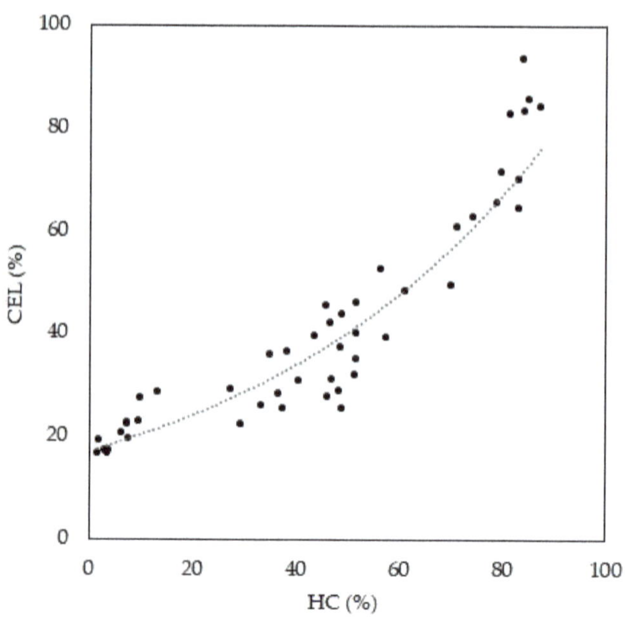

Figure 7. Dispersion graph showing relation between %HC and %CEL at 12 h of enzymatic saccharification.

3.6. *Optimization of Pretreatment Conditions*

During the acid pretreatment of a lignocellulosic material, a set of chemical reactions takes place both in series and in parallel. These reactions generate and/or consume relevant compounds of the 2G-bioethanol production processes. The study of the dynamics (evolution of the concentration) of these compounds both during and after pretreatment operations could provide valuable information about the effects of operational variables (temperature, reaction time and acid concentration) on the global process. However, the nature of the phenomena that take place during the pretreatment means that some of the compounds involved in the reactions may have an opposite effect on the goal of increasing ethanol production. Then, optimal conditions for the overall process cannot be proposed based on optimal individual conditions. Evaluating the models obtained for chemical pretreatment (%HC, furfural, and HMF) and enzymatic hydrolysis (%CEL) in separate ways could be insufficient when searching for optimal operational pretreatment conditions with the goal of maximizing the subsequent ethanol production.

3.6.1. Severity Factor

Since the mid-twentieth century, various severity indices have been proposed to describe pulp and paper industry processes based on correlations between soluble lignins and hemicelluloses and operating conditions, defined by temperature and reaction time [37–39]. This concept has been extended to the study of LB processes with the aim of contributing to both operative aspects and the phenomenological understanding of pretreatment processes. Abatzoglou et al. (1992) [46], developed a modified factor, R_{oH}, that allows for an accounting of the influence of a catalyst on the fractionation of lignocellulosic materials using a complex factor that includes adjustment parameters based on a chemical analysis.

In this work, a modified severity factor (Equation (8)) was adopted in order to identify causal relationships and trends:

$$R_0 = tx\ Ax\ exp\left(\frac{T - T_r}{14.75}\right), \tag{8}$$

where, t is the reaction time, A is the sulfuric acid concentration and T is the temperature of the reaction, whereas Tr is the reference temperature (100 °C).

Figure 8 shows the positive relationship between severity factor $R_{0'}$, the solubilization of structural polymers of SH and the monosaccharide degradation products (furfural and HMF). It can be seen that the solubilization of hemicellulose increases rapidly as the severity of pretreatment increases, becoming almost asymptotic for values greater than 400, with %HC between 80–90 for a wide range of $R_{0'}$ values (Figure 8a). Analysis of the variance of the most severe treatments indicates that there are not significant differences in the %HC (%HC ($R_{0'}$ = 400) = %HC ($R_{0'}$ = 1300) = %HC ($R_{0'}$ = 2500); p-value > 0.34). A similar tendency was observed when a relationship between the glucose release after 12 h of enzymatic hydrolysis and the factor $R_{0'}$ was established (Figure 8b). Moreover, furans production increase slightly in the severity range from 0 to 400 but an exponential increase is observed when more severe conditions were evaluated ([Furans] ($R_{0'}$ = 400) < [Furans] ($R_{0'}$ = 1300) < [Furans] ($R_{0'}$ = 2500); p-value > 0.4×10^{-3}). The analysis suggests that an optimal pretreatment would be performed with 2.5% w/v H_2SO_4 at 125 °C for a duration of 30 min. These conditions correspond to a $R_{0'}$ = 400, whereas a high %HC and low production of furans were obtained at this point.

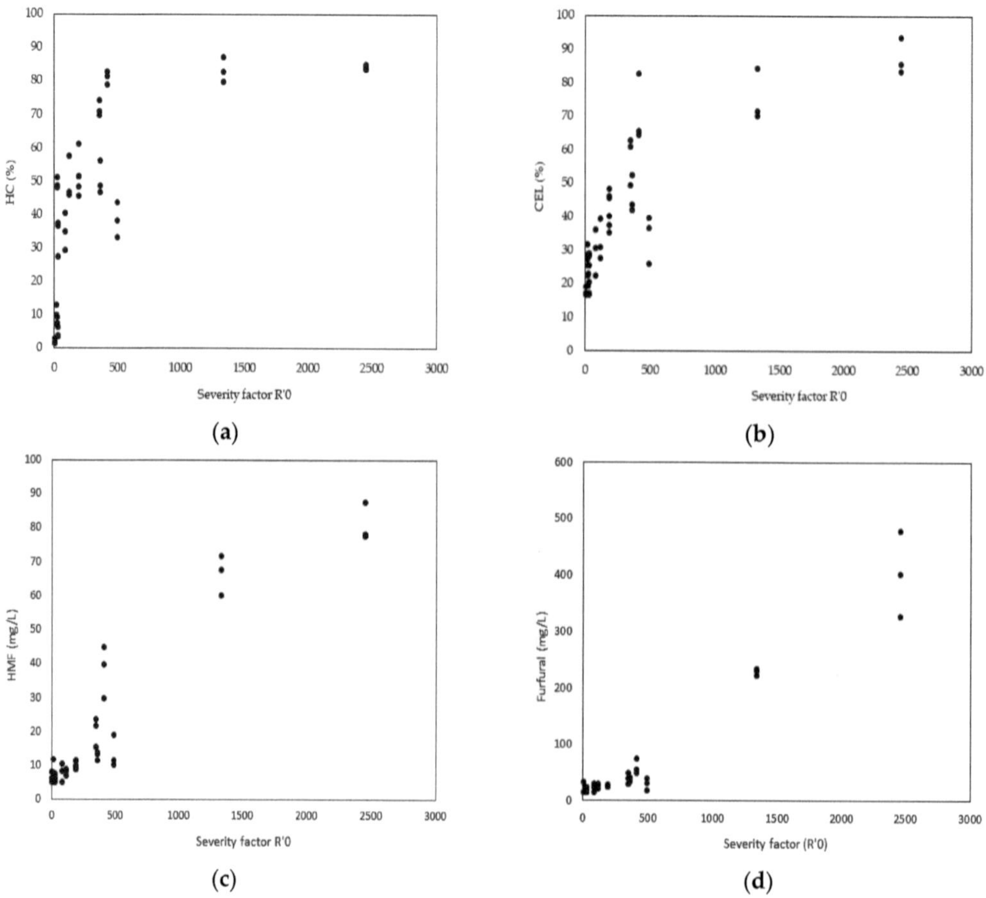

Figure 8. Dispersion graphs showing relations between Severity Factor $R_{0'}$ and %HC (**a**), %CEL (**b**), HMF (**c**) and Furfural (**d**), respectively.

3.6.2. Multiresponse Optimization

Pretreatment optimization was approached as a multi-response optimization problem. Desirability functions, di($\hat{y}$(x)), were constructed using the previously modeled responses (%HC %CEL, furfural and HMF concentration). The optimization of a particular response in the space of conditions (defined by the factors temperature, sulfuric acid concentration and time) was evaluated through the construction of a global desirability index, D = f(di), by combining individual desirability [47,48].

Two optimization problems of identical structure were defined, varying the function to be maximized, %HC or %CEL:

$$\begin{gathered} \text{maximize}\%\text{HC and }\%\text{CEL} \\ \text{minimize [Furfural]} \\ \text{minimize [HMF]} \\ \text{subject to} \begin{cases} 75\ ^\circ\text{C} \leq \text{T} \leq 135 \\ 0.5\ \% \leq \text{A} \leq 2.5\ \% \\ 30\ \text{min} \leq \text{t} \leq 200\ \text{min} \end{cases} \end{gathered} \quad (9)$$

Maximum desirability values (0.83) were found for a sulfuric acid concentration of 2.5% w/v, in the range of 115–125 °C and reaction times between 40–60 min (Figure 9).

This result agrees with the treatments with the maximum %HC and less severity presented in Figure 7, supporting the idea that the severity factor $R_{0'}$ constitutes a good tool for prediction and decision-making.

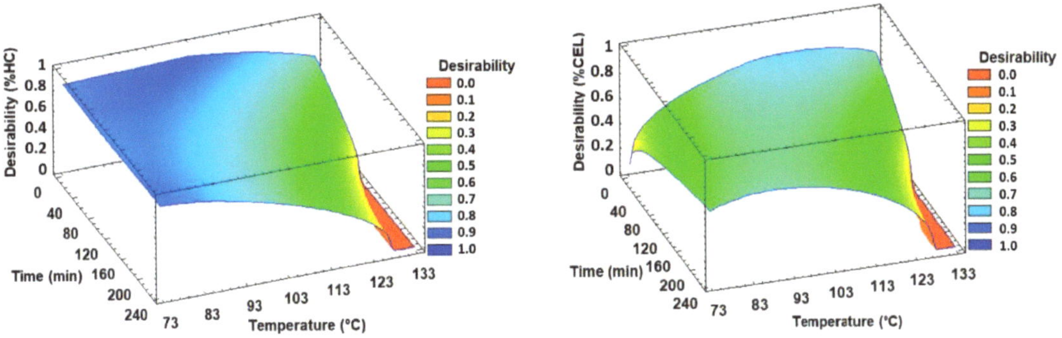

(**a**) (**b**)

Figure 9. Desirability surfaces obtained by resolving the optimization problem described by Equation (9), maximizing %HC (**a**) and %CEL (**b**) at 12 h of enzymatic hydrolysis with sulfuric acid 2.5% *w*/*v*.

3.7. Separate Saccharification and Fermentation

To verify the predicted results, three independent experiments were performed on SHs pretreated at the optimal conditions previously identified (121 °C; 2.5% *w*/*v* Sulfuric acid and 60 min). The pH value of the medium was initially adjusted to 5.00 ± 0.10, and chloramphenicol (1 mg/L) was added to prevent contamination with Gram-positive sporulating microorganisms, which were found in both the raw material and some process supplies. The cellulolytic enzyme complex was added and the reactors were maintained under orbital shaking at 50 °C for 24 h. In all the reactors an average %CEL of 98 ± 2% was obtained.

The values predicted by the models built for %HC, furfural and HMF concentrations, and the values experimentally obtained under optimal conditions selected are shown in Table 4. As can be noted, the models were highly accurate in predicting the responses in different conditions than the ones used for their construction (Table 2).

Table 4. Observed values of response variables defined and predicted values by models built.

Response Variable	Predicted	Observed
%HC	82.07 ± 5.07	90.42 ± 3.90
Furfural (mg/L)	85.06 ± 4.25	97.00 ± 12.07
HMF (mg/L)	80.68 ± 4.03	68.78 ± 10.39

Fermentation in batch mode under anaerobic conditions while a constant temperature of 30 °C was performed. The industrial yeast strain *Saccharomyces cerevisiae* var. Ethanol Red® (Fermentis-Lesaffre, Milwaukee, WI, USA) was used for all the experiments. Because this strain was originally adapted to fast growth in corn-based media, it is the industry standard in 1G-bioethanol production, and all the media were inoculated at 1.00 ± 0.10 g/L with exponential-growth biomass. The yeast culture was previously proliferated under aerobic conditions for 12–18 h. A set of reactors with synthetic medium was included as control. The concentrations of sugar and ethanol were monitored over time, whereas the biomass change was also evaluated in control reactors (Figure 10). Concerning the evaluation of the fermentation performance, some useful parameters were calculated following a previous method [24] and are displayed in Table 5.

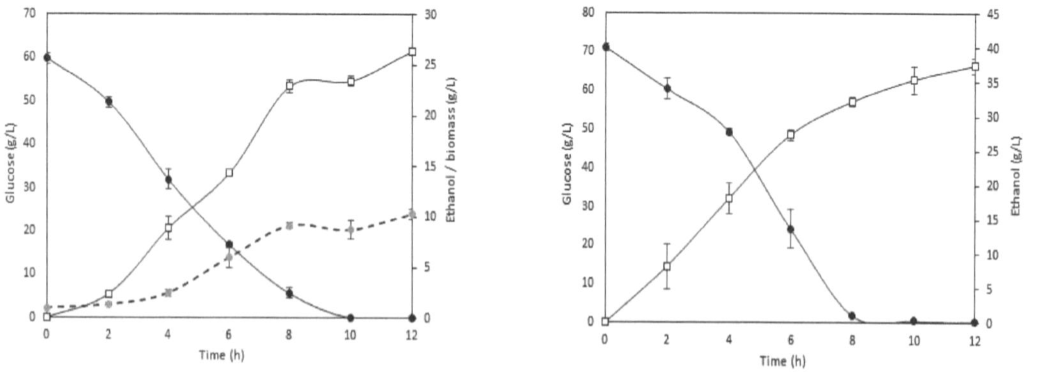

(a) (b)

Figure 10. Yeast performance during batch fermentation experiments on a synthetic medium (**a**), where changes in biomass are indicated by a dashed line, and SH hydrolysates (**b**). The reported values represent the means of three independent experiments.

Table 5. Fermentation parameters for *S. cer*-ER over SH hydrolysates and synthetic medium.

Fermentation Media	$Y_{et/glu}$	$Y_{x/glu}$	r_x (g/L.h)	r_s (g/L.h)	r_p (g/L.h)	Initial Glucose conc. (g/L)	Glucose Consumption (%)	Max. Biomass conc. (g/L)	Max. Ethanol conc. [1] (g/L)
SH Hydrolysates	0.45	-	-	8.65	3.63	69.5	100	-	36.6
Synthetic control medium	0.44	0.15	1.34	8.65	3.63	60.0	100	10.5	26.40

[1] Values for 12 h fermentation.

Sugar was completely depleted in all media in less than 12 h. Despite the presence of potential inhibitory compounds on SH hydrolysates, it is noteworthy that no latency time for sugar consumption was detected (Figure 10b). Fermentation on SH hydrolysates occurs without nutrient supplementation. *S. cerevisiae* is a non-diazotrophic and non-proteolytic organism (i.e., it cannot fix nitrogen and it is incapable of using proteins as a nitrogen source). Thus, the SH hydrolysates should contain readily utilizable sources of nitrogen to support a suitable fermentation. The free amino nitrogen (FAN) content is a measure of assimilable nitrogen, i.e., individual amino acids and small peptides, which can be utilized by microorganisms for cell growth and proliferation. This parameter is frequently used in brewing, particularly as an indicator of good procedure in beer wort preparation. FAN content in the SH hydrolysate previously to fermentation experiments was measured using the ninhydrin method [49], obtaining values of 196.29 $\pm$ 6.03 mg/L. This FAN content is similar to those commonly present in beer wort (approx. 250 mg/L), where subsequent fermentation occurs successfully [50].

Both the profile of glucose consumption and the ethanol production were very similar for the evaluated conditions. Complete glucose starvation occurs between 8–10 h and the maximum ethanol concentration was obtained at 12 h. Ethanol accumulation was slightly different: 36.6 g/L for SH hydrolysate and 26.40 g/L for control media. This result was possibly due to differences in initial glucose content (69.5 $g_{glucose}$/L and 60.0 $g_{glucose}$/L in SH hydrolysates and control media, respectively). The ethanol yield was 0.45 $g_{ethanol}/g_{glucose}$ in experiments on SH hydrolysate, which was close to 90% of the theoretical value (0.51 $g_{ethanol}/g_{glucose}$). Thus, the *S. cerevisiae* var. Ethanol Red emerges as a useful and interesting strain to develop robust processes for the 2G-ethanol industry.

It is of note that present work was focused on pretreatment conditions that maximize the release of glucose from SHs with the minimal accumulation of potential inhibitory compounds. Thus, pentoses were not really considered as substrates for ethanol production. However, for a complete use of the fermentable sugars present in SH hydrolysates, the pentoses derived from hemicellulose also should be metabolized to increase the ethanol concentration, or to produce other value-added products, such as xylitol [51,52]. Because the pentose-metabolization machinery is absent or silent in *Saccharomyces*, it is necessary to use yeast strains naturally capable of consuming pentoses, such as *Spathaspora passalidarum* or *Scheffersomyces stipitis* [53], or genetically modified strains, such as *S. cerevisiae* TMB3400 [54].The recombinant strains *S. cerevisiae* YRH396 and YRH400 were able to ferment soybean and oat hull hydrolysates obtained by pretreatment with dilute sulfuric acid [10]. Total glucose consumption and ethanol yields between 0.34–0.38 $g_{ethanol}/g_{glucose}$ were obtained in media with a significantly higher concentration of HMF (380 mg_{HMF}/L). In another work, *S. cerevisiae* AY-15 was studied in Saccharification and Simultaneous Fermentation (SSF) processes on detoxified and nutrient-supplemented corn stover hydrolysates. The ethanol yields obtained for cellulose was 0.538 $g_{ethanol}/g_{cellulose}$ [55]. Fermentations by *Wickerhamomyces* sp. UFFS-CE-3.1.2 on SH hydrolysates obtained with subcritical water, showed concentrations of acetic acid, HMF and furfural of 3.41, 0.16 and 0.31 g/L, respectively, and a total inhibition of yeast growth. However, *Wickerhamomyces* sp. UFFS-CE-3.1.2 was able to consume 10 g/L of glucose on a $\frac{1}{2}$ dilution of these hydrolysates [56]. Table 6 summarizes results in published works that used soybean hydrolysate or related materials, for comparative purposes.

Table 6. Fermentations parameters obtained in published experiences over different hydrolysates.

Fermentation Media	Pretreatment	$Y_{p/s}$ ($g_{ethanol}/g_{glucose}$)	r_p (g/L.h)	Initial Glucose (g/L)	Final Ethanol (g/L)	Fermentation Time (h)	Ref.
Detoxified corn stover hydrolysate. Nutrient supplementation.	Dilute H_2SO_4	0.538	ND [1]	SSF	37.85	120	[55]
Soybean and oat hulls hydrolysate.	Dilute H_2SO_4	0.34–0.38	NI	60.12 ± 2.03	24.2 ± 2.10	120	[10]
Soybean straw and hulls hydrolysates (SS&HH) supplemented with 10 g/L glucose.	Subcritical water (200 °C)	-	-	10.96	-	96	[56]
SS&HH diluted $\frac{1}{2}$ and supplemented with 10 g/L of glucose.	Hydrolysis with subcritical water (200 °C)	0.557	NI	10.96	6.11 ± 0.11	96	[56]
Soybean hull hydrolysate	Dilute H_2SO_4	0.45	3.63	69.5	36.6 ± 0.50	12	This work

[1] ND: Not determined.

Concerning the utilization of antibiotics in low concentrations in 1G bioethanol production at industrial scale, this is a cost-effective strategy that is frequently used for controlling bacterial contamination, whereas it has vertiginously increased over the past few years [57]. The use of antibiotics in fermentation at industrial scale has been extensively studied as a tool to stop the proliferation of contaminant microorganisms, as well as its impact on yeast performance and ethanol yields. In the fuel ethanol industry, control of bacterial contamination has been achieved by using antibiotics such as penicillin G, streptomycin, tetracycline, virginiamycin or monensin, or mixtures of these antibiotics [58,59]. Lactrol® (virginiamycin) is commonly used in a concentration range of 0.25–2.00 ppm at industrial scale. The enzymatic hydrolysis and fermentation processes were carried out using Lactrol® at 2 ppm as well, obtaining the same results (data not shown) as those exposed next in this section using chloramphenicol. Alternatively, a simultaneous saccharification and fermentation (SSF) strategy is an interesting option for reducing or entirely avoiding the use of antibiotics [60]. In this scenario, yeast is present in the fermentation medium at the initial stage of sugar liberation, and it modifies the medium properties (e.g., pH), displacing other microorganisms such as lactic acid bacteria.

As aforementioned, the pH should be adjusted after preconditioning operations to suitable values for both enzymatic hydrolysis and yeast metabolism. To adjust the pH from 0.7–0.8 to 5.5 of 100 mL of SH hydrolysates with a solid loading of 150 g_{SH} /L, it was necessary to add 2.5 g of NaOH (final concentration of sodium of 14.363 g_{Na}/L).

This represents a quantity of 0.167 g_{NaOH}/g_{SH} or 0.096 g_{Na}/g_{SH}. Conventional strains of *S. cerevisiae* tolerate a sodium chloride concentration of 2 M (45.978 g_{Na}/L). The final concentration of sodium was almost three times smaller than the maximum tolerance concentration values reported in the literature. Moreover, the fermentation performance and evolution of the SH hydrolysates and control (synthetic medium) was very similar for the evaluated conditions. Because the control does not contain sodium added to adjust the pH, it is possible to assume that sodium concentration used was not detrimental for yeast growth and fermentation in the reactors containing hydrolyzed and neutralized SH.

Finally, another compound still remains in fermentation media after ethanol separation: lignin. Although the study of this aromatic biopolymer was outside the scope of the present work, it has the potential to produce various compounds of interest, such as lignin-based hydrogels, surfactants, three-dimensional printing materials, electrodes and fine chemicals through biorefinery activities. Therefore, it is expected to benefit the future circular economy [61]. This aspect will be addressed in future research.

4. Conclusions

This study demonstrates that dilute sulfuric acid, a well-known method, is a suitable option for pretreating a new feedstock in order to develop robust second-generation bioethanol processes. Pretreated hydrolysates were suitable for subsequent enzymatic saccharification, with almost complete recovery of the glucose from the cellulose in 24 h. Neither detoxifying operations nor nutritional supplementation were necessary to ferment the hydrolysates of SH. Additionally, *S. cerevisiae* var. Ethanol Red was able to completely consume the available glucose in less than 12 h with an ethanol yield of 0.45 $g_{ethanol}/g_{glucose}$. Thus, the feasibility of the pretreatment of SH with dilute sulfuric acid was demonstrated.

The developed empirical mathematical models can satisfactorily describe the effects of the thermochemical pretreatment factors on SH hydrolysis. This results in accurate and reliable prediction of hemicellulose solubilization and furans (HMF and furfural) accumulation, as well as the efficiency in the release of glucose during subsequent enzymatic saccharification. Based on these models, it was possible to optimize pretreatment conditions by applying a multi-response methodology. Moreover, the severity factor is an extremely useful tool to guide decision-making in this type of thermochemical processes. Thus, transversal criteria could be included during a process of 2G-bioethanol production.

Last but not least, it is of note that the process has some interesting advantages: (i) absence of latency time to sugar consumption, (ii) short fermentation time (less than 12 h), (iii) high initial glucose content, (iv) absence or minimal inhibitory effect on saccharification and/or fermentation, (v) high ethanol yields and (vi) the fact that SH hydrolysates provide nutrients in sufficient quantities for suitable fermentation. Thus, the thermochemical process evaluated and the SH constitute promising actors for the 2G-ethanol production.

Author Contributions: Conceptualisation: R.N.C.; Methodology: M.G.R., R.J.L. and B.C.B.; Software: M.G.R. and R.J.L.; Data analysis: M.G.R., R.J.L. and R.N.C.; Writing-original draft preparation: M.G.R., R.J.L. and R.N.C.; Writing-review and editing: M.G.R., R.J.L., B.C.B., L.G.S., M.T.B. and R.N.C.; Visualization: M.G.R. and R.J.L.; Supervision: R.N.C. All authors have read and agreed to the published version of the manuscript.

Funding: This research was supported by the Universidad Nacional del Litoral (UNL) via the CAI+D Program (Project 50620190100054LI), Consejo Nacional de Investigaciones Científicas y Técnicas (CONICET) and the Agencia Nacional de Promoción Científica y Tecnológica (ANPCyT) via PICT2018-01458 project.

Institutional Review Board Statement: Not applicable.

Informed Consent Statement: Not applicable.

Data Availability Statement: Not applicable.

Acknowledgments: The collaboration of the undergraduate student G. Delaloye in the realization of some experiments and S. Racca in editing the manuscript is gratefully acknowledged, as well as the permanent support of all GPBIA's members.

Conflicts of Interest: The authors declare no conflict of interest and no competing financial interest.

References

1. Nigam, P.S.; Singh, A. Production of Liquid Biofuels from Renewable Resources. *Prog. Energy Combust. Sci.* **2011**, *37*, 52–68. [CrossRef]
2. Limayem, A.; Ricke, S.C. Lignocellulosic Biomass for Bioethanol Production: Current Perspectives, Potential Issues and Future Prospects. *Prog. Energy Combust. Sci.* **2012**, *38*, 449–467. [CrossRef]
3. Comelli, R.N.; Benzoo, M.T.; Leonardi, R.J.; Seluy, L.G.; Tomassi, A.H.; Isla, M.A. Agro-Industrial Wastewaters as Feedstocks: New Research on Bioethanol Production. In *Agricultural Research Updates*; Nova Science Publishers: Hauppauge, NY, USA, 2023; Volume 43, ISBN 979-8-88697-550-5.
4. Wyman, C.E. (Ed.) *Handbook on Bioethanol: Production and Utilization*; Applied Energy Technology Series; Taylor & Francis: Washington, DC, USA, 1996; ISBN 978-1-56032-553-6.
5. Devi, A.; Singh, A.; Bajar, S.; Pant, D.; Din, Z.U. Ethanol from Lignocellulosic Biomass: An in-Depth Analysis of Pre-Treatment Methods, Fermentation Approaches and Detoxification Processes. *J. Environ. Chem. Eng.* **2021**, *9*, 105798. [CrossRef]
6. Patel, A.; Shah, A.R. Integrated Lignocellulosic Biorefinery: Gateway for Production of Second Generation Ethanol and Value Added Products. *J. Bioresour. Bioprod.* **2021**, *6*, 108–128. [CrossRef]
7. Nishida, V.S.; Woiciechowski, A.L.; Valladares-Diestra, K.K.; Zevallos Torres, L.A.; Vandenberghe, L.P.d.S.; Zandoná Filho, A.; Soccol, C.R. Second Generation Bioethanol Production from Soybean Hulls Pretreated with Imidazole as a New Solvent. *Fermentation* **2023**, *9*, 93. [CrossRef]
8. Amaro Bittencourt, G.; Porto de Souza Vandenberghe, L.; Valladares-Diestra, K.; Wedderhoff Herrmann, L.; Fátima Murawski de Mello, A.; Sarmiento Vásquez, Z.; Grace Karp, S.; Ricardo Soccol, C. Soybean Hulls as Carbohydrate Feedstock for Medium to High-Value Biomolecule Production in Biorefineries: A Review. *Bioresour. Technol.* **2021**, *339*, 125594. [CrossRef]
9. Cassales, A.; de Souza-Cruz, P.B.; Rech, R.; Záchia Ayub, M.A. Optimization of Soybean Hull Acid Hydrolysis and Its Characterization as a Potential Substrate for Bioprocessing. *Biomass Bioenergy* **2011**, *35*, 4675–4683. [CrossRef]
10. Cortivo, P.R.D.; Hickert, L.R.; Hector, R.; Ayub, M.A.Z. Fermentation of Oat and Soybean Hull Hydrolysates into Ethanol and Xylitol by Recombinant Industrial Strains of Saccharomyces Cerevisiae under Diverse Oxygen Environments. *Ind. Crops Prod.* **2018**, *113*, 10–18. [CrossRef]
11. Kumar, A.; Gautam, A.; Dutt, D. Biotechnological Transformation of Lignocellulosic Biomass in to Industrial Products: An Overview. *Adv. Biosci. Biotechnol.* **2016**, *07*, 149–168. [CrossRef]
12. Chiaramonti, D.; Prussi, M.; Ferrero, S.; Oriani, L.; Ottonello, P.; Torre, P.; Cherchi, F. Review of Pretreatment Processes for Lignocellulosic Ethanol Production, and Development of an Innovative Method. *Biomass Bioenergy* **2012**, *46*, 25–35. [CrossRef]
13. Harris, J.F.; Baker, A.J.; Conner, A.H.; Jeffries, T.W.; Minor, J.L.; Pettersen, R.C.; Scott, R.W.; Springer, E.L.; Wegner, T.H.; Zerbe, J.I. *Two-Stage, Dilute Sulfuric Acid Hydrolysis of Wood: An Investigation of Fundamentals*; U.S. Department of Agriculture, Forest Service, Forest Products Laboratory: Madison, WI, USA, 1985; p. FPL-GTR-45.
14. Silverstein, R.A.; Chen, Y.; Sharma-Shivappa, R.R.; Boyette, M.D.; Osborne, J. A Comparison of Chemical Pretreatment Methods for Improving Saccharification of Cotton Stalks. *Bioresour. Technol.* **2007**, *98*, 3000–3011. [CrossRef] [PubMed]
15. Taherzadeh, M.J.; Karimi, K. Acid-Based Hydrolysis Processes for Ethanol from Lignocellulosic Materials: A Review. *BioResources* **2007**, *2*, 472–499.
16. Takeuchi, Y.; Jin, F.; Tohji, K.; Enomoto, H. Acid Catalytic Hydrothermal Conversion of Carbohydrate Biomass into Useful Substances. *J. Mater. Sci.* **2008**, *43*, 2472–2475. [CrossRef]
17. Shahbazi, A.; Zhang, B. Dilute and Concentrated Acid Hydrolysis of Lignocellulosic Biomass. In *Bioalcohol Production*; Elsevier: Amsterdam, The Netherlands, 2010; pp. 143–158, ISBN 978-1-84569-510-1.
18. Tadimeti, J.G.D.; Thilakaratne, R.; Balla, V.K.; Kate, K.H.; Satyavolu, J. A Two-Stage C5 Selective Hydrolysis on Soybean Hulls for Xylose Separation and Value-Added Cellulose Applications. *Biomass Conv. Bioref.* **2020**, *12*, 3289–3301. [CrossRef]
19. Sluiter, A. *Determination of Ash in Biomass: Laboratory Analytical Procedure (LAP)*; Issue Date: 7/17/2005. Technical Report; National Renewable Energy Laboratory: Golden, CO, USA, 2008.
20. Sluiter, A.; Hames, B.; Ruiz, R.; Scarlata, C.; Sluiter, J.; Templeton, D.; Crocker, D. *Determination of Structural Carbohydrates and Lignin in Biomass*; Technical Report; National Renewable Energy Laboratory: Golden, CO, USA, 2012.
21. Eberts, T.J.; Sample, R.H.; Glick, M.R.; Ellis, G.H. A Simplified, Colorimetric Micromethod for Xylose in Serum or Urine, with Phloroglucinol. *Clin. Chem.* **1979**, *25*, 1440–1443. [CrossRef] [PubMed]
22. Rocha, G.J.M.; Silva, F.T.; Curvelo, A.A.S.; Araujo, G.T. A Fast and Accurate Method for Determination of Cellulose and Polyoses by HPLC. In Proceedings of the Fifth Brazilian Symposium on the Chemistry of Lignins and Other Wood Components, Curitiba, Brazil, 31 August–5 September 1997; Sepia Editora e Gráfica LTDA: Curitiba, Brazil, 1997; Volume VI.
23. Comelli, R.N.; Seluy, L.G.; Isla, M.A. Optimization of a Low-Cost Defined Medium for Alcoholic Fermentation—A Case Study for Potential Application in Bioethanol Production from Industrial Wastewaters. *New Biotechnol.* **2016**, *33*, 107–115. [CrossRef]

24. Comelli, R.N.; Seluy, L.G.; Isla, M.A. Performance of Several Saccharomyces Strains for the Alcoholic Fermentation of Sugar-Sweetened High-Strength Wastewaters: Comparative Analysis and Kinetic Modelling. *New Biotechnol.* **2016**, *33*, 874–882. [CrossRef]
25. Eaton, A.D.; Clesceri, L.S.; Greenberg, A.E. *Standard Methods for the Examination of Water and Wastewater*, 21st ed.; American Public Health Association: Washington, DC, USA, 2005; ISBN 978-0-87553-047-5.
26. Isla, M.A.; Comelli, R.N.; Seluy, L.G. Wastewater from the Soft Drinks Industry as a Source for Bioethanol Production. *Bioresour. Technol.* **2013**, *136*, 140–147. [CrossRef]
27. Rojas, M.J.; Siqueira, P.F.; Miranda, L.C.; Tardioli, P.W.; Giordano, R.L.C. Sequential Proteolysis and Cellulolytic Hydrolysis of Soybean Hulls for Oligopeptides and Ethanol Production. *Ind. Crops Prod.* **2014**, *61*, 202–210. [CrossRef]
28. Miron, J.; Yosef, E.; Ben-Ghedalia, D. Composition and in Vitro Digestibility of Monosaccharide Constituents of Selected Byproduct Feeds. *J. Agric. Food Chem.* **2001**, *49*, 2322–2326. [CrossRef]
29. Corredor, D.Y.; Sun, X.S.; Salazar, J.M.; Hohn, K.L.; Wang, D. Enzymatic Hydrolysis of Soybean Hulls Using Dilute Acid and Modified Steam-Explosion Pretreatments. *J Biobased Mater. Bioenergy* **2008**, *2*, 43–50. [CrossRef]
30. Cortivo, P.R.D.; Machado, J.; Hickert, L.R.; Rossi, D.M.; Ayub, M.A.Z. Production of 2,3-Butanediol by *Klebsiella Pneumoniae* BLh-1 and *Pantoea Agglomerans* BL1 Cultivated in Acid and Enzymatic Hydrolysates of Soybean Hull. *Biotechnol. Prog.* **2019**, *35*, e2793. [CrossRef]
31. Toro-Trochez, J.L.; Carrillo-Pedraza, E.S.; Bustos-Martínez, D.; García-Mateos, F.J.; Ruiz-Rosas, R.R.; Rodríguez-Mirasol, J.; Cordero, T. Thermogravimetric Characterization and Pyrolysis of Soybean Hulls. *Bioresour. Technol. Rep.* **2019**, *6*, 183–189. [CrossRef]
32. Robles Barros, P.J.; Ramirez Ascheri, D.P.; Siqueira Santos, M.L.; Morais, C.C.; Ramirez Ascheri, J.L.; Signini, R.; dos Santos, D.M.; de Campos, A.J.; Alessandro Devilla, I. Soybean Hulls: Optimization of the Pulping and Bleaching Processes and Carboxymethyl Cellulose Synthesis. *Int. J. Biol. Macromol.* **2020**, *144*, 208–218. [CrossRef] [PubMed]
33. Jung, Y.H.; Park, H.M.; Kim, D.H.; Park, Y.-C.; Seo, J.-H.; Kim, K.H. Combination of High Solids Loading Pretreatment and Ethanol Fermentation of Whole Slurry of Pretreated Rice Straw to Obtain High Ethanol Titers and Yields. *Bioresour. Technol.* **2015**, *198*, 861–866. [CrossRef] [PubMed]
34. Shiva; Climent Barba, F.; Rodríguez-Jasso, R.M.; Sukumaran, R.K.; Ruiz, H.A. High-Solids Loading Processing for an Integrated Lignocellulosic Biorefinery: Effects of Transport Phenomena and Rheology—A Review. *Bioresour. Technol.* **2022**, *351*, 127044. [CrossRef]
35. Zhang, B.; Liu, X.; Bao, J. High Solids Loading Pretreatment: The Core of Lignocellulose Biorefinery as an Industrial Technology—An Overview. *Bioresour. Technol.* **2023**, *369*, 128334. [CrossRef]
36. Pino, M.S.; Rodríguez-Jasso, R.M.; Michelin, M.; Flores-Gallegos, A.C.; Morales-Rodriguez, R.; Teixeira, J.A.; Ruiz, H.A. Bioreactor Design for Enzymatic Hydrolysis of Biomass under the Biorefinery Concept. *Chem. Eng. J.* **2018**, *347*, 119–136. [CrossRef]
37. Aguilar, D.L.; Rodríguez-Jasso, R.M.; Zanuso, E.; de Rodríguez, D.J.; Amaya-Delgado, L.; Sanchez, A.; Ruiz, H.A. Scale-up and Evaluation of Hydrothermal Pretreatment in Isothermal and Non-Isothermal Regimen for Bioethanol Production Using Agave Bagasse. *Bioresour. Technol.* **2018**, *263*, 112–119. [CrossRef]
38. Schirmer-Michel, Â.C.; Flôres, S.H.; Hertz, P.F.; Matos, G.S.; Ayub, M.A.Z. Production of Ethanol from Soybean Hull Hydrolysate by Osmotolerant Candida Guilliermondii NRRL Y-2075. *Bioresour. Technol.* **2008**, *99*, 2898–2904. [CrossRef]
39. Marzialetti, T.; Valenzuela Olarte, M.B.; Sievers, C.; Hoskins, T.J.C.; Agrawal, P.K.; Jones, C.W. Dilute Acid Hydrolysis of Loblolly Pine: A Comprehensive Approach. *Ind. Eng. Chem. Res.* **2008**, *47*, 7131–7140. [CrossRef]
40. Palmqvist, E.; Hahn-Hägerdal, B. Fermentation of Lignocellulosic Hydrolysates. I: Inhibition and Detoxification. *Bioresour. Technol.* **2000**, *74*, 17–24. [CrossRef]
41. Klinke, H.B.; Thomsen, A.B.; Ahring, B.K. Inhibition of Ethanol-Producing Yeast and Bacteria by Degradation Products Produced during Pre-Treatment of Biomass. *Appl. Microbiol. Biotechnol.* **2004**, *66*, 10–26. [CrossRef] [PubMed]
42. Panagiotou, G.; Olsson, L. Effect of Compounds Released during Pretreatment of Wheat Straw on Microbial Growth and Enzymatic Hydrolysis Rates. *Biotechnol. Bioeng.* **2007**, *96*, 250–258. [CrossRef] [PubMed]
43. Lee, H.-J.; Lim, W.-S.; Lee, J.-W. Improvement of Ethanol Fermentation from Lignocellulosic Hydrolysates by the Removal of Inhibitors. *J. Ind. Eng. Chem.* **2013**, *19*, 2010–2015. [CrossRef]
44. Hoppert, L.; Kölling, R.; Einfalt, D. Synergistic Effects of Inhibitors and Osmotic Stress during High Gravity Bioethanol Production from Steam-Exploded Lignocellulosic Feedstocks. *Biocatal. Agric. Biotechnol.* **2022**, *43*, 102414. [CrossRef]
45. Yoo, J.; Alavi, S.; Vadlani, P.; Amanor-Boadu, V. Thermo-Mechanical Extrusion Pretreatment for Conversion of Soybean Hulls to Fermentable Sugars. *Bioresour. Technol.* **2011**, *102*, 7583–7590. [CrossRef]
46. Abatzoglou, N.; Chornet, E.; Belkacemi, K.; Overend, R.P. Phenomenological Kinetics of Complex Systems: The Development of a Generalized Severity Parameter and Its Application to Lignocellulosics Fractionation. *Chem. Eng. Sci.* **1992**, *47*, 1109–1122. [CrossRef]
47. Derringer, G.; Suich, R. Simultaneous Optimization of Several Response Variables. *J. Qual. Technol.* **1980**, *12*, 214–219. [CrossRef]
48. Carlyle, W.M.; Montgomery, D.C.; Runger, G.C. Optimization Problems and Methods in Quality Control and Improvement. *J. Qual. Technol.* **2000**, *32*, 1–17. [CrossRef]
49. Lie, S. The Ebc-ninhydrin method for determination of free alpha amino nitrogen. *J. Inst. Brew.* **1973**, *79*, 37–41. [CrossRef]

50. Jacob, F.F.; Striegel, L.; Rychlik, M.; Hutzler, M.; Methner, F.-J. Spent Yeast from Brewing Processes: A Biodiverse Starting Material for Yeast Extract Production. *Fermentation* **2019**, *5*, 51. [CrossRef]
51. Hor, S.; Kongkeitkajorn, M.B.; Reungsang, A. Sugarcane Bagasse-Based Ethanol Production and Utilization of Its Vinasse for Xylitol Production as an Approach in Integrated Biorefinery. *Fermentation* **2022**, *8*, 340. [CrossRef]
52. Kaur, S.; Guleria, P.; Yadav, S.K. Evaluation of Fermentative Xylitol Production Potential of Adapted Strains of Meyerozyma Caribbica and Candida Tropicalis from Rice Straw Hemicellulosic Hydrolysate. *Fermentation* **2023**, *9*, 181. [CrossRef]
53. Mouro, A.; dos Santos, A.A.; Agnolo, D.D.; Gubert, G.F.; Bon, E.P.S.; Rosa, C.A.; Fonseca, C.; Stambuk, B.U. Combining Xylose Reductase from Spathaspora Arborariae with Xylitol Dehydrogenase from Spathaspora Passalidarum to Promote Xylose Consumption and Fermentation into Xylitol by Saccharomyces Cerevisiae. *Fermentation* **2020**, *6*, 72. [CrossRef]
54. Wahlbom, C.; Vanzyl, W.; Jonsson, L.; Hahnhagerdal, B.; Otero, R. Generation of the Improved Recombinant Xylose-Utilizing TMB 3400 by Random Mutagenesis and Physiological Comparison with CBS 6054. *FEMS Yeast Res.* **2003**, *3*, 319–326. [CrossRef]
55. Chen, Y.; Dong, B.; Qin, W.; Xiao, D. Xylose and Cellulose Fractionation from Corncob with Three Different Strategies and Separate Fermentation of Them to Bioethanol. *Bioresour. Technol.* **2010**, *101*, 6994–6999. [CrossRef]
56. Vedovatto, F.; Bonatto, C.; Bazoti, S.F.; Venturin, B.; Alves, S.L., Jr.; Kunz, A.; Steinmetz, R.L.R.; Treichel, H.; Mazutti, M.A.; Zabot, G.L.; et al. Production of Biofuels from Soybean Straw and Hull Hydrolysates Obtained by Subcritical Water Hydrolysis. *Bioresour. Technol.* **2021**, *328*, 124837. [CrossRef]
57. Aquarone, E. Penicillin and Tetracycline as Contamination Control Agents in Alcoholic Fermentation of Sugar Cane Molasses. *Appl. Microbiol.* **1960**, *8*, 263–268. [CrossRef]
58. Hynes, S.H.; Kjarsgaard, D.M.; Thomas, K.C.; Ingledew, W.M. Use of Virginiamycin to Control the Growth of Lactic Acid Bacteria during Alcohol Fermentation. *J. Ind. Microbiol. Biotechnol.* **1997**, *18*, 284–291. [CrossRef]
59. Peng, J.; Zhang, L.; Gu, Z.-H.; Ding, Z.-Y.; Shi, G.-Y. The Role of Nisin in Fuel Ethanol Production with Saccharomyces Cerevisiae: Nisin and Fuel Ethanol Production. *Lett. Appl. Microbiol.* **2012**, *55*, 128–134. [CrossRef] [PubMed]
60. Chohan, N.A.; Aruwajoye, G.S.; Sewsynker-Sukai, Y.; Gueguim Kana, E.B. Valorisation of Potato Peel Wastes for Bioethanol Production Using Simultaneous Saccharification and Fermentation: Process Optimization and Kinetic Assessment. *Renew. Energy* **2020**, *146*, 1031–1040. [CrossRef]
61. Sethupathy, S.; Murillo Morales, G.; Gao, L.; Wang, H.; Yang, B.; Jiang, J.; Sun, J.; Zhu, D. Lignin Valorization: Status, Challenges and Opportunities. *Bioresour. Technol.* **2022**, *347*, 126696. [CrossRef] [PubMed]

fermentation

Article

Heterologous Expression and Biochemical Characterization of a Thermostable Endoglucanase (*Mt*EG5-1) from *Myceliophthora thermophila*

Wenyuan Zhou [1,2,3,†], Sheng Tong [2,3,†], Farrukh Raza Amin [2,3,4], Wuxi Chen [2,3], Jinling Cai [1,*] and Demao Li [2,3,*]

1 Tianjin Key Laboratory of Brine Chemical Engineering and Resource Eco-Utilization, College of Chemical Engineering and Materials Science, Tianjin University of Science & Technology, Tianjin 300457, China; zhouwy@tib.cas.cn

2 Tianjin Key Laboratory for Industrial Biological System and Bioprocessing Engineering, Tianjin Institute of Industrial Biotechnology, Chinese Academy of Sciences, Tianjin 300308, China; tongs@tib.cas.cn (S.T.); chen_wx@tib.cas.cn (W.C.)

3 National Innovation Centre for Synthetic Biology, Tianjin 300308, China

4 Department of Chemistry, COMSATS University Islamabad, Park Road, Tarlai Kalan, Islamabad 45550, Pakistan

* Correspondence: jinlingcai@tust.edu.cn (J.C.); li_dm@tib.cas.cn (D.L.); Tel.: +86-(022)-84861993 (D.L.)

† These authors contributed equally to this work.

Abstract: Thermophilic endoglucanases have become of significant interest for effectively catalyzing the hydrolysis of cellulose. *Myceliophthora thermophila* is an ideal source of thermophilic enzymes. Interestingly, different hosts differently express the same enzymes. In this study, we successfully overexpressed endoglucanase (*Mt*EG5-1) from *M. thermophila* in the methylotrophic yeast, *Pichia pastoris* GS115, via electroporation. We found that purified *Mt*EG5-1 exhibited optimum activity levels at pH 5 and 70 °C, with 88% thermal stability after being incubated at 70 °C for 2 h. However, we observed that purified *Mt*EG5-1 had a molecular weight of 55 kDa. The Km and Vmax values of purified *Mt*EG5-1 were approximately 6.11 mg/mL and 91.74 μmol/min/mg at 70 °C (pH 5.0), respectively. Additionally, the optimum NaCl concentration of purified *Mt*EG5-1 was found to be 6 g/L. Furthermore, we observed that the activity of purified *Mt*EG5-1 was significantly enhanced by Mn^{2+} and was inhibited by K^+. These results indicated that *Mt*EG5-1 expressed by *P. pastoris* GS115 is more heat-tolerant than that expressed by *A. niger* and *P. pastoris* X33. These properties of *Mt*EG5-1 make it highly suitable for future academic research and industrial applications.

Keywords: endoglucanase; enzymatic activity; *Pichia pastoris* GS115; *Myceliophthora thermophila*

Citation: Zhou, W.; Tong, S.; Amin, F.R.; Chen, W.; Cai, J.; Li, D. Heterologous Expression and Biochemical Characterization of a Thermostable Endoglucanase (*Mt*EG5-1) from *Myceliophthora thermophila*. *Fermentation* **2023**, *9*, 462. https://doi.org/10.3390/fermentation9050462

Academic Editors: Miguel Ladero and Victoria E. Santos

Received: 22 March 2023
Revised: 8 May 2023
Accepted: 10 May 2023
Published: 12 May 2023

1. Introduction

Currently, 88% of global primary energy use is still derived from fossil fuels [1]. However, due to their depleting reserves and serious environmental impact, it is imperative to find sustainable and renewable alternatives to fossil energy sources [2]. Cellulose is considered to be one of the best alternatives to fossil energy, as it is a renewable energy source that is abundantly available in many forms, such as corn straw and bagasse [3]. Nevertheless, the widespread use of cellulose has been limited by the high energy consumption and pollution associated with chemical and physical pretreatment methods [4]. Fortunately, the emergence of synthetic biology provides novel methods to address this problem. Cellulose can be efficiently and gently degraded by cellulase, providing more opportunities to apply it in a wide range of industrial processes.

Cellulose is a macromolecular polysaccharide composed of repeating glucose units with β-1,4-glycosidic bonds [5]. The complete conversion of cellulose to glucose requires the combined action of three enzymes: cellobiohydrolases (CBHs, EC3.2.1.9), endoglucanases

(EGs, EC 3.2.1.4), and β-glucosidases (BGLs, EC 3.2.1.21) [6,7]. The principle of the cellulase synergistic decomposition of cellulose is shown in Figure 1. EGs, as the most important rate-limiting enzyme in cellulose degradation, focus on the hydrolysis of β-1, 4-glucosidic glucan linkages [8]. This result in the conversion of long cellulase chains into short chains, thereby increasing the amount of reducing ends exposed in the cellulose chain [9]. Consequently, EGs are not only a prerequisite for the quick initial liquefaction of biomass, but they are also a key biocatalyst for cellulose biodegradation [10]. Increasing the system temperature for cellulose degradation may reduce the risk of contamination with mesophilic microorganisms and increase the reaction rate [11]. In addition, thermostable enzymes can be transferred and stored more easily than mesophilic enzymes can [12]. Therefore, it is necessary to screen for thermostable EGs and apply them to cellulose degradation.

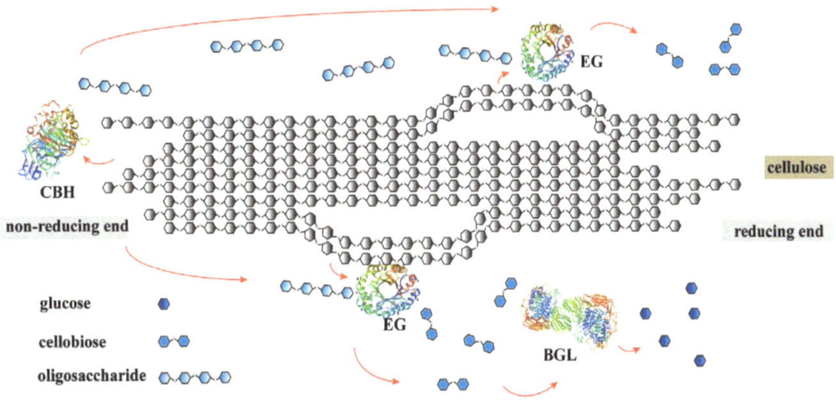

Figure 1. The synergistic effect of cellulase for cellulose decomposition [6,7]. CBH: cellobiohydrolase, EG: endoglucanase, BGL: beta-glucosidase. The red arrows indicate the sequence of cellulase synergy required for the complete degradation of cellulose into monosaccharides.

Myceliophthora thermophila is a thermophilic fungus with a good cellulose-degrading ability. Its cellulase products are generally recognized as safe [13]. Additionally, EGs possess a high catalytic efficiency and thermal stability, making them ideal candidates for industrial cellulose degradation. However, when cellulase is overexpressed in *M. thermophila*, several endogenous enzymes, such as xylanase [14], are overexpressed, which complicates the isolation of cellulose and makes it difficult to study the enzymatic properties and structure of the enzyme.

Pichia pastoris is a methylotrophic strain that is widely used as a potent heterologous protein expression system, owing to its high cell density and low levels of endogenous protein expression [11]. Furthermore, the expressed protein always exhibits higher stability for glycosylation and other posttranslational modifications [15]. Consequently, *P. pastoris* is becoming increasingly popular for academic or industrial applications. Although *Mt*EG5-1 has been successfully overexpressed in *Aspergillus niger* and *P. pastoris* X33, the results indicate that *Mt*EG5-1 has poor thermal stability and a higher molecular weight [16,17]. In contrast, an expression system constructed using *P. pastoris* GS115 as the host strain can produce a thermally stable *Mt*EG5-1.

In this study, we used the enzyme obtained from *M. thermophila* and analyzed some enzymatic properties of *Mt*EG5-1 overexpressed in *P. pastoris* GS115. Therefore, this work provides a thermo-acid-stable enzyme for the utilization of cellulose resources.

2. Materials and Methods

2.1. Strains and Culture Conditions

M. thermophila ATCC 42464 and the *P. pastoris* GS115 strain were obtained at the Tianjin Institute of Industrial Biotechnology. *M. thermophila* was routinely grown in/on

GY broth/agar (30 g/L glucose and 8 g/L yeast extract) at 45 °C. *P. pastoris* GS115 was grown in/on YPD broth/agar (10 g/L yeast extract, 20 g/L peptone, 20 g/L glucose, and 15 g/L agarose) at 30 °C. *Escherichia coli* DH5α (Biomed, Beijing, China) was used for gene cloning and vector construction, and it was grown in/on LB broth/agar (10 g/L NaCl, 10 g/L peptone, and 5 g/L yeast extract) at 37 °C.

2.2. Chemicals

The SE Seamless Cloning and Assembly Kit was purchased from Beijing Zoman Biotechnology Co., Ltd. (Beijing, China). All restriction endonucleases were purchased from Guangzhou Chuangrong Biotechnology Co., Ltd. (Guangzhou, China). The pre-stained protein marker and the DNA maker were purchased from Beijing Yishan Huitong Technology Co., Ltd. (Beijing, China). PCR primers were prepared by QingKe Biotechnology Company (Beijing, China). The DNA agarose recycling kit and the plasmid extraction kits were purchased from Omega (Norcross, GA, USA). The total RNA Extraction Kit was purchased from Shenzhen Yibaishun Technology Co., Ltd. (Shenzhen, China). SuperScript II reverse transcriptase was purchased from Thermo Fisher Scientific (Wilmington, MA, USA). All chemicals used were of analytical grade.

2.3. Bioinformatics Analysis of *Mt*EG5-1

The evolutionary location of *Mt*EG5-1 was analyzed using MEGA 7.0 software [18]. The signal peptide of *Mt*EG5-1 was predicted using the Signal P online website [19]. The tertiary structure of *Mt*EG5-1 was predicted using the SWISS-MODEL online website and PyMOL software [20].

2.4. Expression Vector Construction

M. thermophila was inoculated into GY liquid medium at 45 °C and 180 rpm for 24 h with shock culture. After filtration, mycelium was collected and stored in liquid nitrogen. The total RNA of *M. thermophila* was extracted using the total RNA Extraction Kit, and cDNA was obtained using reverse transcriptase. The entire coding region of *mtEG5* in *M. thermophila* was amplified via PCR from cDNA as a template using the primer *Eco*RI-*mtEG5*-F/R. The PCR program was set up with an initial denaturation step of 95 °C for 5 min, followed by 38 cycles of 95 °C for 10 s, 55 °C for 15 s, and 72 for 60 s, and a final extension at 72 °C for 10 min. The PCR product was purified using the glue recovery kit according to the manufacturer's protocol. The *mtEG5* PCR product with the *Eco*RI fragment was ligated to the *Eco*RI-digested and dephosphorylated vector, pPIC9K, at a 5:1 M ratio of insert: vector. The recombinant plasmid was transformed into *E. coli* DH5α. The correct vector was verified by sequencing after verification by performing colony PCR with the primers Y*mtEG5*-F/R. All the primers are listed in Table 1.

Table 1. Primers were used in this study.

Primers	Sequence (5′-3′) Restriction Italic/Underlined	Restriction Sites
*Eco*RI-*mtEG5*-F	CTGAAGCTTACGTAGAATTC [a] GGTCCGTGGCAGCAATGTGG	*Eco*RI [a]
*Eco*RI-*mtEG5*-R	CGGCCGCCCTAGGGAATTC [a] TTA*ATGGTGATGGTGATGATG* [b] CGGCAAGTACTTCTTCAAGA	*Eco*RI [a]
Y*mtEG5*-F	CGATGTTGCTGTTTTGCCAT	
Y*mtEG5*-R	GGTTACAAATAAAAAAGTAT	

[a] The *Eco*RI restriction site is underlined. [b] The HIS label is marked in italics.

2.5. Transformation and Selection of Positive Transformants

The *Pme*I-linearized pPIC9K-*mtEG5* plasmid and pPIC9K plasmid were transformed into the *P. pastoris* GS115 strain via electroporation (BTX, ECM630 at 1.5 kV, 200 Ω, 25 μF, and 5–10 ms with a 0.2 cm cuvette) [21]. After the transformation, 100 μL cuvette contents was spread on minimal dextrose agar plates (0.2 mg/mL biotin, 20 g/L glucose, 15 g/L agarose, and 13.4 g/L yeast nitrogen base, with ammonium sulfate and without amino acids) and incubated at 30 °C for 36 h. As a negative control, the plasmid of linearized pPIC9K was transferred onto the plates. Positive transformants were screened with YPD medium containing 1.75 mg/mL G418 and confirmed via PCR. One transformer, *P. pastoris* GS115, expressing the highest recombinant *Mt*EG5-1 activity level was selected for the analysis of recombinant enzymatic properties.

2.6. Fermentation of MtEG5-1 in P. pastoris GS115

A positive colony of *P. pastoris* GS115 expressing *Mt*EG5-1 was cultured in 100 mL of buffered glycerol-complex medium (10 g/L yeast extract, 20 g/L peptone, 13.4 g/L yeast nitrogen base with ammonium sulfate and without amino acids, 0.2 mg/mL biotin, 100 mM (pH 6.0) potassium phosphate, and 1.0% glycerol) grown at 30 °C and 180 pm for 48 h. The yeast cells were resuspended in 100 mL of buffered methanol-complex medium via the natural sinking method (10 g/L yeast extract, 20 g/L peptone, 13.4 g/L yeast nitrogen base with ammonium sulfate and without amino acids, 0.2 mg/mL biotin, 100 mM (pH 6.0) potassium phosphate, and 1.0% methanol). The suspension was grown in a 1 L baffled flask at 30 °C and 200 rpm for 48 h and supplemented with 1% methanol every 24 h for induction. To determine the time course of *Mt*EG5-1 expression in *P. pastoris* GS115, the fermented supernatant was collected daily (0–7 d), and the total *Mt*EG5-1 concentration (BCA Protein Assay Reagent) and the activity of *Mt*EG5-1 were analyzed [22]. Three replicates of each experiment were performed to obtain accurate data.

2.7. Enzymatic Assay of MtEG5-1

The activity of *Mt*EG5-1 was determined via the DNS method [23], which is based on the generation of reducing sugar ends from carboxymethyl cellulose (CMC) as a substrate and its reaction with dinitro salicylic acid. The activity of *Mt*EG5-1 was determined in an assay using 1% CMC in 50 mM citrate buffer (pH 4.8) at 50 °C for 30 min, and then the reaction was stopped with the addition of 3 mL of DNS reagent. The resulting mixture was placed in boiling water for 5 min for color development, and the amount of reducing ends was determined by measuring absorbance at 540 nm. As a negative control, the enzyme was replaced by the pPIC9K vector in *P. pastoris*. An endoglucanase unit (IU) is defined as the amount of enzyme that generates the equivalent to 1 μmol of glucose per minute in the abovementioned reaction conditions. Cell-free supernatant (CFS) was extracted from a 1% methanol-induced *P. pastoris* GS115 converter for 1–7 days, enzyme activity was measured daily, and 2 mL of CFS was collected and stored at −20 °C. Finally, CFS was run through SDS-PAGE from 1 to 7 days.

2.8. Activity Was Determined via Congo-Red Staining

P. pastoris GS115 transformant cultures were centrifuged at 12,000 rpm for 10 min, and the clarified CFS obtained after passing it through a 0.2 μm filter was used as the source of extracellular enzymes for the agar well diffusion assay. A total of 1% agarose gel supplemented with 1 g/L CMC and Congo-red was cast on a plate, and wells were made [24]. After adding 100 μL of CFS to each well, the CMC plate was incubated at 50 °C for 1 h. Gels were stained for 1 h, followed by de-staining with 1 mol/L NaCl for another 10 min (they were washed until there was no color).

2.9. Scanning Electron Microscopy (SEM) Analysis

As a natural lignocellulose material, corn straw has a dense and stable structure, which inhibits the decomposition and utilization of polysaccharide, thereby reducing the treatment efficiency and greatly increasing the production cost of industrial ethanol and other products [25]. To remove lignin and solubilize hemicellulose to break down the natural barrier of lignocellulose and increase the contact area with enzymes and microorganisms, corn straw pretreatment is necessary, which can greatly improve the availability of corn straw [26]. Dilute acid pretreatment can break the structure of lignocellulose and expose cellulose, thus providing hydrolysis sites for cellulase [27]. Zhang et al. pretreated rice straw with HCl, and the cellulose content of the pretreated straw increased from 38.10 to 52.54~56.09% [28]. In the study, corn straw was cut into long strips of 0.5 × 1.0 cm, pretreated with 2 mol/L HCl at 28 °C for 24 h, washed with distilled water until neutral, and dried at 60 °C. Pretreated corn straw was reacted with 1.5 mL of crude enzyme solution and sterile water in a water bath at 70 °C for 7 days. At the same time, untreated corn straw was also reacted with 1.5 mL sterile water in a water bath at 70 °C for 7 days as a blank control. The samples were sputter coated with platinum nanoparticles in an ion-sputter coater before viewing via SEM at 10 mm and 3 kV acceleration voltage and 1000× magnification [29].

2.10. Purification of Recombinant MtEG5-1 from P. pastoris GS115

In the process of protein purification, liquids used were filtered through a 0.22 μm filter membrane and ultrasonicated for more than 15 min. *Mt*EG5-1 was purified via Ni^{2+} affinity chromatography [30]. The system was balanced with 3–5 times more of the volume of Tris-HCl (pH 7.0). The supernatant after centrifugation was taken and slowly allowed to flow through the Ni column at a flow rate of 1 mL/min. The protein was eluted with the same buffer at a flow rate of 5 mL/min, the system pressure was no more than 1 MPa, and 2 mL of eluent was collected in each tube using AKTA Purifier 10. The weakly bound hetero proteins were washed with 10 mM imidazole, and the proteins were eluted with 60 mM imidazole. The target protein was eluted with 400 mM imidazole. The purified recombinant protein *Mt*EG5-1 was run using an SDS-PAGE gel.

2.11. Enzyme Characterization

The optimum pH for *Mt*EG5-1 was determined via the standard assay at different pH values (3–11) using either 50 mM sodium citrate buffer at pH 2–5, 50 mM phosphate buffer at pH 6–8, 100 mM Tris-HCl buffer at pH 9, or 50 mM $NaHCO_3$-NaOH buffer at pH 9–11. On the other hand, the stability at different pH conditions was determined after the incubation of *Mt*EG5-1 in the abovementioned buffers at 4 °C for 24 h and the measurement of the remaining activity after the standard assay. The optimal temperature of *Mt*EG5-1 was estimated by using the standard assay procedure at varying temperatures (30–90 °C) for 2 h. The highest activity level obtained under standard conditions was defined as 100% activity. To determine temperature stability, 0.38 mg of purified *Mt*EG5-1 was incubated at various temperatures (50, 60, and 70 °C) for different time intervals, and the residual activity level was measured via the standard assay procedure. The thermostability of *Mt*EG5-1 was tested at an optimal pH value. The optimal NaCl concentration of *Mt*EG5-1 was estimated via the standard assay procedure at NaCl final concentrations of 1–10 g/L at 50 °C for 2 h. The influence of metal ions on purified *Mt*EG5-1 activity was determined by incorporating metal ions such as 0.33 mol/L Na_2SO_4, $MgSO_4$, $MnSO_4$, $CuSO_4$, K_2SO_4, $FeSO_4$, and $NiSO_4$ solutions during the standard assay procedure; however, the reaction duration was 2 h.

2.12. Enzymatic Kinetics of MtEG5-1

CMC (0.2–2.0%) was used as a substrate to measure enzyme kinetics under standard conditions (50 mM sodium citrate buffer, pH 5.0, 70 °C). Michaelis–Menten constants (*Km*) and the maximum velocity (*Vm*) of purified *Mt*EG5-1 were determined by measuring the rate of CMC hydrolysis using a Lineweaver—Burk plot, as shown in Equation (1), where *V* is the reaction rate and [*S*] is the substrate concentration [31].

$$\frac{1}{V} = \frac{Km}{Vm} \times \frac{1}{[S]} + \frac{1}{Vm}, \tag{1}$$

3. Results and Discussion

3.1. Bioinformatics Analysis of MtEG5-1 and Construction of pPIC9K-mtEG5

The *mtEG5* gene (NCBI: XP_003659014.1) was obtained from *M. thermophila* ATCC 42464, which is also known as *Sporotrichum thermophile* ATCC 42464, *Thielavia heterothallica* ATCC 42464, and *Thermothelomyces thermophilus* ATCC 42464 (https://www.uniprot.org/taxonomy/573729 (accessed on 8 May 2023)). This species can grow at 45–55 °C [32]. The enzyme expressed by the *mtEG5* gene in *P. pastoris* GS115 was named *Mt*EG5-1 in this study. *Mt*EG5-1 (UniProtKB: G2Q5D8) has 389 amino acids and an 18-aa N-terminal signal peptide, as depicted in Figure 2a. Carbohydrate-active enzymes are classified based on their protein sequence and structure [33]. *Mt*EG5-1 belongs to glycoside hydrolase family 5 (GH5), one of the largest families among all GHs. The two catalytic residues in GH5 enzymes are both glutamic acids: one is a nucleophile and the other is a catalytic proton donor [34]. The carbohydrate-active enzyme database (http://cazy.org (accessed on 8 May 2023)) lists 53 sub-families, with approximately 80% of all known GH5 sequences falling into one of these subfamilies [35]. *Mt*EG5-1 belongs to the GH5-1 subfamily. The 3D structure of *Mt*EG5-1 was predicted to be a classical $(\alpha/\beta)_8$ TIM-barrel, which was connected to eight parallel β-strands and eight parallel α-helices through αβ or βα loops [33] (Figure 2b). These loops are typically found on the protein surface and play a key role in the interactions between the catalytic core and substrate [36,37]. During protein folding, the $(\alpha/\beta)_8$ barrel protein provides stability for hydrogen bonds and non-covalent interactions between amino acid residues to resist the local tendency for unfolding [38]. A mutation in these loops can alter their flexibility and improve the biochemical properties of cellulase [39]. The neighbor-joining method was used to further analyze the evolutionary position of *Mt*EG5-1, and we found that it shares 99% identity with homologous proteins from *Precursor Humicola insolens* (Figure 2c). Finally, the expression vector pPIC9K-*mtEG5* was constructed following the described procedure (Figure 2d).

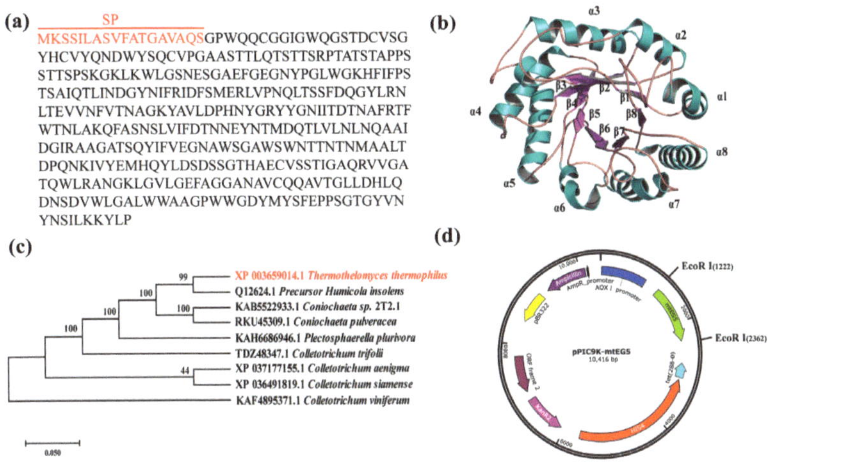

Figure 2. Bioinformatics analysis of *Mt*EG5-1. (**a**) The number and location of *Mt*EG5-1's signal peptide according to the Signal P 5.0 online website. (**b**) Homology-based (3D) model showing the barrel $(\alpha/\beta)_8$ fold in *mtEG5*. The labels α1-α8 are the eight parallel α-helices located on the outside of the TIM-barrel, and the labels β1-β8 are the eight parallel β-strands located on the inside of the TIM barrel. (**c**) Neighbor-joining phylogenetic tree constructed using *Mt*EG5-1's amino acid sequence along with other cellulase sequences retrieved from the NCBI database using MEGA 7.0 software. The digit at each branch point represents the percentage of bootstrap support calculated from 1000 replicates. The deduced amino acid sequence of *Mt*EG5-1 exhibited maximum homology with the endoglucanase of *P. Humicola insolens*. (**d**) Construction map of the recombinant expression vector pPIC9K-*mtEG5* in *P. pastoris* GS115 obtained using SnapGene 6.0.2 software.

3.2. Selection of *P. pastoris* GS115 Positive Transformants and Heterologous Expression

The sequencing of the *mtEG5* gene fragment revealed that it consists of 1174 base pairs. In order to identify positive transformants, eight random samples were selected, all of which showed the same band expansion as the target band in PCR amplification. Based on this result, transformants 1, 3, 6, and 8, which produced wide and bright bands, were selected for further experiments, as depicted in Figure 3a. Four transformants were then tested every 24 h, and with *Mt*EG5-1 activity induced by 1% methanol, and positive transformants with high copy numbers were selected for subsequent experiments. No corresponding bands were amplified in the CK control group. The hydrolysis of cellulose was observed in cellulose plates inoculated with the CFS of *Mt*EG5-1 culture, indicating the production of *Mt*EG5-1 by *P. pastoris* GS115 transformants. Cellulose hydrolysis was detected on the first day, indicating that *Mt*EG5-1 is capable of decomposing cellulose, as shown in Figure 3b. Daily detection of the CFS of cultured *P. pastoris* GS115 via 12% SDS-PAGE revealed a protein with a molecular weight of approximately 55 kDa, compared to that of wild-type yeast. The protein bands were clear from day 1 to day 7, indicating that *Mt*EG5-1 had good stability (Figure 3c). The concentration of *Mt*EG5-1 was 0.95 g/L on day 1, and the highest concentration was observed on day 5 at 1.15 g/L. The activity level of *Mt*EG5-1 was 133.76 IU/mL on day 1, with the highest activity level being observed on day 5 at 162.39 IU/mL. Furthermore, the enzyme activity level was maintained at above 100 IU/mL from days 1–7, providing further evidence of the stability of *Mt*EG5-1, as shown in Figure 3d.

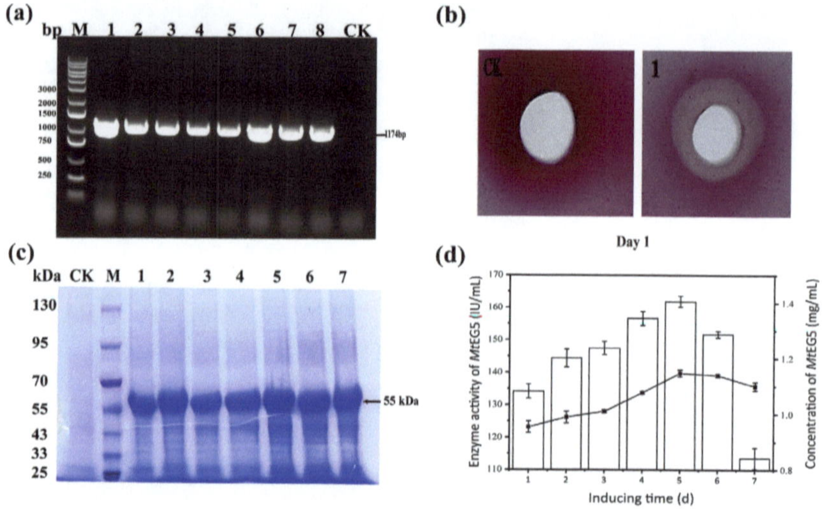

Figure 3. Positive transformants were screened for molecular identification and activity detection. (**a**) *P. pastoris* GS115 transformants of *mtEG5* verified by PCR. Lane M, marker 10,000 kb; Lanes 1–8, different *mtEG5* colonies; Lane 9, pPIC9K plasmid. (**b**) The CFS of *P. pastoris* transformed with pPIC9K-*mt*EG5 was stained with Congo-red on the CMC plate for the first day. The CK well was supplemented with 100 μL of CFS from pPIC9K yeast transformation, while well 1 was supplemented with 100 μL of CFS from pPIC9K-*mtEG5* yeast transformation. (**c**) The expression of *Mt*EG5-1 induced for different times was detected via 12% SDS-PAGE. Lane M, molecular weight marker; Lane CK, fermentation broth of *P. pastoris* transformed with pPIC9K induced by methanol; Lanes 1–7, fermentation broth of *P. pastoris* transformed with pPIC9K-*mtEG5* induced by methanol for the first day to the seventh day. Twenty microliters of pretreated sample were loaded in every lane. (**d**) The expression of *Mt*EG5-1 on different induction days was measured via the BCA method, and the activity of *Mt*EG5-1 on different induction days was measured via the DNS method.

3.3. Scanning Electron Microscopy of Hydrolyzed Corn Straw

We further studied whether the crude enzyme solution of *Mt*EG5-1 has an obvious ability to degrade corn straw. SEM can help to identify morphological changes in pretreated lignocellulosic materials [40]. SEM was used to compare the morphological changes in pretreated corn straw samples before and after catalysis by *Mt*EG5-1. Prior to pretreatment, the fiber surface of corn straw was smooth, with a dense and rigid structure and only a few cracks (Figure 4a). However, after the HCl pretreatment, the fiber of corn straw broke and disconnected at many points, which exposed the cellulose fibrils partially to the surface, thereby increasing the size of the contact site of cellulose with cellulase (Figure 4b). In contrast to the previous two images, the surface of the corn straw decomposed as a result of the use of *Mt*EG5-1 and displayed more fibrous nematocytic filaments, as demonstrated in Figure 4c. The findings indicate that *Mt*EG5-1 decomposed the pretreated corn straw. Nevertheless, future research is required to determine the specific decomposition capacity.

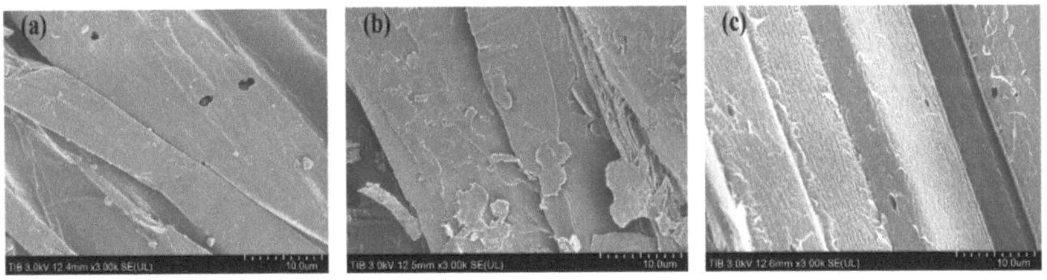

Figure 4. Analysis of the results of scanning electron microscopy. (**a**) Corn straw without any treatment. (**b**) Corn straw pretreated with 2 mol/L HCl for 24 h. (**c**) The CFS of 1.5 mL crude enzyme solution was mixed with the pretreated corn straw at 70 °C for seven days.

3.4. Characterization of Purified MtEG5-1

In this study, the purified recombinant, *Mt*EG5-1, demonstrates a clear band in the SDS-PAGE results, with an approximate size of 55 kDa (Figure 5a), which differs from the theoretical value of 43 kDa. Typically, cellulases are glycosylated enzymes [41]. Enzymatic modifications can increase the actual molecular weight beyond the theoretical value, as demonstrated by Karnaouri et al., who obtained a purified recombinant enzyme with a molecular weight of 83 kDa by using a purified protein expressed by yeast; deglycosylation reduced the molecular weight to 66 kDa [42].

Figure 5b, and c demonstrates that the recombinant enzyme reaches its maximum activity at an optimum pH of 5 and a temperature of 70 °C. After incubation in different pH buffers for 24 h, *Mt*EG5-1 was most stable at pH 5 and least stable at pH 11, with approximately 30% at pH 2. Therefore, *Mt*EG5-1 is an acidic and highly stable protein at pH 4–6 (Figure 5d). Furthermore, *Mt*EG5-1 also exhibited good thermal stability, retaining a relative activity level of above 88% at all temperatures until it was incubated for 2 h. Subsequently, the relative activity levels decreased to approximately 97, 85, and 72% at temperatures of 50, 60 and 70 °C, respectively, after 3 h of incubation (Figure 5e). The glycosylation of the protein may contribute to its thermal stability. Moreover, the optimization of glycosylation sites may serve as an effective and feasible strategy to enhance the enzymatic activity and thermostability [43]. In addition, the relative activity of *Mt*EG5-1 incubated with NaCl (1–10 g/L) at 50 °C for 2 h was above 80%, and the optimal salt concentration of *Mt*EG5-1 was 6 g/L (Figure 5f).

Metal ions have been found to cause structural changes in enzymes [44], resulting in alterations to their activity. This study investigates the effect of Na^+, Mg^{2+}, Mn^{2+}, Cu^{2+}, K^+, Fe^{2+}, and Ni^{2+} on the activity level of *Mt*EG5-1, which was incubated at 50 °C for 2 h. The results showed that Mn^{2+} increased the activity level of *Mt*EG5-1 by 142.6%, while K^+ decreased its activity level by 76.1% (Table 2). These results indicate that *Mt*EG5-1 has remarkable tolerance to harsh conditions, such as high salt concentrations, high temperatures, and low pH, making it an ideal candidate for cellulose degradation.

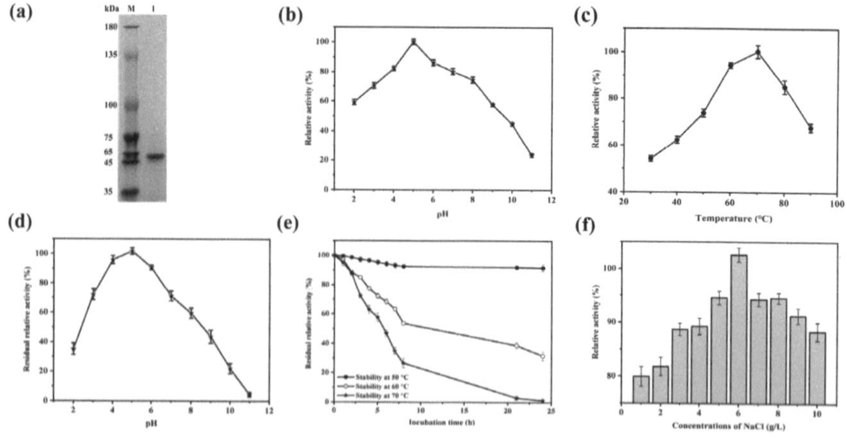

Figure 5. Effects of different factors on the activity of purified *Mt*EG5-1. (**a**) Purified *Mt*EG5-1 expression was detected via 12% SDS-PAGE. Lane M, molecular weight marker; Lane 1, a purified *Mt*EG5-1 protein. (**b**) Effects of assay pH on *Mt*EG5-1 activity. The standard assay was used, except the reaction pH is shown in the graph. (**c**) Effects of assay temperature on *Mt*EG5-1 activity. The standard assay was used, except the reaction temperature is shown in the graph. (**d**) Effects of pH on *Mt*EG5-1 stability. The enzyme was incubated for 24 h at 4 °C, and the pH values are indicated, followed by an activity assay under standard conditions. (**e**) Effects of temperature on *Mt*EG5-1 stability. The enzyme was incubated for the times plotted at the temperatures indicated and at pH 5 for 24 h, followed by an activity assay under standard conditions (black circle is 50 °C, white circle is 60 °C, and black pentagram is 70 °C). (**f**) Effects of NaCl concentration on *Mt*EG5-1 activity. The enzyme was incubated for 2 h at 50 °C and various NaCl concentrations indicated, followed by an activity assay under standard conditions.

Table 2. Effect of various metal ions on *Mt*EG5-1 activity. The standard assay was used, except the reaction duration was 2 h.

Metal Ion	Relative Activity (%)	Metal Ion	Relative Activity (%)
None	100	Cu^{2+}	100.2 ± 2.9
Na^{+}	86.1 ± 2.4	K^{+}	76.1 ± 1.6
Mg^{2+}	97.4 ± 1.3	Fe^{2+}	123.3 ± 3.8
Mn^{2+}	142.6 ± 2.5	Ni^{2+}	87.6 ± 2.2

Interestingly, the experimental results differ from those reported in another published paper. While *Mt*EG5-1 was successfully expressed in *P. pastoris* X33 [17] and *A. niger* [16], the properties of *Mt*EG5-1 expressed by *P. pastoris* GS115, as displayed in Table 3, were dissimilar. The optimal temperature and pH of *Mt*EG5-1 expressed by the three strains were 70 °C and 5, respectively. However, the relative activity levels of the enzyme expressed by *A. niger*, *P. pastoris* X33, and *P. pastoris* GS115 were determined at different temperatures and pH 5 levels. The results showed that the relative enzyme activity levels were still above 80% in the temperature ranges of 60–70 °C, 60–75 °C, and 60–80 °C. The relative activity levels of the enzyme expressed by *P. pastoris* X33 and *P. pastoris* GS115 were measured at 4 °C and different pH values for 24 h, while the relative activity level of the enzyme expressed by *A. niger* was measured at 37 °C and different pH values for 2 h. The results showed that the relative enzyme activity levels were still above 80% in the pH ranges of 4–6, 4–6, and 5.5–6.5. Notably, although there was no significant difference in the optimal temperature and pH of the expressed *Mt*EG5-1 among the three hosts, *Mt*EG5-1 exhibited a greater tolerance to high temperatures than the enzymes expressed by *P. pastoris* X33 and *A. niger* did. For instance, the thermal stability of *Mt*EG5-1 expressed by *P. pastoris* GS115 remained at the 88% residual activity level after 2 h at 70 °C, compared to the 25% residual activity level of the enzyme expressed in *P. pastoris* X33 after 2 h at 65 °C and the complete inactivation of the enzyme expressed in *A. niger* after 2 h at 60 °C. Altogether, the enzyme expressed by *P. pastoris* GS115 was resistant to a wider range of temperatures than the enzyme expressed by *A. niger* and *P. pastoris* X33 was. Moreover, the molecular weight of *Mt*EG5-1 expressed by *P. pastoris* X33 was 75 kDa, while the molecular weight of *Mt*EG5-1 expressed by *P. pastoris* GS115 was smaller (55 kDa). Generally, larger molecular weight proteins are less stable as they are more susceptible to denaturation as a result of chemical and physical factors [45,46]. Although *A. niger*, *P. pastoris* X33, and *P. pastoris* GS115 expressed the enzyme encoded by the *mtEG5* gene, the enzymatic properties and molecular weight of the expressed enzymes were different. Future analysis of the crystal structure of *Mt*EG5-1 could help to explain its characteristics. Therefore, the diverse expression of *Mt*EG5-1 in different hosts requires further investigation. The microbial cell factory constructed in our study can be applied more widely in the industry. As the properties of *Mt*EG5-1 were further verified and improved, this study provides an effective strategy for the utilization of cellulose resources.

Table 3. Differences between *Mt*EG5-1 in this experiment and other experiments.

	Karnaouri [17]	Tambor [16]	This Study
The optimum temperature	70	70	70
The optimum pH	5–6	6	5
Source of *mtEG5* gene	Genomic DNA	RNA to cDNA	RNA to cDNA
Host strain	*P. pastoris* X33	*A. niger*	*P. pastoris* GS115
The expression of the vector	pPICZαC	pDONR201 and pGBFIN-GTW	pPIC9K
The molecular weight of *Mt*EG5-1	75 kDa	–	55 kDa
The relative activity of *Mt*EG5-1 was above 80% at different temperatures (pH 5) for 2 h	60–75	60–70	60–80
Comparison of the stability of *Mt*EG5-1	25% residual activity after 2 h at 65 °C	inactivity after 2 h at 60 °C	88% residual activity after 2 h at 70 °C
The relative activity of *Mt*EG5-1 was above 80% at different pH for 24 h at 4 °C	4–6	5.5–6.5 (2 h, 37 °C)	4–6
Maximum level obtained *Mt*EG5-1 expression	0.98 g/L	>0.3 g/L	1.15 g/L

3.5. Enzyme Kinetic Parameters of Purified MtEG5-1

For different carbon source substrates, GH5 family cellulases have different substrate binding manners [47]. They are known to hydrolyze cellulose and non-cellulose substrates, usually acting on β-1,4 linkages [48]. In most cases, CMC is one of the best substrates for fungal cellulase [49]. For example, Ma et al. showed that Cel-5A was highly effective at hydrolyzing CMC [50]. By studying the cellulose decomposition activity of endoglucanase caused by bacterial flora, Salehi et al. suggested that endoglucanase could decompose all three types of substrates, including CMC, cellulose, and Avicel. The results showed that CMC was the best substrate for the purified enzyme, which was considered to have 100% activity. There was no significant difference between the specificity of the enzyme in the presence of CMC and cellulose. However, the activity levels recorded were 90% in the presence of cellulose and 35% in the presence of Avicel [51]. Based on the Lineweaver—Burk double inverse method, the values of Km and Vmax for purified *Mt*EG5-1 were approximately 6.11 mg/mL and 91.74 µmol/min/mg at 70 °C (pH 5), respectively (Figure 6). The k_{cat} value was found to be 84.94/s.

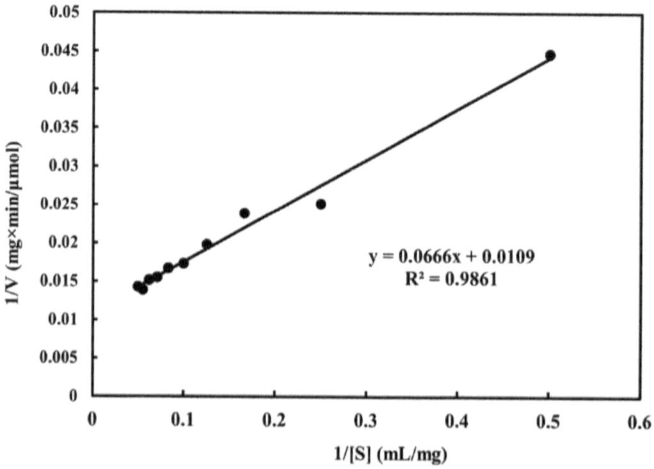

Figure 6. Lineweaver—Burk plot for *Mt*EG5-1 with CMC as substrate.

4. Conclusions

In this study, *Mt*EG5-1 was successfully overexpressed in *P. pastoris* GS115, and the enzymatic properties were characterized. Recombinant protein, *Mt*EG5-1, was stable in extreme environments, such as an acidic pH (4–6), a high temperature (60–80 °C), and the presence of salt (6 g/L), in our study. *Mt*EG5-1 expressed by *P. pastoris* GS115 is more tolerant to high temperatures than the enzyme expressed by *P. pastoris* X33 and *A. niger*. These unique characteristics suggest that *Mt*EG5-1 has a lot of potential for practical applications under extreme environmental conditions, including varying pH levels, osmolarities, and temperatures.

Author Contributions: Conceptualization, D.L.; methodology, W.Z. and S.T.; formal analysis, W.Z. and W.C.; investigation, W.Z.; validation, W.Z.; writing—original draft, W.Z., S.T.; writing—review and editing, S.T., F.R.A., J.C. and D.L.; conceptualization, S.T.; project administration, S.T., W.C., J.C. and D.L.; supervision, J.C. and D.L.; funding acquisition, D.L. All authors have read and agreed to the published version of the manuscript.

Funding: This work was supported by the Strategic Pilot Science and Technology Project of the Chinese Academy of Sciences (XDA28030301). Many thanks to Xu Jian for his valuable suggestions for this article.

Institutional Review Board Statement: Not applicable.

Informed Consent Statement: Not applicable.

Data Availability Statement: All data are included in the manuscript.

Conflicts of Interest: The authors declare no conflict of interest.

References

1. Correa, D.F.; Beyer, H.L.; Fargione, J.E.; Hill, J.D.; Possingham, H.P.; Thomas-Hall, S.R.; Schenk, P.M. Towards the implementation of sustainable biofuel production systems. *Renew. Sustain. Energy Rev.* **2019**, *107*, 250–263. [CrossRef]
2. Zhou, Z.; Liu, D.; Zhao, X. Conversion of lignocellulose to biofuels and chemicals via sugar platform: An updated review on chemistry and mechanisms of acid hydrolysis of lignocellulose. *Renew. Sustain. Energy Rev.* **2021**, *146*, 111169. [CrossRef]
3. Xia, Z.; Li, J.; Zhang, J.; Zhang, X.; Zheng, X.; Zhang, J. Processing and valorization of cellulose, lignin and lignocellulose using ionic liquids. *J. Bioresour. Bioprod.* **2020**, *5*, 79–95. [CrossRef]
4. Tan, J.; Li, Y.; Tan, X.; Wu, H.; Li, H.; Yang, S. Advances in Pretreatment of Straw Biomass for Sugar Production. *Front. Chem.* **2021**, *9*, 696030. [CrossRef]
5. Kadowaki, M.A.S.; Higasi, P.; de Godoy, M.O.; Prade, R.A.; Polikarpov, I. Biochemical and structural insights into a thermostable cellobiohydrolase from *Myceliophthora thermophila*. *FEBS J.* **2018**, *285*, 559–579. [CrossRef] [PubMed]

6. Himmel, M.E.; Ruth, M.F.; Wyman, C.E. Cellulase for commodity products from cellulosic biomass. *Curr. Opin. Biotechnol.* **1999**, *10*, 358–364. [CrossRef]
7. Yang, W.; Fan, H.; Zhou, M.; Zhou, Z.; Yan, L.; Ju, X.; Li, L. Synergistic effect of ionic liquid and surfactant for enzymatic hydrolysis of lignocellulose by *Paenibacillus* sp. LLZ1 cellulase. *Biomass Bioenergy* **2020**, *142*, 105760. [CrossRef]
8. Hamalainen, V.; Barajas Lopez, J.D.; Berlina, Y.; Alvarez Rafael, R.; Birikh, K. New thermostable endoglucanase from *Spirochaeta thermophila* and its mutants with altered substrate preferences. *Appl. Microbiollogy Biotechnol.* **2021**, *105*, 1133–1145. [CrossRef]
9. Singh, B. *Myceliophthora thermophila* syn. *Sporotrichum thermophile*: A *thermophilic* mould of biotechnological potential. *Crit. Rev. Biotechnol.* **2016**, *36*, 59–69. [CrossRef]
10. Chen, X.T.; Li, W.G.; Ji, P.; Zhao, Y.; Hua, C.Y.; Han, C. Engineering the conserved and noncatalytic residues of a thermostable beta-1,4-endoglucanase to improve specific activity and thermostability. *Sci. Rep.* **2018**, *8*, 2954. [CrossRef]
11. Javanmard, A.S.; Matin, M.M.; Bahrami, A.R. Polycistronic cellulase gene expression in *Pichia pastoris*. *Biomass Convers. Biorefinery* **2021**, *7*, 1–15. [CrossRef]
12. Kumar, S.; Nussinov, R. How do thermophilic proteins deal with heat? *Cell. Mol. Life Sci. CMLS* **2001**, *58*, 1216–1233. [CrossRef] [PubMed]
13. Da Rosa-Garzon, N.G.; Laure, H.J.; Rosa, J.C.; Cabral, H. Valorization of agricultural residues using *Myceliophthora thermophila* as a platform for production of lignocellulolytic enzymes for cellulose saccharification. *Biomass Bioenergy* **2022**, *161*, 106452. [CrossRef]
14. Wang, J.; Wu, Y.; Gong, Y.; Yu, S.; Liu, G. Enhancing xylanase production in the thermophilic fungus Myceliophthora thermophila by homologous overexpression of Mtxyr1. *J. Ind. Microbiol. Biotechnol.* **2015**, *42*, 1233–1241. [CrossRef]
15. Saxena, S.; Shrivastava, S.; Arora, R.; Hussain, S.; Jena, S.C.; Kumar, M.; Vasu, R.K.; Srivastava, S.; Sharma, P.; Kumar, N.; et al. Development of Real-Time PCR Assays for Detecting Matrix Metalloproteinases-2 & 9 Over-expression in Canine Mammary Tumours. *Adv. Anim. Vet. Sci.* **2016**, *4*, 342–345. [CrossRef]
16. Tambor, J.H.; Ren, H.; Ushinsky, S.; Zheng, Y.; Riemens, A.; St-Francois, C.; Tsang, A.; Powlowski, J.; Storms, R. Recombinant expression, activity screening and functional characterization identifies three novel endo-1,4-beta-glucanases that efficiently hydrolyse cellulosic substrates. *Appl. Microbiol. Biotechnol.* **2012**, *93*, 203–214. [CrossRef] [PubMed]
17. Karnaouri, A.; Muraleedharan, M.N.; Dimarogona, M.; Topakas, E.; Rova, U.; Sandgren, M.; Christakopoulos, P. Recombinant expression of thermostable processive MtEG5 endoglucanase and its synergism with MtLPMO from *Myceliophthora thermophila* during the hydrolysis of lignocellulosic substrates. *Biotechnol. Biofuels* **2017**, *10*, 2–17. [CrossRef] [PubMed]
18. Malik, Y.S.; Sircar, S.; Bhat, S.; Sharun, K.; Dhama, K.; Dadar, M.; Tiwari, R.; Chaicumpa, W. Emerging novel coronavirus (2019-nCoV)—Current scenario, evolutionary perspective based on genome analysis and recent developments. *Vet. Q.* **2020**, *40*, 68–76. [CrossRef]
19. Reynolds, S.M.; Käll, L.; Riffle, M.E.; Bilmes, J.A.; Noble, W.S. Transmembrane Topology and Signal Peptide Prediction Using Dynamic Bayesian Networks. *PLOS Comput. Biol.* **2008**, *4*, e1000213. [CrossRef]
20. Biasini, M.; Bienert, S.; Waterhouse, A.; Arnold, K.; Studer, G.; Schmidt, T.; Kiefer, F.; Cassarino, T.G.; Bertoni, M.; Bordoli, L.; et al. SWISS-MODEL: Modelling protein tertiary and quaternary structure using evolutionary information. *Nucleic Acids Res.* **2014**, *42*, 252–258. [CrossRef]
21. Karbalaei, M.; Rezaee, S.A.; Farsiani, H. Pichia pastoris: A highly successful expression system for optimal synthesis of heterologous proteins. *J. Cell. Physiol.* **2020**, *235*, 5867–5881. [CrossRef] [PubMed]
22. Hill, H.D.; Straka, J.G. Protein determination using bicinchoninic acid in the presence of sulfhydryl reagents. *Anal. Biochem.* **1988**, *170*, 203–208. [CrossRef] [PubMed]
23. Miller, G.L. Use of Dinitrosalicylic Acid Reagent for Determination of Reducing Sugars. *Anal. Chem.* **1959**, *31*, 426–428. [CrossRef]
24. Gavande, P.V.; Basak, A.; Sen, S.; Lepcha, K.; Murmu, N.; Rai, V.; Mazumdar, D.; Saha, S.P.; Das, V.; Ghosh, S. Functional characterization of thermotolerant microbial consortium for lignocellulolytic enzymes with central role of *Firmicutes* in rice straw depolymerization. *Sci. Rep.* **2021**, *11*, 3032. [CrossRef]
25. Sun, N.; Rahman, M.; Qin, Y.; Maxim, M.L.; Rodríguez, H.; Rogers, R.D. Complete dissolution and partial delignification of wood in the ionic liquid 1-ethyl-3-methylimidazolium acetate. *Green Chem.* **2009**, *11*, 646–655. [CrossRef]
26. Sorn, V.; Chang, K.-L.; Phitsuwan, P.; Ratanakhanokchai, K.; Dong, C.-D. Effect of microwave-assisted ionic liquid/acidic ionic liquid pretreatment on the morphology, structure, and enhanced delignification of rice straw. *Bioresour. Technol.* **2019**, *293*, 121929. [CrossRef]
27. Sahoo, D.; Ummalyma, S.B.; Okram, A.K.; Pandey, A.; Sankar, M.; Sukumaran, R.K. Effect of dilute acid pretreatment of wild rice grass (*Zizania latifolia*) from Loktak Lake for enzymatic hydrolysis. *Bioresour. Technol.* **2018**, *253*, 252–255. [CrossRef] [PubMed]
28. Zhang, W.; Liu, J.; Wang, Y.; Sun, J.; Huang, P.; Chang, K. Effect of ultrasound on ionic liquid-hydrochloric acid pretreatment with rice straw. *Biomass Convers. Biorefinery* **2020**, *11*, 1749–1757. [CrossRef]
29. Ghorbani, S.; Eyni, H.; Bazaz, S.R.; Nazari, H.; Asl, L.S.; Zaferani, H.; Kiani, V.; Mehrizi, A.A.; Soleimani, M. Hydrogels Based on Cellulose and its Derivatives: Applications, Synthesis, and Characteristics. *Polym. Sci. Ser. A* **2018**, *60*, 707–722. [CrossRef]
30. Dalal, S.; Raghava, S.; Gupta, M. Single-step purification of recombinant green fluorescent protein on expanded beds of immobilized metal affinity chromatography media. *Biochem. Eng. J.* **2008**, *42*, 301–307. [CrossRef]
31. Lindmo, T.; Boven, E.; Cuttitta, F.; Fedorko, J.; Bunn, P.A., Jr. Determination of the immunoreactive function of radiolabeled monoclonal antibodies by linear extrapolation to binding at infinite antigen excess. *J. Immunol. Methods* **1984**, *72*, 77–89. [CrossRef] [PubMed]

32. Turner, P.; Mamo, G.; Karlsson, E.N. Potential and utilization of thermophiles and thermostable enzymes in biorefining. *Microb. Cell Factories* **2007**, *6*, 1–23. [CrossRef] [PubMed]
33. Wierenga, R. The TIM-barrel fold: A versatile framework for efficient enzymes. *FEBS Lett.* **2001**, *492*, 193–198. [CrossRef] [PubMed]
34. Aspeborg, H.; Coutinho, P.M.; Wang, Y.; Brumer, H.; Henrissat, B. Evolution, substrate specificity and subfamily classification of glycoside hydrolase family 5 (GH5). *BMC Evol. Biol.* **2012**, *12*, 1–16. [CrossRef] [PubMed]
35. Singh, S.; Kumar, K.; Nath, P.; Goyal, A. Role of glycine 256 residue in improving the catalytic efficiency of mutant endoglucanase of family 5 glycoside hydrolase from *Bacillus amyloliquefaciens* SS35. *Biotechnol. Bioeng.* **2020**, *117*, 2668–2682. [CrossRef]
36. Tu, T.; Pan, X.; Meng, K.; Luo, H.; Ma, R.; Wang, Y.; Yao, B. Substitution of a non-active-site residue located on the T3 loop increased the catalytic efficiency of endo-polygalacturonases. *Process Biochem.* **2016**, *51*, 1230–1238. [CrossRef]
37. Zhai, X.; Amyes, T.L.; Richard, J.P. Role of loop-clamping side chains in catalysis by triosephosphate isomerase. *J. Am. Chem. Soc.* **2015**, *137*, 15185–15197. [CrossRef]
38. Gromiha, M.M.; Pujadas, G.; Magyar, C.; Selvaraj, S.; Simon, I. Locating the stabilizing residues in (alpha/beta)8 barrel proteins based on hydrophobicity, long-range interactions, and sequence conservation. *Proteins* **2004**, *55*, 316–329. [CrossRef]
39. Badieyan, S.; Bevan, D.R.; Zhang, C. Study and design of stability in GH5 cellulases. *Biotechnol. Bioeng.* **2012**, *109*, 31–44. [CrossRef]
40. Silva, T.P.; de Albuquerque, F.S.; Nascimento Ferreira, A.; Santos, D.; Santos, T.V.D.; Meneghetti, S.M.P.; Franco, M.; Luz, J.; Pereira, H.J.V. Dilute acid pretreatment for enhancing the enzymatic saccharification of agroresidues using a Botrytis ricini endoglucanase. *Biotechnol. Appl. Biochem.* **2022**, *70*, 184–192. [CrossRef]
41. Beckham, G.T.; Dai, Z.; Matthews, J.F.; Momany, M.; Payne, C.M.; Adney, W.S.; Baker, S.E.; Himmel, M.E. Harnessing glycosylation to improve cellulase activity. *Curr. Opin. Biotechnol.* **2012**, *23*, 338–345. [CrossRef] [PubMed]
42. Karnaouri, A.; Topakas, E.; Antonopoulou, I.; Christakopoulos, P. Genomic insights into the fungal lignocellulolytic system of *Myceliophthora thermophila*. *Front. Microbiol.* **2014**, *5*, 281. [CrossRef] [PubMed]
43. Han, C.; Wang, Q.Q.; Sun, Y.X.; Yang, R.R.; Liu, M.Y.; Wang, S.Q.; Liu, Y.F.; Zhou, L.F.; Li, D.C. Improvement of the catalytic activity and thermostability of a hyperthermostable endoglucanase by optimizing N-glycosylation sites. *Biotechnol. Biofuels* **2020**, *13*, 11–30. [CrossRef] [PubMed]
44. Schiffmann, R.; Heine, A.; Klebe, G.; Klein, C.D. Metal ions as cofactors for the binding of inhibitors to methionine aminopeptidase: A critical view of the relevance of in vitro metalloenzyme assays. *Angew. Chem. Int. Ed.* **2005**, *44*, 3620–3623. [CrossRef]
45. Shoichet, B.K.; Baase, W.A.; Kuroki, R.; Matthews, B.W. A relationship between protein stability and protein function. *BIochemistry* **1995**, *92*, 452–456. [CrossRef]
46. Deller, M.C.; Kong, L.; Rupp, B. Protein stability: A crystallographer's perspective. *Acta Crystallogr. Sect. F* **2016**, *72*, 72–95. [CrossRef]
47. Bianchetti, C.M.; Brumm, P.; Smith, R.W.; Dyer, K.; Hura, G.L.; Rutkoski, T.J.; Phillips Jr, G.N. Structure, dynamics, and specificity of endoglucanase D from *Clostridium cellulovorans*. *J. Mol. Biol.* **2013**, *425*, 4267–4285. [CrossRef]
48. Henrissat, B. A classification of glycosyl hydrolases based on amino acid sequence similarities. *Biochem. J.* **1991**, *280*, 309–316. [CrossRef]
49. Niyonzima, F.N. Detergent-compatible fungal cellulases. *Folia Microbiol.* **2021**, *66*, 25–40. [CrossRef]
50. Ma, L.; Aizhan, R.; Wang, X.; Yi, Y.; Shan, Y.; Liu, B.; Zhou, Y.; Lu, X. Cloning and characterization of low-temperature adapted GH5-CBM3 endo-cellulase from *Bacillus subtilis* 1AJ3 and their application in the saccharification of switchgrass and coffee grounds. *AMB Express* **2020**, *10*, 42. [CrossRef]
51. Salehi, M.E.; Asoodeh, A. Extraction, Purification, and Biochemical Characterization of an Alkalothermophilic Endoglucanase from Bacterial Flora in Gastrointestinal Tract of Osphranteria coerulescens Larvae. *Waste Biomass Valorization* **2022**, *14*, 1251–1265. [CrossRef]

fermentation

Article

Biotechnological Production of Fumaric Acid by *Rhizopus arrhizus*—Reaching Industrially Relevant Final Titers

Anja Kuenz *, Laslo Eidt and Ulf Prüße

Thünen Institute of Agricultural Technology, 38116 Braunschweig, Germany; lasloeidt@gmx.de (L.E.); ulf.pruesse@thuenen.de (U.P.)
* Correspondence: anja.kuenz@thuenen.de

Abstract: Fumaric acid is used in various areas of the chemical industry due to its functional groups. For example, it is used in the polymer industry to produce unsaturated polyester resins, which are nowadays mostly produced from fossil raw materials. With regard to sustainable biotechnological fumaric acid production, the main challenge is to develop a cost-effective and robust fermentation process with industrially relevant final titers, productivities and yields. For biotechnological fumaric acid production, mainly fungi of the genus *Rhizopus* are used, which require very complex and challenging morphology control. The aim of this work is the development of an effective biotechnological fumaric acid production process with *R. arrhizus* NRRL 1526. Significant insights into the morphology control of the fungus and optimization of production characteristics were obtained, and a final titer of 86.3 g/L fumaric acid was achieved in a batch cultivation, with a yield of 0.67 g/g and a productivity of 0.60 g/(L·h). In addition, a fed-batch process was developed, in which the production phase was extended, and a maximum final titer of 195.4 g/L fumaric acid was achieved. According to current knowledge, this value is the highest final concentration of fumaric acid produced using biotechnology.

Keywords: fumaric acid; *Rhizopus arrhizus*; process optimization; fermentation

Citation: Kuenz, A.; Eidt, L.; Prüße, U. Biotechnological Production of Fumaric Acid by *Rhizopus arrhizus*—Reaching Industrially Relevant Final Titers. *Fermentation* **2023**, *9*, 588. https://doi.org/10.3390/fermentation9070588

Academic Editor: Zhihua Liu

Received: 23 May 2023
Revised: 14 June 2023
Accepted: 20 June 2023
Published: 23 June 2023

1. Introduction

Fumaric acid (FA) is a naturally occurring unsaturated dicarboxylic acid with diverse fields of application. Due to its double bond and two carboxylic groups, FA is used as comonomer for polymerization and esterification reactions, resulting in unsaturated polyester resins and alkyd resins. In the pulp industry, FA is applied as acidic tackifier for the production of rosin paper [1–3]. Additionally, FA is used as acidulant and nutritional additive in the food and feed sector [1,4,5]. More recently, derivatives of FA like fumaric acid esters have been identified to have potential biomedical applications, such as multiple sclerosis and psoriasis treatment [6–8]. All these different fields of application lead to a growing demand for FA, which is currently fulfilled exclusively through the chemical synthesis of petroleum-derived maleic acid anhydride. However, as petroleum prices are rising and the emphasis on low carbon footprint production strategies is increasing, there is a renewed interest in the biotechnological production of FA, which was operational during the 1940s by Pfizer, but was discontinued due to the more economical method of petrochemical-based synthesis [4].

For the microbial production of FA, multiple fungal strains have been identified as natural overproducers, including *Rhizopus* species (e.g., *arrhizus* and *oryzae*) as the most promising productions strains [9–12]. In fungal metabolism, the substrate glucose, after conversion into pyruvate, is converted into fumaric acid via two different metabolic pathways: (1) in the tricarboxylic acid (TCA) cycle, which is present in all eukaryotic organisms and occurs in the mitochondria and (2) in the reductive TCA cycle, which occurs in the cytosol. The reductive TCA cycle is responsible for the overproduction of fumaric acid in filamentous fungi, and requires CO_2 fixation [1,13].

Depending on the metabolic pathway, different theoretical yields are calculated. (1) The formation of fumaric acid via the TCA cycle allows a theoretical yield of a maximum of 1 mol of fumaric acid/mol of glucose consumed, or 0.64 $g_{FA}/g_{glucose}$. (2) Reductive carboxylation allows a maximum yield of 2 moles of fumaric acid/mole of glucose consumed, or 1.29 $g_{FA}/g_{glucose}$. In practice, however, a lower yield is expected, since an exclusive course of reductive carboxylation would lead to an energy deficit due to ATP consumption during CO_2 fixation. Therefore, in the fermentative production of fumaric acid, the oxidative citric acid cycle is also active to maintain the energy balance [14–16].

To reach high product titer, productivity and product yield via submerged fermentations, the control of fungal morphology is one of the most challenging tasks within the bioprocess. In general, the formation of fungal biomass can be distinguished as clumps, disperse filaments or pellets [1,17]. Taking the specific production potential into account, clump and filamentous morphologies show a high tendency to grow on cultivation equipment like bioreactor internals and walls [18–20]. This growth behavior causes a poor oxygen supply and leads to the unintentional production of ethanol [13]. In contrast, the formation of small spherical pellets promotes oxygen mass transfer due to a lower overall medium viscosity [13,15,21]. In recent decades, a wide variety of scientific research has been published in the field of morphological control, focusing on the influence of different cultivation parameters. Identified factors affecting the growth behavior and the production performance include physical and chemical parameters like the cultivation system itself, inoculum size, working volume, agitation, aeration, pH and temperature. The second group of factors are medium-related parameters such as nutrient supply, the use of complex medium components and the concentration of metal ions [18]. The concentration of provided nitrogen in particular affects the relationship between additional biomass growth and the accumulation of FA during the production phase [3]. All the previously mentioned parameters influence the final type of morphology and necessitate a targeted optimization of each process stage during the development of a promising strategy for biotechnological FA production.

Comparing the numerous published approaches, Ling and Ng (1989) described in a patent of Du Pont a fermentation procedure with a regulated oxygen concentration of 80% during the production phase [12]. This strategy allowed the production of 135.3 g/L FA with a productivity of 1.77 g/(L·h) and a yield of 1.04 g/g. To date, this fermentation represents the best biotechnological production of FA with the highest final titer. Furthermore, later published studies stayed remarkably far behind these results [1,13,17,22], making the fermentative procedure of Ling and Ng (1989) [12] a highly interesting template for further investigations [23].

To control the morphology and, therefore, achieve reproducible and comparable cultivations, the effect of the carbon to nitrogen (C/N) ratio was investigated. Based on this batch fermentation, a specific feeding strategy of glucose and ammonium was developed to extend the production phase and reach industrially relevant final titers of FA.

2. Materials and Methods

2.1. Microorganism and Inoculum Preparation

Rhizopus arrhizus NRRL 1526 was obtained from the Agricultural Research Service Culture Collection (Peoria, IL, USA). For experimental usage, the strain was stored in the form of spores in 50% glycerol at −80 °C (stock culture). For inoculum preparation, a stock culture was spread on agar plates (medium A) at 32 °C, containing 4 g/L glucose, 10 mL/L glycerol, 6 g/L lactose, 0.6 g/L urea, 0.4 g/L KH_2PO_4, 1 mL/L corn steep liquid, 1.6 g/L tryptone/peptone, 0.3 g/L $MgSO_4 \times 7 H_2O$, 0.088 g/L $ZnSO_4 \times 7 H_2O$, 0.25 g/L $FeSO_4 \times 7 H_2O$, 0.038 g/L $MnSO_4 \times H_2O$, 0.00782 g/L $CuSO_4 \times 5 H_2O$, 40 g/L NaCl, 0.4 g/L KCl and 30 g/L agar-agar. After six days of sporulation, the spores were suspended by adding 0.9 % (*w*/*w*) NaCl solution. The resulting spore solution was stored at 4 °C until inoculation. The spore concentration was determined using a counting chamber (Thoma) and a Zeiss microscope (Axioplan, Carl Zeiss AG, Oberkochen, Germany). Chemicals were either purchased from Merck KGaA (Darmstadt, Germany), Carl Roth GmbH and Co. KG (Karlsruhe, Germany) or from Sigma Aldrich (St. Louis, MO, USA) in an appropriate purity for biochemistry.

2.2. Pre-Culture Conditions

For pre-culture, 500 mL shaking flasks (unbaffled) with 50 g/L $CaCO_3$ (precipitated, ≥99%, VWR) were sterilized (20 min at 121 °C) then 100 mL of sterile fermentation medium B was added. The composition of medium B was 130 g/L glucose, 1.2 g/L $(NH_4)_2SO_4$, 0.3 g/L KH_2PO_4, 0.4 g/L $MgSO_4 \times 7\ H_2O$, 0.044 g/L $ZnSO_4 \times 7\ H_2O$ and 0.0075 g/L $FeCl_3 \times 6\ H_2O$. All components of medium B were prepared separately in stock solutions and heat-sterilized; the iron solution was sterile-filtered. For inoculation, spore suspension was added to ensure an initial concentration of 1×10^5 spores/mL. Pre-cultures were carried out at 34 °C and 200 rpm in a rotary shaking flask incubator for 24 h.

2.3. Batch Fermentations

Batch cultivations were studied in 500 mL shaking flasks (unbaffled), each containing 50 g/L $CaCO_3$ and 90 mL medium B at 34 °C, in a rotary incubator at 200 rpm. Then, 10 mL of the 24 h seed culture was transferred into the fermentation medium for inoculation. The solubility of FA at room temperature is very low, 5–7 g/L [22]. $CaCO_3$ is used as neutralizing agent, so calcium fumarate precipitates due to low solubility [17]. Next, 2 mL well-mixed samples were taken periodically, diluted with a 5% (*w*/*w*) HCl solution and heated to 80 °C to remove excessive $CaCO_3$ and resolve precipitated calcium fumarate. After 2 d of fermentation, 20 g/L $CaCO_3$ was added under sterile conditions to maintain a pH value of approximately 6.0. Within all shaking flask cultivations, the weight loss due to evaporation was balanced by the addition of deionized water. All cultivations were performed in duplicate.

2.4. Fed-Batch Fermentations

To prolong the phase of FA production, batch fermentations in shaking flasks were extended by the addition of 40 g/L or 80 g/L glucose (solid) before reaching a glucose concentration of less than 30 g/L. In addition, whenever the cultivation showed decreasing FA productivity, 0.6 g/L or 1.2 g/L $(NH_4)_2SO_4$ (200 g/L stock solution) was transferred into the broth. Depending on the over production of organic acids, portions of 20 g/L $CaCO_3$ were added periodically for pH control. Within all shaking flask cultivations, the weight loss due to evaporation was balanced by the addition of deionized water. All cultivations were performed in duplicate.

2.5. Analytical Methods

Glucose, FA, other organic acids (malic acid and succinic acid) and ethanol were quantified through high-performance liquid chromatography (HPLC) with an HPX-87H ion-exclusion column (300 × 7.8 mm) (Bio-Rad, Hercules, CA, USA), a refraction index (RI) detector (Shodex RI-101, Shōwa Denkō, Tokyo, Japan) and ultraviolet (UV) detector (LaChrom Elite L-2400, Hitachi, Tokyo, Japan) at 250 nm. The column was tempered at 40 °C and eluted with a 5 mM H_2SO_4 solution at a flow rate of 0.6 mL/min.

The concentration of the cation ammonium and the anion phosphate were determined through ion chromatography (IC) (Dionex ICS-100 IC, Thermo Fisher Scientific Inc., Sunnyvale, CA, USA). Cations were quantified with an IonPac CS16 column (cation-IC, 5 × 250 mm) and a suppressor CSRS-500 (4 mm) at 88 mA and a temperature of 40 °C, and eluted with 30 mM methlysulfonic acid at a flow rate of 1 mL/min. Anions were quantified with an IonPac AG11-HC (cation-IC, 4 × 250) and a suppressor ASRS-300 (4 mm) at 68 mA and a temperature of 40 °C, and eluted with 25 mM sodium hydroxide at a flow rate of 1 mL/min. For measurement, samples were diluted with ultrapure water, filtered using a nylon syringe filter (pore size 0.22 µm), and a sample volume of 2 mL was transferred to a Dionex Polyvial with a filter cap (Thermo Fisher Scientific Inc., Sunnyvale, CA, USA).

External cation and anion multi-element standards (Carl Roth GmbH + Co. KG, Karlsruhe, Germany) were measured for each series of measurements to calibrate the IC system.

The morphology of *R. arrhizus* NRRL 1526 was documented with a phase-contrast microscope (Axioplan, Carl Zeiss AG, Oberkochen, Germany) and processed with the software Analysis (Analysis 5.0, Soft Imaging System GmbH).

The yield is calculated from the product concentration of fumaric acid relative to the concentration of substrate consumed up to that cultivation time and is described by the product yield coefficient (Equation (1)).

$$Y_{P/S} = c_{FAt}/(c_{S0} - c_{St}), \quad (1)$$

where:

$Y_{P/S}$—product yield coefficient (g/g);
c_{FAt}—fumaric acid concentration at time t (g/L);
c_{S0}—initial concentration of the substrate (g/L);
c_{St}—substrate concentration at time t (g/L).

Productivity is calculated from the quotient of the concentration of fumaric acid formed and the cultivation period required for this (Equation (2)). Here, the maximum productivity is defined as the productivity that assumes the highest value between two sampling points.

$$P = (c_{Fat} - c_{FA0})/\Delta t, \quad (2)$$

where:

P—productivity (g/(L·h));
c_{FAt}—fumaric acid concentration at time t (g/L);
c_{FA0}—initial concentration of fumaric acid (g/L);
Δt—cultivation period (h).

3. Results and Discussion

Based on the procedure of Ling and Ng (1989) as a benchmark, an effective manufacturing process for the biotechnological production of fumaric acid was investigated in the present study [12].

Cultivation with the process strategy of Ling and Ng (1989) allowed the production of 29.5 g/L fumaric acid within 6 d in shake flasks. Under complete glucose consumption, this corresponds to a yield of 0.26 g/g, with a productivity of 0.21 g/(L·h). As by-products, malic and succinic acid with, in sum, a maximum of 5.7 g/L after 5 d were detected. The main by-product produced within the first two days of cultivation time was ethanol, formed at a concentration of 18.1 g/L. The accumulation of ethanol here indicates oxygen-limiting conditions within the cultivation system. The biomass of *R. arrhizus* NRRL 1526 was predominantly present as dense pellets with a diameter of >1 mm in the main culture. This form of compact biomass creates potentially anaerobic conditions inside the pellets that favor ethanol formation while negatively affecting fumaric acid production [24]. Comparable to the findings in the literature, a direct transfer of Ling and Ng's (1989) cultivation strategy and their published results of 135.3 g/L FA with a productivity of 1.77 g/(L·h) and a yield of 1.04 g/g was not possible.

Therefore, the influence of different cultivation parameters on morphology, reproducibility, productivity and yield was investigated in the following experiments. Subsequently, a process strategy was developed on the basis of the findings obtained.

3.1. Optimizing Cultivation Parameters

- Influence of process steps

According to Ling and Ng (1989), three individual process steps (spore production, preculture and main culture) and glucose as the carbon source were used for FA production [12]. To date, no cultivation conditions have been observed to induce and favor submerged spore production of *R. arrhizus* NRRL 1526. Therefore, spores were produced via the surface method using agar plates.

Within the overall process, during the pre-culture, the germination of spores and building up biomass take place. Various cultivation parameters, such as the size of shake flasks, the shaking speed, the cultivation temperature or the composition of the medium, have a major influence on the morphology of the pre-culture and, thus, also on the cultivation result of the main culture [17,21,25,26]. The morphology of *Rhizopus* sp. can be divided into three different growth forms: clumps, pellets and mycelium. The production behavior of FA differs significantly depending on the corresponding growth form of the fungus [1]. Which form of morphology is formed within the cultivation system depends largely on the process parameters described above. However, these influencing factors cannot be considered alone. Rather, the morphology results from a complex interaction of all cultivation parameters. This fact not only limits the comparability of different research studies with each other, but also makes it difficult to identify generally valid regularities in the targeted control of morphology [3,25].

Regarding the potential to accumulate fumaric acid, various publications exist that achieved more effective production of FA with either mycelium or pellet morphology. For example, Rhodes et al. (1962) and Papadaki et al. (2017) realized better FA production using mycelium morphology than pellets [26,27]. Opposing results were obtained in studies by Liao et al. (2007) and Zhou et al. (2011) [21,28]. The contradictory results of these publications thus indicate that a generally valid preference of a certain growth form is not possible. Rather, when developing an effective cultivation system for the production of FA, it makes sense to identify specifically occurring problems and optimization potentials in the context of morphology and, if necessary, to counteract them specifically by controlling the morphology.

In order to perform an optimization of the actual fermentation (main culture) despite this complex interplay of different influencing factors, the main culture was directly inoculated with the spore suspension. By omitting the pre-culture, the influence of process changes on the production of fumaric acid could be directly investigated. The direct spore inoculation of the main culture also allowed an increase in the general reproducibility of cultivations.

- Influence of spore concentration

To investigate the influence of spore concentration, cultivations were inoculated with different amounts of spores (1×10^4, 1×10^5, 1×10^6 spores/mL). Cultivations with concentrations $<1 \times 10^5$ spores/mL were found to cause clump morphology and, thus, were not suitable for fumaric acid production. Above a concentration of 1×10^5 spores/mL, *R. arrhizus* NRRL 1526 grew as loose mycelium, and the production of approximately 50 g/L of fumaric acid was observed after a duration of 7 d, with yields of 0.44 and 0.43 g/g, respectively. A comparison of this result with the literature data shows that the use of higher spore concentrations is quite common in cultivations for the production of fumaric acid. For example, Riscaldati et al. (2000), Zhou et al. (2011) and Das et al. (2015) used a spore concentration of 1×10^6 spores/mL to inoculate the pre-culture [21,29,30]. At 1×10^7 spores/mL, Fu et al. (2010), Ding et al. (2011) and Gu et al. (2014) used the highest spore concentration reported in the literature for inoculating cultivations [31–33]. A further increase in spore concentration during cultivation inoculation did not improve fumaric acid production in this work. Thus, a spore concentration of 1×10^5 spores/mL was used for subsequent cultivations with direct spore inoculation.

- Influence of corn steep liquor

According to Ling and Ng (1989), different amounts of corn steep liquor were used for preparing the different culture media (sporulation: 1 mL/L; germination: 0.5 mL/L; production: 0.5 mL/L) for FA production [12]. Corn steep liquor is a complex medium component that can differ significantly in composition depending on the batch and manufacturer. It contains different concentrations of peptides, amino acids, vitamins, nucleotides and various trace elements [34]. Thus, by using two different batches of corn steep liquor and different concentrations in the range of 0.1 to 1 mL/L, corn steep liquor resulted in different morphologies and, thus, a different production of fumaric acid could also be observed. Alternatively, different concentrations of yeast extract (0.1 until 1 g/L) were used in cultivations. However, all cultivations using corn steep liquor or yeast extract yielded lower concentrations of fumaric acid compared to cultivations without the addition of a complex medium component. The morphology obtained exhibited almost exclusively mycelial morphology. In the literature, several publications have already described biomass and fumaric acid production using minimal media [35–37]. For example, in Wang et al. (2013), using pre-culture and main culture without complex medium components, cultivation was able to produce 56.5 g/L of fumaric acid, with yields of 0.71 g/g and productivities of 0.67 g/(L·h) [36].

Therefore, in the adapted cultivation strategy, the addition of complex medium components was completely omitted and, thus, a precisely defined minimal medium was used.

- Influence of tartaric acid

Tartaric acid is a saturated dicarboxylic acid. In the cultivation medium of Ling and Ng (1989), 0.0075 g/L tartaric acid was used [12]. To investigate the principle necessity of tartaric acid for the production of fumaric acid, cultivation with 0.0075 g/L (2S,3R)-tartaric acid and cultivation without the addition of tartaric acid were performed. Cultivation with *R. arrhizus* NRRL 1526 with and without the addition of tartaric acid demonstrated that a concentration of 0.0084 g/L (2S,3R)-tartaric acid × H_2O had no effect on the fermentation process. An identical concentration of fumaric acid of approximately 50.5 g/L, with a yield of 0.41 g/g and productivity of 0.31 g/(L·h), was obtained in both preparations after 7 d. Thus, the addition of tartaric acid was omitted in subsequent studies.

- Influence of calcium carbonate

To regulate the pH value, 50 g/L calcium carbonate was added to the system in previous cultivations. Due to the optimizations made and the resulting increase in the production of fumaric acid, the calcium carbonate is completely consumed during cultivation. This was recognizable by a drop in the pH value as the production of fumaric acid progressed. To counteract a drop in pH occurring too early on, which negatively influences the formation of fumaric acid, a threefold feed of 10 g/L calcium carbonate was performed after 3, 4 and 5 d. This successfully stabilized the pH at a value of 6, thus realizing an improvement in fumaric acid production. In addition, the calcium carbonate particle size and shape have been shown to be very important parameters for the building of a proper morphology and, therefore, have direct effects on the FA production, productivity and yield [16]. Depending on the particle size of $CaCO_3$, different morphologies of *R. arrhizus* NRRL 1526 occurred. Using $CaCO_3$ (precipitated, ≥99%, VWR), which has the smallest particle size, the best production of fumaric acid was obtained for loose mycelial morphology [38].

- Influence of ammonium and phosphate

For biomass formation, 1.8 g/L $(NH_4)_2SO_4$ is used as the nitrogen source in the main culture medium, according to Ling and Ng (1989) [12]. To optimize the amount of growth-relevant ammonium, different initial concentrations of 0.3–3.6 g/L $(NH_4)_2SO_4$ were used. Using 0.3 g/L $(NH_4)_2SO_4$, only a low growth of *R. arrhizus* NRRL 1526 could be detected, which only allowed a very slow production of fumaric acid. In contrast, at concentrations of 3.6 g/L, excessive biomass formation was documented, leading to clump formation. At initial concentrations between 0.6 g/L and 1.8 g/L ammonium sulfate, an increasing production of by-products (e.g., ethanol) could be shown under mycelial morphology. This resulted in a decrease in the yield of fumaric acid with increasing concentrations of ammonium sulfate. With regard to the effective production of fumaric acid, an initial concentration of 1.2 g/L $(NH_4)_2SO_4$ showed an ideal compromise in terms of productivity and yield. Therefore, for further cultivations, 1.2 g/L $(NH_4)_2SO_4$ was used in the fermentation medium.

Besides ammonium, phosphate is another component of the main culture medium that significantly influences the biomass growth of *R. arrhizus* NRRL 1526. Cultivations with 0.05–0.60 g/L KH_2PO_4 were performed to identify the ideal initial concentration. This series of experiments showed that the best cultivation result in terms of final titer, yield and productivity was achieved with 0.30 g/L KH_2PO_4. Thus, the concentration of KH_2PO_4 determined according to Ling and Ng (1989) was not changed.

- Influence of initial glucose concentration

So far, an initial glucose concentration of 130 g/L has been used, according to Ling and Ng (1989) [12]. However, a comparison with the literature shows that the use of lower concentrations such as 80 g/L glucose is quite common [39,40]. In addition to being a carbon and energy source and being used for the production of fumaric acid, glucose is also used for the production of biomass. With regard to an optimal ratio of built-up biomass and the potential to produce fumaric acid, the C/N ratio thus plays a decisive role in cultivation [32,41].

Therefore, cultivations with different initial concentrations of glucose (40–200 g/L) were carried out. In all cultivations, a drop in the pH value was prevented by adding additional calcium carbonate. Furthermore, the medium contained a concentration of 1.20 g/L $(NH_4)_2SO_4$. The resulting different C/N ratios are shown comparatively in Table 1. In addition, this table contains an overview of the morphology achieved and the corresponding cultivation time until complete conversion of the glucose.

Table 1. Cultivations with different initial concentrations of glucose (40–200 g/L) using *R. arrhizus* NRRL 1526 with direct spore inoculation (1×10^5 spores/mL) and optimized medium (100 mL of medium B) in 500 mL shake flasks without baffles at 34 °C and 200 rpm, and 50 g/L $CaCO_3$ with addition of 10 g/L $CaCO_3$ after 3, 4 and 5 d.

Experiment		1	2	3	4	5	6	7
Initial glucose	(g/L)	40	60	80	100	130	160	200
C/N-ratio	(g_C/g_N)	63	94	126	157	204	252	314
Morphology	-	clumps	mycelium	mycelium	mycelium	mycelium	mycelium	mycelium
Duration	(d)	2	3	4	6	7	11	15

In relation to the morphology obtained, the growth of *R. arrhizus* NRRL 1526 in the form of loose mycelial flakes was observed in a concentration range of 60–200 g/L initial glucose. In contrast, in the cultivation approach with the lowest glucose concentration of 40 g/L, clumps formed, which are unsuitable for the production of fumaric acid. Taking into account the C/N ratio of 63 g_C/g_N used in this approach, a critical minimum of approximately 60–90 g_C/g_N can thus be identified, which should not be fallen short of in order to form the preferred mycelial morphology. In the cultivations described in the subsection "Influence of ammonium and phosphate", the formation of mycelium was observed at a glucose concentration of 130 g/L and 1.80 g/L $(NH_4)_2SO_4$ (C/N ratio of 136 g_C/g_N). In contrast, the cultivation batch with 130 g/L glucose and 3.60 g/L $(NH_4)_2SO_4$ (C/N ratio of 68 g_C/g_N) showed compact clumps. Thus, based on these results, the influence of the C/N ratio on the growth form could be confirmed and, thus, a targeted control of the morphology in the cultivation system used here could be made possible.

In order to investigate the influence of the initial glucose concentration on the production of fumaric acid, the results of the individual cultivation approaches were shown in Figure 1 after complete consumption of glucose.

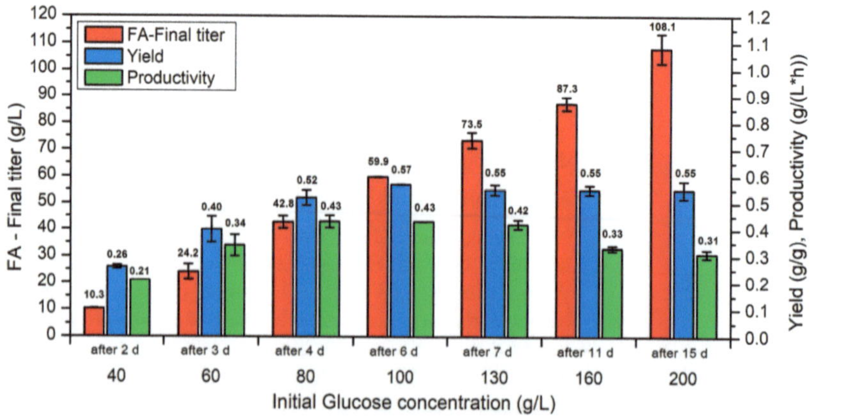

Figure 1. Cultivations with different initial concentrations of glucose (40–200 g/L) using *R. arrhizus* NRRL 1526 with direct spore inoculation (1×10^5 spores/mL) and optimized medium (100 mL of medium B) in 500 mL shake flasks without baffles at 34 °C and 200 rpm, and 50 g/L $CaCO_3$ with addition of 10 g/L $CaCO_3$ after 3, 4 and 5 d.

With increasing initial glucose concentration, more product is formed, and with 200 g/L glucose, the highest final concentration of 108.1 g/L FA was achieved in this set of experiments (Figure 1). However, this tendency could not be confirmed in terms of yield and productivity. Rather, with regard to these two process parameters, an optimum range of 80–130 g/L initial glucose was identified. Thus, the cultivation approach with 100 g/L glucose showed the highest yield (0.57 g/g). At initial glucose concentrations higher than 100 g/L, however, only a minimally lower yield of 0.55 g/g was detected. With regard to the achieved productivities, the highest productivities were detected in the range of 80–130 g/L glucose, with 0.43 and 0.42 g/(L·h), respectively. A further increase in the initial glucose concentration leads to lower productivities. This showed that FA concentrations above 100 g/L are possible, but that a fed-batch strategy should be investigated.

3.2. Direct Spore Inoculation Compared to Pre-Culture Inoculation with Optimized Parameters

Based on the previously described optimizations (Section 3.1), the cultivation process shown in Figure 2 was developed.

In the optimized cultivation with direct spore inoculation, a lag phase caused by spore germination of 1 day without fumaric acid production was observed. A final titer of 73.5 g/L fumaric acid was achieved after the present concentration of glucose was used up after approximately 7 d. This corresponds to a yield of 0.55 g/g and a total productivity of 0.42 g/(L·h). A positive aspect is the reduced accumulation of ethanol, which can be attributed to the improved oxygen supply of the biomass.

With regard to the most efficient cultivation process possible for the production of fumaric acid, the cultivation strategy described above with direct spore inoculation still has potential for optimization. For example, due to the initial germination of the spores, no production of fumaric acid takes place within the first 24 h of the main culture. In order to recognize the effects and influences during optimization, the production strategy using spores as an inoculum of the main cultivation has advantages. For an optimized process, on the other hand, the use of a pre-culture has its advantages, e.g., with regard to the key parameters of productivity and yield. Therefore, the use of the previously omitted pre-culture as a process step was reconsidered and further developed. This involves growing biomass (pre-culture) within 1 d, using (10% ($v_{pre\text{-}culture}/v_{main\text{-}culture}$)) as the inoculum. For this purpose, the optimized medium B was used in the pre-culture as well as in the subsequent main culture (Figure 3).

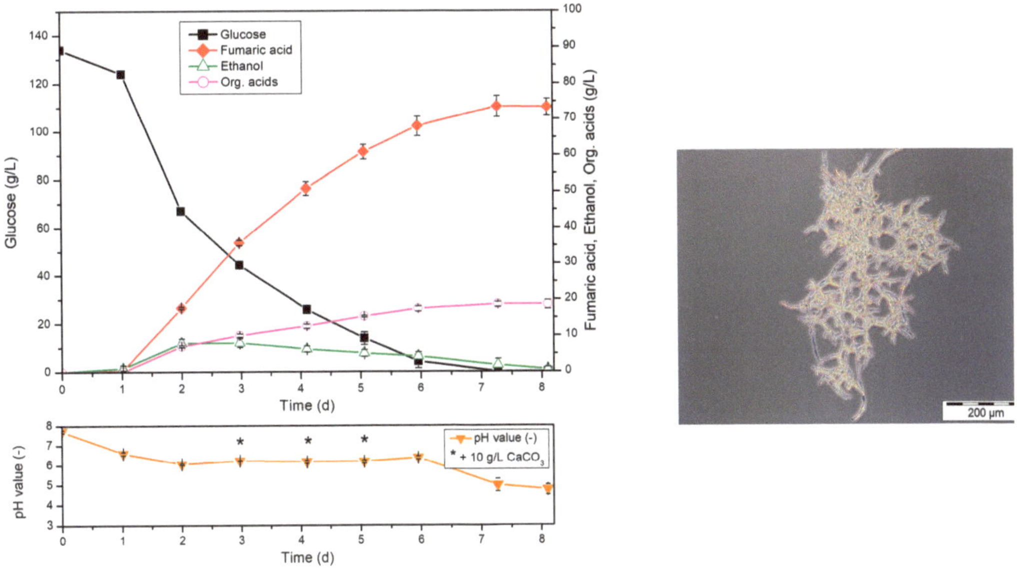

(a) (b)

Figure 2. (**a**) Cultivation of *R. arrhizus* NRRL 1526 with direct spore inoculation (1×10^5 spores/mL) and optimized medium (100 mL of medium B) in 500 mL shake flasks without baffles at 34 °C and 200 rpm, and 50 g/L $CaCO_3$ with addition of 10 g/L $CaCO_3$ after 3, 4 and 5 d (left). (**b**) Morphology of *R. arrhizus* NRRL 1526 cultured with direct spore inoculation after 3 d (right).

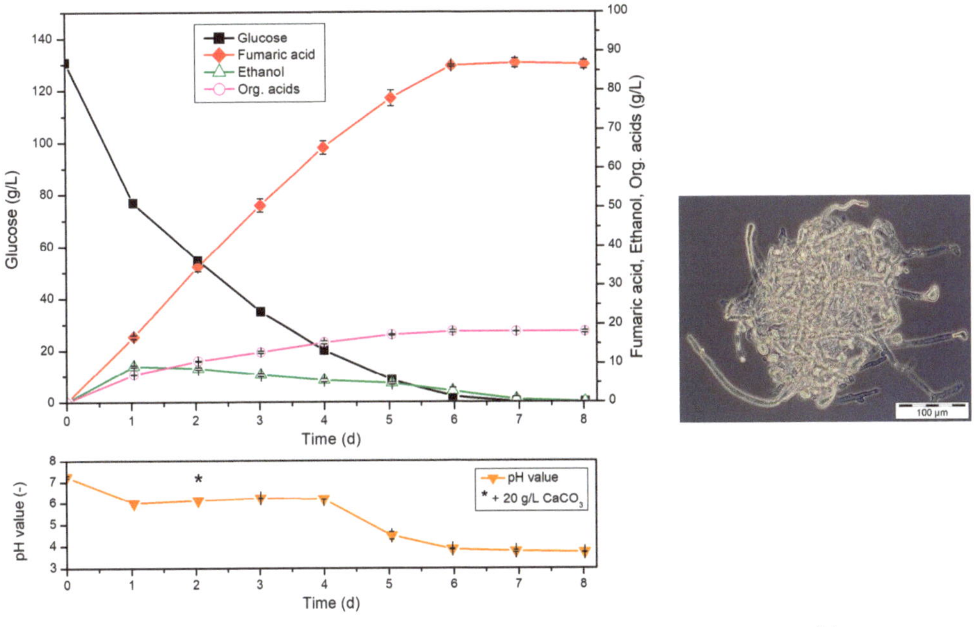

(a) (b)

Figure 3. (**a**) Cultivation of *R. arrhizus* NRRL 1526 with 10 % (v/v) pre-culture and optimized medium B in 500 mL shake flasks without baffles at 34 °C and 200 rpm, and 50 g/L $CaCO_3$ with addition of 20 g/L $CaCO_3$ after 2 d (left). (**b**) Morphology of *R. arrhizus* NRRL 1526 using pre-culture after 3 d (right).

Due to the outsourced cultivation of the biomass (pre-culture), a direct production of fumaric acid could be detected in this cultivation. With an almost linear increase in fumaric acid, a final titer of 86.3 g/L fumaric acid was reached after 6 d at almost complete conversion of glucose. This corresponds to a yield of 0.67 g/g, with a total productivity of 0.60 g/(L·h). Thus, compared to cultivation with direct spore inoculation, the lag phase occurring there was successfully transferred to the pre-culture. Since only 10 % (*v*/*v*, based on the main culture volume) is required for this, the efficiency of fumaric acid production was successfully improved. In addition, the final titer of fumaric acid as well as the yield and productivity show significantly increased values, which confirm the success of this cultivation strategy. With regard to the yield, it should be noted that a yield higher than 0.64 g/g can no longer be explained exclusively by the oxidative synthesis pathway. Thus, the yield of 0.67 g/g identified here indicates at least partial production of fumaric acid by means of reductive carboxylation [13,16].

Since the production of fumaric acid in this cultivation was completely stopped by the consumption of the supplied amount of glucose, no conclusion can be made about a possible final titer of fumaric acid at this point. However, a comparison of this final titer and especially of the yield with the literature shows that the optimized process conditions and achieved morphology are very well suited for efficient biotechnological production of fumaric acid using *R. arrhizus* NRRL 1526 (Table 2).

Table 2. Overview of FA production and yield based on glucose as substrate in batch cultivations using *Rhizopus* spp.; STR—stirred tank reactor, SF—shaking flask.

Year	Strain	System	FA (g/L)	Yield (g/g)	Source
1962	*R. arrhizus* NRRL 2582	STR	90	0.7	[27]
1989	*R. arrhizus* NRRL 1526	STR	130	1.0	[12]
2000	*R. arrhizus* NRRL 1526	STR	38	0.3	[29]
2002	*R. oryzae*	STR	37	0.5	[42]
2010	*R. oryzae*	STR	56	0.7	[39]
2022	*R. arrhizus* NRRL 1526	SF	65	0.46	[16]
2023	*R. arrhizus* NRRL 1526	SF	86	0.67	Present study

3.3. Fed-Batch Cultivation

In order to further increase the final titer of FA, the addition of glucose during the cultivation process was investigated. For this purpose, different process strategies involving an additional dosage of a N-source and $CaCO_3$ were applied, which are described in more detail in the following subsections.

3.3.1. Glucose Feed

The absolute amount of glucose to be converted into FA was increased during cultivation by adding glucose several times. Glucose was added in the form of solids and in an amount leading to an additional 40 g/L in the cultivation medium (Figure 4). The regular addition of calcium carbonate also prevented a drop in pH throughout the cultivation process.

Under continuous consumption of glucose within 14 d, the total added amount of glucose was not completely metabolized. During the cultivation time, the glucose consumption rate decreased, and a glucose concentration of 55.0 g/L was still detected at the end of cultivation. With respect to the production of fumaric acid, a final titer of 103.2 g/L was achieved after 14 d with the formation of mycelial morphology. This corresponds to a yield of 0.52 g/g, with a productivity of 0.31 g/(L·h). During the cultivation time, the FA production rate decreased and this prevented effective production of fumaric acid.

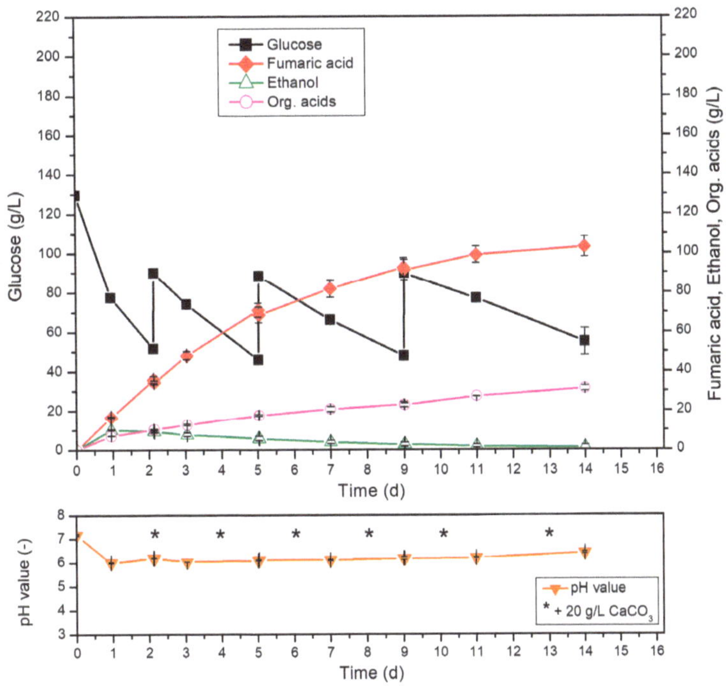

Figure 4. Fed-batch cultivation of *R. arrhizus* NRRL 1526 with 10% (*v*/*v*) pre-culture and optimized medium B (130 g/L initial glucose) in 500 mL shake flasks without baffles at 34 °C and 200 rpm, and 50 g/L $CaCO_3$ with addition of 40 g/L glucose after 2, 5 and 9 d and addition of 20 g/L $CaCO_3$ after 2, 4, 6, 8, 10 and 13 d.

Overall, it was demonstrated that glucose can be metabolized even if it is added subsequently during cultivation. However, without further optimization, this process strategy is not suitable for effective FA production, constant glucose consumption rates and constant FA production or FA concentrations higher than 100 g/L.

3.3.2. Glucose and Ammonium Feed

Neither the addition of glucose nor the use of a high initial concentration of glucose provided an effective cultivation system. In particular, the decrease in productivity with advancing cultivation time proved to be problematic. A possible cause for this observation was identified as aging or damage to the biomass formed with increasing cultivation time. In this context, it was hypothesized that the formation of new biomass in an advanced cultivation period could potentially improve productivity. Thus, a process strategy combining the addition of glucose, the usage of a high initial concentration of glucose and the addition of ammonium during cultivation, which, in principle, enables a further growth phase, was investigated (Figure 5).

During the first 7 days, the usage of a high initial glucose concentration (200 g/L initial glucose) allowed a constant production of 107.0 g/L fumaric acid without any intervention in the cultivation system. At this point, at which approximately 30 g/L glucose remained, glucose and $(NH_4)_2SO_4$ were added in the form of solids and in amounts leading to an additional 80 g/L glucose and 1.2 g/L $(NH_4)_2SO_4$ in the cultivation medium. In the following two days, only a small amount of FA was formed under increasing substrate consumption rate. However, ethanol was formed during this period, and its concentration reached a maximum of 14.1 g/L after 8 d. Therefore, the addition of ammonium was omitted during the second addition of 80 g/L glucose (after 9 days). As a result of this adjustment, only low glucose consumption and low FA production were detected between the ninth and tenth day of cultivation. To counteract this trend, a concentration of 0.6 g/L $(NH_4)_2SO_4$ was added to the cultivation system after 10 d, 12 d and 14 d. Through this process strategy, a complete consumption of the total amount of glucose was documented after 16 d under increased glucose consumption. In relation to the production of FA, a maximum titer of 195.4 g/L could be identified after a cultivation time of 15 d. This corresponds to a yield of 0.54 g/g, with a total productivity of 0.54 g/(L·h). It was shown that very high productivities are possible at late cultivation times and that effective production of fumaric acid was still possible even at concentrations of >150 g/L fumaric acid. For example, maximum productivities of 0.72 g/(L·h) (13–14 d) and 0.74 g/(L·h) (14–15 d) were detected at the end of cultivation.

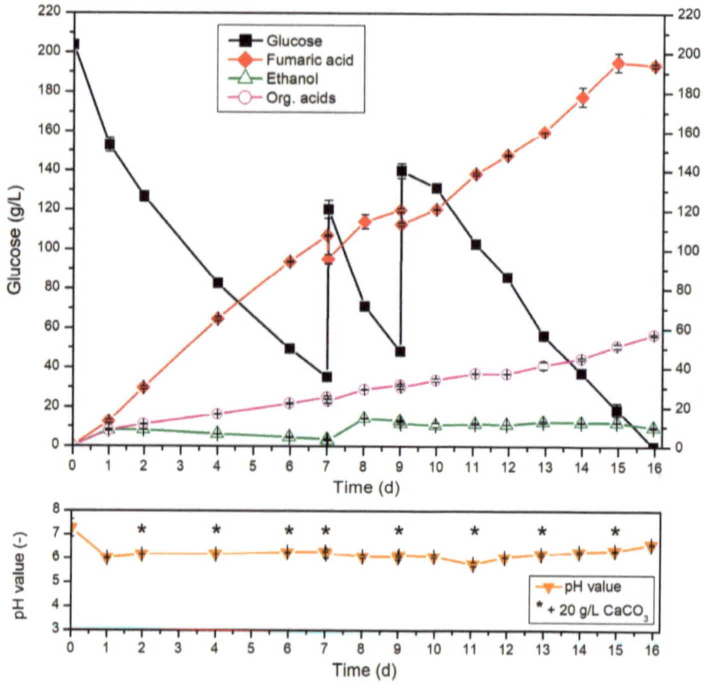

Figure 5. Fed-batch cultivation of *R. arrhizus* NRRL 1526 with 10% (*v*/*v*) pre-culture and optimized medium B (200 g/L initial glucose) in 500 mL shake flasks without baffles at 34 °C and 200 rpm, and 50 g/L $CaCO_3$ with addition of 80 g/L glucose after 7 and 9 d, addition of 1.2 g/L $(NH_4)_2SO_4$ after 7 d, addition of 0.6 g/L $(NH_4)_2SO_4$ after 10, 12 and 14 d and addition of 20 g/L $CaCO_3$ after 2, 4, 6, 7, 9, 11, 13 and 15 d.

Related to the obtained morphology of *R. arrhizus* NRRL 1526, the formation of mycelium was also documented during this cultivation. With increasing cultivation time, the aggregation of individual mycelial flakes into larger agglomerates was observed.

Fumaric acid is currently produced petrochemically. It has a high potential to be produced using biotechnological processes and it has been designated by the US Department of Energy (DOE) as one of the top 12 valuable chemicals to be produced from biomass [43].

This requires the development of a fermentation process that allows specific morphology control of the production strain used and, at the same time, enables effective production of fumaric acid. Biotechnologically produced fumaric acid, in a multi-feedstock biorefinery, could contribute to the development of the circular economy and the bioeconomy [44,45].

The results of this work successfully demonstrated that effective biotechnological fumaric acid production with a high final titer of FA is possible using the provided process strategy. The combined addition of glucose and ammonium provides a promising possibility to build up new biomass during cultivation and, thus, to produce FA constantly over long cultivation periods. As shown, this resulted in very high final titers of 195.4 g/L FA. To our knowledge, this is the highest biotechnologically produced concentration of FA (Table 2). Furthermore, the only slight decrease in yield and productivity compared to the reference cultivation (without the addition of glucose and ammonium, Figure 3) illustrates the high potential of this cultivation approach.

4. Conclusions

In this work, different fermentation strategies were developed on the basis of the substrate glucose and with the production strain *Rhizopus arrhizus* NRRL 1526, and the respective production behavior was characterized. The patented cultivation procedure of Ling and Ng (1989) served as a template. Based on this, an initial cultivation strategy could be developed on a shake flask-scale, consisting of direct spore inoculation of the main culture. By omitting a pre-culture, a significantly simplified process could be investigated, which allowed systematic medium and process optimization. Significant insights regarding morphology control could be gained. For example, the use of calcium carbonate with a smaller particle size, which was used to regulate pH, allowed the formation of individual mycelial flakes, instead of the unsuitable clump morphology or growth in the form of large and compact pellets. In addition, by varying the C/N ratio used, medium optimization identified a critical minimum of about 90 gC/gN, which should not be fallen below to avoid lumps. Deviating from the starting point, the optimized fermentation medium did not contain any complex medium constituents such as corn steep liquor or yeast extract, the use of tartaric acid was also omitted and a reduced concentration of growth-relevant ammonium sulfate was used. Based on these optimizing cultivation parameters, a different cultivation strategy could be developed on a shake flask-scale.

In another cultivation strategy, the germination of the spores and, thus, the first cultivation of biomass were carried out in a pre-culture. In contrast to cultivation with direct spore inoculation, the occurrence of a lag phase could thus be successfully outsourced to the pre-culture. This had a positive effect on the overall balance of the production process and resulted in a final titer of 86.3 g/L fumaric acid, with a yield of 0.67 g/g and an overall productivity of 0.60 g/(L·h).

Furthermore, the maximum possible final titer of fumaric acid was determined on the shake flask-scale. For this purpose, the cultivation strategy was modified and extended by adding additional glucose and ammonium during the main culture. Depending on the absolute amount of substrate used, the production phase could be successfully extended to a duration of 15 d. This allowed a final concentration of fumaric acid to be determined. As a result, a final titer of fumaric acid of 195.4 g/L could be achieved. These values are by far the highest final titers of fumaric acid achieved through fermentation that have been published, according to the best of our knowledge.

Author Contributions: Conceptualization, L.E., A.K. and U.P.; methodology, L.E.; investigation, L.E.; data curation, L.E.; writing—original draft preparation, A.K.; writing—review and editing, L.E., A.K. and U.P.; supervision, A.K. and U.P.; project administration, U.P. and A.K.; funding acquisition, A.K. and U.P. All authors have read and agreed to the published version of the manuscript.

Funding: This work was funded by the German Federal Ministry of Food and Agriculture, following a decision of the German Bundestag, via the Agency of Renewable Resources (Grant No. 22029515).

Institutional Review Board Statement: Not applicable.

Informed Consent Statement: Not applicable.

Data Availability Statement: All data supporting this paper results are included in this document.

Conflicts of Interest: The authors declare no conflict of interest. The funders had no role in the design of the study; in the collection, analyses, or interpretation of data; in the writing of the manuscript or in the decision to publish the results.

References

1. Roa Engel, C.A.; Straathof, A.J.J.; Zijlmans, T.W.; Gulik, W.M.; van der Wielen, L.A.M. Fumaric acid production by fermentation. *Appl. Microbiol. Biotechnol.* **2008**, *78*, 379–389. [CrossRef] [PubMed]
2. Tsao, G.T.; Cao, N.J.; Du, J.; Gong, C.S. Production of multifunctional organic acids from renewable resources. In *Recent Progress in Bioconversion of Lignocellulosics*; Springer: Berlin/Heidelberg, Germany, 1999; pp. 243–280.
3. Das, R.K.; Brar, S.K.; Verma, M. Chapter 8—Fumaric acid: Production and application aspects. In *Platform Chemical Biorefinery*; Elsevier: Amsterdam, The Netherlands, 2016; pp. 133–157.
4. Goldberg, I.; Rokem, J.S.; Pines, O. Organic acids: Old metabolites, new themes. *J. Chem. Technol. Biotechnol.* **2006**, *81*, 1601–1611. [CrossRef]
5. McGinn, S.; Beauchemin, K.; Coates, T.; Colombatto, D. Methane emissions from beef cattle: Effects of monensin, sunflower oil, enzymes, yeast, and fumaric acid. *J. Anim. Sci.* **2004**, *82*, 3346–3356. [CrossRef] [PubMed]
6. Weißert, R. Multiple Sklerose-Risiken und Nutzen der neuen antiinflammatorischen Substanzen. *J. für Neurol. Neurochir. und Psychiatr.* **2014**, *16*, 95–101.
7. Smith, D. Fumaric acid esters for psoriasis: A systematic review. *Ir. J. Med. Sci.* **2017**, *186*, 161–177. [CrossRef]
8. Das, R.K.; Brar, S.K.; Verma, M. Recent advances in the biomedical applications of fumaric acid and its ester derivatives: The multifaceted alternative therapeutics. *Pharmacol. Rep.* **2016**, *68*, 404–414. [CrossRef] [PubMed]
9. Waksman, S.A. Process for the Production of Fumaric Acid. US Patent 2,326,986 (to Merck & Co., Inc. and Pfizer & Co., Inc.), 1943.
10. Lubowitz, H.R.; La, R.E.G. Fumaric Acid Fermentation Process. US Patent 2,861,922 (to National Distillers and Chemical Corporation), 1958.
11. Goldberg, I.; Stieglitz, B. Fermentation Process for Production of Carboxylic Acids. US Patent 4,564,594 (to E. I. Du Pont de Nemours and Company, Wilmington, Del.), 1986.
12. Ling, L.B.; Ng, T.K. Fermentation Process for Carboxylic Acids. US Patent 4,877,731 (to E. I. Du Pont de Nemours and Company, Wilmington, Del.), 1989.
13. Xu, Q.; Li, S.; Huang, H.; Wen, J. Key technologies for the industrial production of fumaric acid by fermentation. *Biotechnol. Adv.* **2012**, *30*, 1685–1696. [CrossRef]
14. Kenealy, W.; Zaady, E.; du Preez, J.C.; Stieglitz, B.; Goldberg, I. Biochemical aspects of fumaric acid accumulation by *Rhizopus arrhizus*. *Appl. Environ. Microbiol.* **1986**, *52*, 128–133. [CrossRef]
15. Zhang, K.; Zhang, B.; Yang, S.T. Production of citric, itaconic, fumaric and malic acids in filamentous fungal fermentations. In *Bioprocessing Technologies in Biorefinery for Sustainable Production of Fuels, Chemicals and Polymers*; John Wiley & Sons: Hoboken, NJ, USA, 2013; pp. 375–397.
16. Martin-Dominguez, V.; Cabrera, P.I.A.; Eidt, L.; Pruesse, U.; Kuenz, A.; Ladero, M.; Santos, V.E. Production of Fumaric Acid by *Rhizopus arrhizus* NRRL 1526: A Simple Production Medium and the Kinetic Modelling of the Bioprocess. *Fermentation* **2022**, *8*, 64. [CrossRef]
17. Martin-Dominguez, V.; Estevez, J.; Ojembarrena, F.D.B.; Santos, V.E.; Ladero, M. Fumaric Acid Production: A Biorefinery Perspective. *Fermentation* **2018**, *4*, 33. [CrossRef]
18. Byrne, G.S.; Ward, O.P. Growth of *Rhizopus arrhizus* in fermentation media. *J. Ind. Microbiol.* **1989**, *4*, 155–161. [CrossRef]
19. Kosakai, Y.; Soo Park, Y.; Okabe, M. Enhancement of L(+)-lactic acid production using mycelial flocs of *Rhizopus oryzae*. *Biotechnol. Bioeng.* **1997**, *55*, 461–470. [CrossRef]
20. Ilica, R.A.; Kloetzer, L.; Galaction, A.I.; Caşcaval, D. Fumaric acid: Production and separation. *Biotechnol. Lett.* **2018**, *41*, 1–11. [CrossRef] [PubMed]
21. Zhou, Z.; Du, G.; Hua, Z.; Zhou, J.; Chen, J. Optimization of fumaric acid production by *Rhizopus delemar* based on the morphology formation. *Bioresour. Technol.* **2011**, *102*, 9345–9349. [CrossRef]
22. Sebastian, J.; Hegde, K.; Kumar, P.; Rouissi, T.; Brar, S.K. Bioproduction of fumaric acid: An insight into microbial strain improvement strategies. *Crit. Rev. Biotechnol.* **2019**, *39*, 817–834. [CrossRef]
23. Eidt, L. Nutzung Nachwachsender Rohstoffe für die Biotechnologische Produktion von Fumarsäure. Ph.D. Thesis, Technical University of Braunschweig, Braunschweig, Germany, 2021. [CrossRef]
24. Roa Engel, C.A.; van Gulik, W.M.; Marang, L.; van der Wielen, L.A.M.; Straathof, A.J.J. Development of a low pH fermentation strategy for fumaric acid production by *Rhizopus oryzae*. *Enzyme Microb. Technol.* **2011**, *48*, 39–47. [CrossRef]
25. Papagianni, M. Fungal morphology and metabolite production in submerged mycelial processes. *Biotechnol Adv* **2004**, *22*, 189–259. [CrossRef]
26. Papadaki, A.; Androutsopoulos, N.; Patsalou, M.; Koutinas, M.; Kopsahelis, N.; Castro, A.M.d; Papanikolaou, S.; Koutinas, A.A. Biotechnological production of fumaric acid: The effect of morphology of *Rhizopus arrhizus* NRRL 2582. *Fermentation* **2017**, *3*, 33. [CrossRef]
27. Rhodes, R.A.; Lagoda, A.A.; Misenheimer, T.J.; Smith, M.L.; Anderson, R.F.; Jackson, R.W. Production of fumaric acid in 20-liter fermentors. *Appl. Microbiol.* **1962**, *10*, 9–15. [CrossRef] [PubMed]
28. Liao, W.; Liu, Y.; Frear, C.; Chen, S. A new approach of pellet formation of a filamentous fungus—*Rhizopus oryzae*. " *Bioresour. Technol.* **2007**, *98*, 3415–3423. [CrossRef]
29. Riscaldati, E.; Moresi, M.; Federici, F.; Petruccioli, M. Direct ammonium fumarate production by *Rhizopus arrhizus* under phosphorous limitation. *Biotechnol. Lett.* **2000**, *22*, 1043–1047. [CrossRef]

30. Das, R.K.; Brar, S.K.; Verma, M. Valorization of egg shell biowaste and brewery wastewater for the enhanced production of fumaric acid. *Waste Biomass Valorization* **2015**, *6*, 535–546. [CrossRef]
31. Fu, Y.Q.; Li, S.; Chen, Y.; Xu, Q.; Huang, H.; Sheng, X.Y. Enhancement of fumaric acid production by *Rhizopus oryzae* using a two-stage dissolved oxygen control strategy. *Appl. Biochem. Biotechnol.* **2010**, *162*, 1031–1038. [CrossRef] [PubMed]
32. Ding, Y.; Li, S.; Dou, C.; Yu, Y.; Huang, H. Production of fumaric acid by *Rhizopus oryzae*: Role of carbon–nitrogen ratio. *Appl. Biochem. Biotechnol.* **2011**, *164*, 1461–1467. [CrossRef] [PubMed]
33. Gu, S.; Xu, Q.; Huang, H.; Li, S. Alternative respiration and fumaric acid production of *Rhizopus oryzae*. *Appl. Microbiol. Biotechnol.* **2014**, *98*, 5145–5152. [CrossRef] [PubMed]
34. Klotz, S. Biotechnisch Hergestellte D-Milchsäure—Substitution von Hefeextrakt Durch Agrarische Rohstoffhydrolysate. Ph.D. Thesis, Technical University of Braunschweig, Braunschweig, Germany, 2017. [CrossRef]
35. Yu, S.; Huang, D.; Wen, J.; Li, S.; Chen, Y.; Jia, X. Metabolic profiling of a *Rhizopus oryzae* fumaric acid production mutant generated by femtosecond laser irradiation. *Bioresour. Technol.* **2012**, *114*, 610–615. [CrossRef]
36. Wang, G.; Huang, D.; Qi, H.; Wen, J.; Jia, X.; Chen, Y. Rational medium optimization based on comparative metabolic profiling analysis to improve fumaric acid production. *Bioresour. Technol.* **2013**, *137*, 1–8. [CrossRef] [PubMed]
37. Das, R.K.; Brar, S.K.; Verma, M. Application of calcium carbonate nanoparticles and microwave irradiation in submerged fermentation production and recovery of fumaric acid: A novel approach. *RSC Adv.* **2016**, *6*, 25829–25836. [CrossRef]
38. Eidt, L.; Kuenz, A.; Prüße, U. Biotechnologische Produktion von Fumarsäure: Prozessoptimierung und Kontrolle der Morphologie. *Chem. Ing. Tech.* **2018**, *90*, 1272. [CrossRef]
39. Fu, Y.; Xu, Q.; Li, S.; Huang, H.; Chen, Y. A novel multi-stage preculture strategy of *Rhizopus oryzae* ME-F12 for fumaric acid production in a stirred-tank reactor. *World J. Microbiol. Biotechnol.* **2009**, *25*, 1871–1876. [CrossRef]
40. Xu, Q.; He, S.; Jiang, L.; Li, S.; Wen, J.; Guan, R.; Huang, H. Extractive fermentation for fumaric acid production by *Rhizopus oryzae*. *Sep. Sci. Technol.* **2017**, *52*, 1512–1520. [CrossRef]
41. Swart, R.M.; Ronoh, D.K.; Brink, H.; Nicol, W. Continuous Production of Fumaric Acid with Immobilised *Rhizopus oryzae*: The Role of pH and Urea Addition. *Catalysts* **2022**, *12*, 82. [CrossRef]
42. Zhou, Y.; Du, J.; Tsao, G.T. Comparison of fumaric acid production by *Rhizopus oryzae* using different neutralizing agents. *Bioprocess. Biosyst. Eng.* **2002**, *25*, 179–181. [PubMed]
43. Werpy, T.; Petersen, G. *Top Value Added Chemicals from Biomass Volume I—Results of Screening for Potential Candidates from Sugars and Synthesis Gas Energy Efficiency and Renewable Energy*; PNLL: Richland, WA, USA, 1992.
44. Esteban, J.; Ladero, M. Food waste as a source of value-added chemicals and materials: A biorefinery perspective. *Int. J. Food Sci.Technol.* **2018**, *53*, 1095–1108. [CrossRef]
45. Di Lorenzo, R.D.; Serra, I.; Porro, D.; Branduardi, P. State of the Art on the Microbial Production of Industrially Relevant Organic Acids. *Catalysts* **2022**, *12*, 234. [CrossRef]

fermentation

Article

Lactic Acid Production from Cow Manure: Experimental Process Conditions Analysis

Ricard Garrido [1], Víctor Falguera [2], Omar Pérez Navarro [3], Amanda Acosta Solares [3] and Luisa F. Cabeza [1,*]

[1] GREiA Research Group, Universitat de Lleida, Pere de Cabrera s/n, 25001 Lleida, Spain; ricard.garrido@udl.cat
[2] AKIS International, 25171 Albatàrrec, Spain; v.falguera@akisinternational.com
[3] Department of Chemical Engineering, Faculty of Chemistry and Pharmacy, Central University "Marta Abreu" of Las Villas, Santa Clara 54830, Cuba; omarpnavrr69@gmail.com (O.P.N.); aasolares@uclv.cu (A.A.S.)
* Correspondence: luisaf.cabeza@udl.cat; Tel.: +34-973-003576

Abstract: The production of cow manure far exceeds the quantity that can be utilized in primary applications such as fertilizer or for the generation of biogas. As a result, alternative value-added applications are being investigated. The purpose of this study is to evaluate the production of lactic acid, using cow manure as the raw material. The methodology involved the implementation of thermochemical pretreatment for the cow manure, followed by simultaneous saccharification and fermentation for lactic acid production. Response surface methodology based on a central composite design was employed to analyze the simultaneous saccharification and fermentation process. The factorial design of the experiments was carried out with three factors, cow manure concentration, temperature, and enzyme concentration, with 80 $g \cdot L^{-1}$, 50 °C, and 212.5 $IU/gCM_{Dry\ Matter}$ as central point values, respectively. Following the addition of *Bacillus coagulans* DSM2314 inoculum to enzymatically hydrolyzed cow manure at pH 5.0, after a 24 h period the concentration of lactic acid was recorded at 13.65 $g \cdot L^{-1}$, with a conversion efficiency of 33.1%. Studies were conducted until 48 h to analyze time impact. Characterization studies for native cow manure and that pretreated using acid reagent were conducted. Sugar content and by-product formation were analyzed, resulting in 23.24 $g \cdot L^{-1}$ of sugar remaining as the maximum after fermentation, while low values of furfural (1.04 $g \cdot L^{-1}$), 5-hydroxymethylfurfural (1.35 $g \cdot L^{-1}$), and acetic acid (1.45 $g \cdot L^{-1}$) were found. Optimal conditions were calculated at 24 and 48 h with R software, obtaining the lactic acid, with yields of 13.4 $g \cdot L^{-1}$, 36.28% (for 24 h) and 15.27 $g \cdot L^{-1}$, 32.76% (for 48 h), respectively. Experimental and statistical studies of enzymatic hydrolysis and fermentation stated that cow manure was a feasible substrate for the production of lactic acid.

Keywords: bioplastic; lactic acid; response surface methodology; optimization; cow manure

Citation: Garrido, R.; Falguera, V.; Pérez Navarro, O.; Acosta Solares, A.; Cabeza, L.F. Lactic Acid Production from Cow Manure: Experimental Process Conditions Analysis. *Fermentation* **2023**, *9*, 604. https://doi.org/10.3390/fermentation9070604

Academic Editors: Miguel Ladero and Victoria E. Santos

Received: 27 May 2023
Revised: 18 June 2023
Accepted: 26 June 2023
Published: 27 June 2023

1. Introduction

In the culinary, pharmaceutical, cosmetic, and textile industries, lactic acid (LA) is employed as an acidulant and a preservative [1]. In the baking industry, it serves as a precursor in the manufacturing of emulsifiers such as stearoyl-2-lactylates. It performs a wide range of tasks, including flavoring, regulating pH, acting as an acidulant, enhancing the microbiological quality, fortifying minerals, and extending shelf life [2].

Lactic acid can be produced either by chemical synthesis or by the fermentation of renewable carbohydrates. It is possible to produce LA by using biomass as a source of carbohydrates. Lactic acid is an organic acid that occurs naturally and serves as the primary metabolic intermediate in the majority of organisms, including in people and anaerobic prokaryotes [3]. It is classified as Generally Recognized As Safe (GRAS) for general purpose food additives by the United States Food and Drug Administration (USFDA) [4]. Due to the high cost of product recovery and purification, as well as the expensive base materials, production costs are high. So that it can be produced economically, it is essential to find low-cost raw materials for lactic acid fermentation. Typically, lactose, maltose, or glucose are employed in its production [5]. Lactic acid is in high demand as a raw material for the synthesis of poly(lactic acid) (PLA) due to the recent increase in interest in the creation of biodegradable plastic [6]. To produce highly crystalline PLA, which results in the polymer's high strength and chemical and heat resistant qualities, optically pure lactic acid is required [7].

The most prevalent type of agricultural waste is cow manure, which is also a lignocellulosic substance [8]. Enzymatic hydrolysis into fermentable sugars could successfully disrupt the treated lignocellulosic fraction [9]. *Bacillus coagulans* DSM 2314 is a fascinating strain to use for manufacturing lactic acid from lignocellulose using a Simultaneous Saccharification and Fermentation (SSAF) technique [10]. With conversion efficiencies above 90 wt.%, it can homoferment glucose and xylose. Furthermore, *B. coagulans* has a high productivity, from 2.5 to 3 $g \cdot L^{-1} \cdot h^{-1}$ of lactic acid. It can thrive in surroundings that are slightly acidic and it is a moderate thermophile with an ideal growth temperature of about 50 °C, which is comparable to the ideal circumstances for commercial enzyme combinations like GC220 (Genencor, Denmark) and CTeC2 (Novozymes, Denmark). However, byproducts of processed lignocellulose can hinder the growth of *Bacillus coagulans* DSM 2314 [11].

Byproducts are produced in every pretreatment techniques. Phenolic compounds, furans, and tiny organic acids have been recognized as the three main groups of byproducts. Based on the quantities present and their inhibitory effects, these byproducts may block the fermentation that leads to the production of biochemicals, reducing productivity, growth, and occasionally the yield of the microorganisms in these processes [12].

The process from cow manure to lactic acid consists of raw material milling, acid pretreatment, and SSAF, as reviewed in [13]. While physical pretreatments involve size reduction and steam explosion, chemical pretreatments involve changing the structure of biomass with solvents that stimulate the breakdown of cellulose, hemicellulose, and lignin [14]. To convert the majority of lignocellulose into dextrose, a fermentable sugar, amylolytic enzymes such as amylase and glucoamylase must first hydrolyze it twice. The first stage is typically rapidly finished at high temperatures (between 90 and 130 °C), and the second stage is typically finished at lower temperatures after an extended saccharification to dextrose process. For many years, this technology has been used on an industrial basis. Industrial enzyme manufacturers like Novozymes and Genencor, for example, offer highly developed, effective, and reasonably priced enzymes for this process. This procedure yields dextrose, which can be used to ferment lactic acid [15].

The bioconversion of carbohydrate materials to lactic acid can be considerably enhanced by combining the microbial fermentation of the resulting sugars and the enzymatic hydrolysis of the carbohydrate substrates into a single phase, known as SSAF [16]. Enzymatic hydrolysis should progress considerably faster when fermentation and enzymatic hydrolysis are combined in an SSAF process, because the microbe can directly absorb the monomerized sugars, reducing product inhibition. Consequently, an SSAF process' processing time can be significantly reduced [17].

Presently, there are no available tests for using cow manure as a feedstock for lactic acid production. This paper describes the conversion of cow manure into lactic acid through pretreatment and efficient enzymatic hydrolysis and fermentation. Furthermore, this study determined the most common composition of cow manure, inhibitors, byproducts, and LA production performance. Cow manure could be effectively disrupted by enzymatic hydrolysis into fermentable sugars to produce lactic acid. Taken together, the research has implications for all cow farms as it is the first attempt to investigate the potential utilization of cow manure for lignocellulosic–lactic acid in combination with lignocellulosic enzyme production, which could serve as a reference for improving bovine waste economics. The aim of this study was to reach lactic acid productivity in SSAF experiments using cow manure, similar to what has been reported for fermentations using high-grade sugars or lignocellulose as feedstock.

2. Materials and Methods

2.1. Raw Material (Cow Manure)

Raw material was collected in cattle fattening stables of a farm located in Lleida (Spain), with straw and compound feeding regime. Collection was carried out inside the stable and in the manure heap, collecting a total of nine samples in each section. Samples were subjected to drying at 55 °C for 72 h in a SELECTA oven (DIGITRONIC-TFT) and ground to an average diameter of 1 mm in a MOULINEX fruit grinder in 50 g portions for 5 s. Dried and milled cow manure was subjected to experimental steps and analysis.

2.2. Cow Manure Analysis

Cellulose, hemicellulose, and lignin content were determined using an Ankrom 200 fiber analyzer. Samples were weighted in a Mettler-Toledo balance, model XS204. Acid Detergent Fiber (ADF) analysis, Neutral Detergent Fiber (NDF), and Crude Fiber Analysis (CFA) were performed. The ADF reagent consisted of 20 g of cetyl trimethylammonium bromide (CTAB) with 1 L 1.00 N H_2SO_4 previously standardized. The NDF reagent consisted of 30 g sodium dodecyl sulfate, 18.61 g ethylenediaminetetraacetic disodium salt (dehydrate), 6.81 g sodium borate, 4.56 g sodium phosphate dibasic (anhydrous), and 10 mL of triethylene glycol to 1 L distilled water. pH was controlled from 6.9 to 7.1. CFA reagent consisted of sulfuric acid (72% by weight), and reagent grade H_2SO_4 diluted to a specific gravity of 1634 g·L^{-1} at 20 °C (24.00 *N*) by adding 1200 g H_2SO_4 to 350 mL H_2O in a 1 L of mono chloracetic acid (MCA) volumetric flask with cooling. Solution was standardized to 1634 g·L^{-1} at 20 °C specific gravity by removing solution and adding H_2O or H_2SO_4 as required. ADF and NDF were performed on 0.5 mg of milled and dried samples, sealed in a filter bag F57 from Ankrom. After analysis, samples were rinsed with tap water and dried at 105 °C for 2 h. After that, ADF, NDF, and CFA were performed. Lignin content was determined directly by CFA. Cellulose was determined as ADF minus CFA. Finally, to account for hemicellulose content, NDF minus ADF and CFA was calculated. Ash [18] and humidity [19] values were obtained using gravimetric analysis.

2.3. Microorganism

Bacillus coagulans DSM 2314 was acquired as freeze-dried stock from the German collection of microorganisms and cell cultures (Deutsche Sammlung von Mikroorganismen und Zellkulturen, Braunschweig, Germany).

Strain was grown in culture medium, TrypticSoy Broth (TSB), Scharlau 02-200-500, batch 132425. The composition was (per liter of medium) 17 g casein peptone, 3 g soy peptone, 5 g sodium chloride, 2.5 g dipotassium phosphate, 2.5 g dextrose, 1000 mL deionized water, and 7.3 ± 0.2 pH ready-to-use. The incubation temperature was 50 °C and time 24 ± 3 h. Optical density (absorbance) was measured by means of culture medium plate count. Plate count growth parameters medium consisted of Tryptic Soy Agar (TSA) and biocardiagnostics BK047HA, batch 0016009. The composition (per liter of medium) was 15 g tryptone, 5 g papaic of digested soybean meal, 5 g sodium chloride, 15 g bacteriological agar, 1000 mL deionized water, and 7.3 ± 0.2 pH ready-to-use. A 300 mL quantity of TSB medium in borosilicate glass bottles with non-hermetic closure was constantly shaken at 150 rpm in an orbital incubator. Peptone saline medium was used to correct dilution. The composition (per liter of medium) was 1 g tryptone, 8.5 g sodium chloride, 1000 mL deionized water, and 6.8 ± 0.2 pH ready-to-use. It was sterilized at 121 °C for 15 min.

2.4. Process Flow

The process of producing acid lactic from cow manure consisted of cow manure collection; cow manure mixing, drying and grinding/sieving; chemical pretreatment; and simultaneous saccharification and co-fermentation process, as shown in Figure 1.

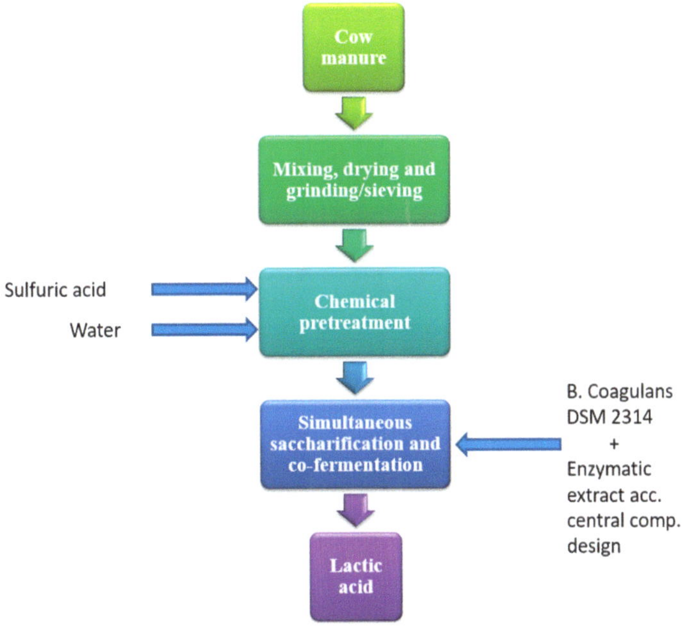

Figure 1. Process flow for lactic acid production.

The chemical hydrolysis pretreatment was carried out with 96% sulfuric acid (Panreac), using 0.5 wt.% of acid on cattle manure on a dry basis, which has been used successfully in similar lignocellulosic raw materials. Dilute acid pretreatment has low requirements, minimizes possible environmental impacts in the proposal, and its effectiveness has been demonstrated in the hydrolysis of cellulose and hemicellulose [11]. The process was carried out in a glass reactor with an effective capacity of 5 L, treating 2.5 L of cow manure suspension in deionized water. It was stirred at 250 min^{-1} using a propeller stirrer with the geometrical diameter relation of 0.25 reactor diameter. It was operated isothermally at 90 °C for 120 min, by circulating thermal oil P20.275.50 from a HUBERT thermostat through the jacket. The process gases were recovered through a reflux condenser with water circulation. The acid hydrolysate was subjected to vacuum filtration. The amount of glucose, galactose, mannose, xylose and arabinose, acetic acid, furfural and 5-hydroxymethylfurfural (HMF) was determined from the filtered liquid. The fluid was subjected to drying at 55 °C for 72 h in a SELECTA oven (DIGITRONIC-TFT).

The cow manure mixture from the chemical pretreatment was subjected to SSAF in a glass reactor with an effective capacity of 3 L, treating 0.7 L. It was stirred at 150 min^{-1} using a propeller stirrer with 0.25 diameter agitator/reactor relation. It was operated isothermally. Temperature regulation was carried out by circulating deionized water from a SELECTA thermostat (DIGITERM 200) through the jacket. Pre-saccharification was carried out for 18 h. At the beginning, the acid hydrolysate pH was adjusted to 5 ± 0.1 by adding a 6.25 $mol \cdot L^{-1}$ NaOH suspension, and 75% of the enzymatic cocktail SAE0020 corresponding to each run was added. The SAE0020 enzyme cocktail with enzyme activity of 1000 IU/g was supplied by SIGMA-ALDRICH. After 18 h of presaccharification, the pH of the medium was adjusted to 5.8 ± 0.1 by adding a suspension of NaOH at 6.25 $mol \cdot L^{-1}$. The rest of the enzymatic cocktail corresponding to each run was added and supplemented with 1 $g \cdot L^{-1}$ of KCl, 1 $g \cdot L^{-1}$ of Na_2HPO_4, 1.25 $g \cdot L^{-1}$ of NH_4Cl, 3 $g \cdot L^{-1}$ of yeast extract, 5 $g \cdot L^{-1}$ of glucose, and 10 $g \cdot L^{-1}$ of casein. The enriched medium was inoculated with a 5 vol.% suspension of *Bacillus coagulans* DSM 2314 and non-active aeration was maintained for 48 h of fermentation under isothermal conditions corresponding to the temperature of each experiment.

2.4.1. Experimental Design

Response surface methodology (RSM) is an approach that brings several benefits to traditional one-variable-at-a-time optimization, including the ability to generate a large amount of data from a small number of tests, and the ability to assess how the interaction between factors affects the answer [20]. The application of RSM as an experimental design allows the extraction of complex information, while reducing costs associated with labor, supplies, and time [21].

The experimental design was generated by R language [22] with the RSM package [23]. It consisted of a 2^k full factorial design for the factorial portion of a central composite design (CCD), four central points in the cube, and 3 factors (k), with 6 axial points at a rotational distance of $\alpha = 1.682$, for a total of 18 runs. The quadratic model was selected for predicting the optimal point, and is expressed as Equation (1):

$$Y = b0 + b1X1 + b2X2 + b3X3 + b11X12 + b22X22 + b33X32 + b12X1X2 + b13X1X3 + b23X2X3 \quad (1)$$

where Y represents response variables (LA, productivity, yield); b0 is the interception coefficient; b1, b2, and b3 are the linear terms; b11, b22, and b33 are the quadratic terms; and X1, X2, and X3 represent the variables studied.

As experimental factors, the solids concentration in the reaction medium subjected to SSAF was considered and expressed as CM ($gCM \cdot L^{-1}$), the concentration of enzymatic cocktail SAE0020 was expressed as E ($IU \cdot gCM_{Dry\ Matter}{}^{-1}$) (DM), and temperature was expressed as T (°C). Table 1 shows a summary of independent variables, ranges, and levels.

Table 1. Experimental range and levels of independent process variables.

Independent Variables	Symbol	Range and Levels				
		$-\alpha$	-1	0	1	$+\alpha$
Cow manure ($g \cdot L^{-1}$)	X1	46.36	60	80	100	113.63
Enzyme ($IU \cdot gCM_{DM}{}^{-1}$)	X2	141.02	170	212.5	255	283.97
Temperature (°C)	X3	41.59	45	50	55	58.41

2.4.2. Analysis of Monomeric Sugars, Lactic Acid, and Byproducts

The equipment used to analyze the sugar content [24–27] was a Waters chromatograph, model Acquity UPLC binary, a CORTECS C18+ 1.6 μm, 2.1 mm × 100 mm column and a Waters Mass spectrometry, Xevo TQS model. To prepare pattern samples, a solution of each aldose (mannose, glucose, galactose, arabinose, and xylose) was separately prepared, together with a 500 ppm (μg/mL) sample of glucose 13C6 (internal standard–IS1), as reference. To analyze it, 100 μL of 3-methyl-1-phenyl-2-pyrazoline-5-one solution and 200 μL of the 25.9% ammonium hydroxide solution were mixed together in each Eppendorf tube. This was stirred and heated at 70 °C for 40 min, then cooled at room temperature and 200 μL of formic acid added to each to neutralize. It was filtered with a 0.22 μ hydrophilic polytetrafluoroethylene (PTFE) sample and injected into the mass spectrometer. The material from the SSAF was subjected to vacuum filtration. Glucose, galactose, mannose, xylose and arabinose, acetic acid and lactic acid content were measured from the filtered liquid. The fluid was subjected to drying at 55 °C for 72 h in a SELECTA oven (DIGITRONIC-TFT). Sugar content was analyzed after chemical pretreatment to check its effectiveness, and after SSAF to check the remaining quantity which was not used by microorganism.

To analyze the lactic acid and byproducts (acetic acid, ethanol, furfural, and 5-HMF), gas chromatography with Flame Ionization Detector (FID) was used. The reactives used were MilliQ water and phosphoric acid. The patterns used were lactic acid, acetic acid, ethanol, furfural, and 5-HMF, respectively. The samples were filtered with a 3 mL aliquot, 1:3 dilutions with 0.35% phosphoric acid solution, using a gas chromatography vial to inject. The equipment for the lactic acid analysis was an HP-5MS UI 30 m × 0.25 mm × 0.25 μm column chromatography. The configuration parameters were an injection split of 1:50; a flux of 1 mL/min He; an injector temperature of 250 °C; an oven ramp with an initial temperature of 65 °C for 1 min, 20 °C/min ramp, ending at 290 °C for 5 min; and the flame ionization detector using an injector temperature of 290 °C. The equipment for the by-products' analysis was ZB-FFAP 30 m × 0.25 mm × 0.25 µm column chromatography. The configuration parameters were an injection split of 1:10; a flux of 1 mL/min H_2; an injector temperature of 230 °C; an oven ramp with an initial temperature of 70 °C for 1 min, 15 °C/min ramp up to 200 °C, and a ramp of 30 °C until 240 °C for 6 min; and an FID injector temperature of 240 °C.

2.5. Experimental Results Calculation

LA production refers to the grams of LA produced per liter of substrate. LA yield (Y) was calculated as the ratio between lignocellulosic material included in CM and LA production. LA productivity (P) was calculated as LA production per hour. Sugar production was determined as the total quantity of sugars (mannose, glucose, galactose, arabinose, and xylose) in diluted acid and SSAF (after 24 and 48 h), minus glucose added.

3. Results and Discussion

3.1. Results

The resulting parameters of the central composite design, specified per experimental set, are summarized in Table 2.

Table 2. Central composed design parameters.

Run	Codified Variables			Experimental Levels		
	X1	X2	X3	Cow Manure ($g{\cdot}L^{-1}$)	Enzyme ($IU{\cdot}gCM_{DM}{}^{-1}$)	Temperature (°C)
1	1	−1	−1	100	170	45
2	1	1	1	100	255	55
3	1	−1	1	100	170	55
4	0	0	0	80	212.5	50
5	1.682	0	0	113.636	212.5	50
6	−1	1	1	60	255	55
7	1	1	−1	100	255	45
8	−1	−1	1	60	170	55
9	−1	−1	−1	60	170	45
10	−1.682	0	0	46.364	212.5	50
11	−1	1	−1	60	255	45
12	0	1.682	0	80	283.976	50
13	0	0	1.682	80	212.5	58.41
14	0	0	0	80	212.5	50
15	0	−1.682	0	80	141.024	50
16	0	0	0	80	212.5	50
17	0	0	0	80	212.5	50
18	0	0	−1.682	80	212.5	41.59

Homogenized samples were analyzed to determine initial humidity, cellulose, hemicellulose, and lignin content. Results are shown in Table 3.

Table 3. Cow manure lignocellulosic material analysis.

Sample	Hemicellulose (%)	Cellulose (%)	Lignin (%)	Ash (%)	Humidity (%)
Manure heap A *	27.3	24.5	3.8	17.69	2.32
Manure heap B *	26.3	25	3.8	18.85	2.5
Farm 1A	28.3	23.4	3.5	18.04	2.32
Farm 1B	30.2	21.7	3.1	17.43	2.11
Farm 2A	28.4	23.6	4.1	17.13	2.76
Farm 2B	28.3	22.9	5.3	16.5	2.23
Average	28.1	23.5	3.9	17.6	2.4

* A and B distinguish different locations.

The components analyzed represent approximately 75% of the total. The remaining percentage was not analyzed, and nutrients and other components were left out of this study, as the focus was on Table 3 components. The average portion of hemicellulose plus cellulose obtained was 51.6%, which means that cow manure has a biodegradable potential for obtaining value-added products. This result, higher than reported in studies to obtain bioethanol from this raw material [8], was associated with a high proportion of lignocellulosic fiber in cow diets. According to the cow manure composition, proximity to other lignocellulosic residues used for same purpose (such as sugarcane bagasse [11] and corn forage [28]) converts this material in a novel and attractive alternative for this (or similar) purpose. A reported average lignin content of only 3.9% was lower than that studied in lignocellulosic materials [29]. Therefore, it does not constitute a significant barrier in the development of these processes. These results allowed the evaluation of the production of lactic acid from cow manure as a raw material, also considering the values obtained of sugars and inhibitory compounds in the stages of the process. Nevertheless, possible barriers should be studied, such as inhibition by product, to improve the current production.

Determinations of sugar concentrations in the liquid fraction of the substrate showed a dependence on the experimental operating conditions. Glucose and xylose were found in high concentrations after chemical pretreatment and also after SSAF. The same behavior was observed in pretreated sugarcane bagasse sugar analysis papers [30]. This means there was a sugar fraction which had not been consumed by the microorganism.

During the pretreatment stage with acid at a high temperature, cow manure content was considered for the central composite design runs. The summary of the analysis of the sugar content after the solubilization of the substrate and the release of sugar oligomeric polymers and monosaccharides is shown in Table 4, as a percentage of lignocellulosic fraction. Xylose and glucose concentrations showed a dependence on the experimental conditions. The highest xylose rating was achieved with 48 h SSAF, high cow manure content, high temperature, and high enzyme value (the same as glucose), as shown in Table 5. Higher sugar concentrations were obtained at higher temperatures, combined with high concentrations of cow manure.

Table 4. Sugar content per experimental run after chemical pretreatment.

Cow Manure (g/L)	Mannose %	Glucose %	Galactose %	Arabinose %	Xylose %	Total %
46.4	0.08	10.09	6.36	18.63	48.30	44.56
60	0.06	4.32 - 4.39	3.02 - 3.44	10.66 - 11.14	22.75 - 25.06	22.36 - 23.10
80	0.04 - 0.53	5.85 - 52.89	3.55 - 6.40	8.84 - 15.68	26.35 - 31.77	25.53 - 47.19
100	0.04 - 0.32	8.93 - 47.84	4.23 - 4.48	9.53 - 10.34	28.58 - 31.10	29.12 - 45.00
113.6	0.22	8.34	4.50	11.04	28.62	27.98

Table 5. Sugar content per experimental run after SSAF process.

Run	Mannose ($g·L^{-1}$)	Glucose ($g·L^{-1}$)	Galactose ($g·L^{-1}$)	Arabinose ($g·L^{-1}$)	Xylose ($g·L^{-1}$)	Total ($g·L^{-1}$)
1	0.48	21.23	1.35	2.82	11.59	32.47
2	0.61	40.53	2.09	3.4	11.68	53.31
3	0.29	24.76	1.37	2.36	8.23	32.01
4	0.31	12.9	2.32	2.2	9.7	22.43
5	0.55	29.07	3.91	3.4	11.64	43.57
6	0.31	15.86	0.75	1.9	6.96	20.78
7	0.45	20.81	1.13	2.34	9.65	29.38
8	0.19	12.89	0.55	1.49	5.79	15.91
9	0.21	15.29	0.11	1.52	6.77	18.9
10	0.2	7.21	1.13	1.52	5.91	10.97
11	0.21	17.25	0.12	0.93	9.15	22.66
12	0.42	18.37	3.21	2.41	9.33	28.74
13	0.27	24.66	1.33	2.45	8.07	31.78
14	0.29	12.43	2.33	2.14	8.53	20.72
15	0.41	25.26	2.98	2.71	10.15	36.51
16	0.48	27.15	1.29	2.88	11.16	37.96
17	0.37	18.15	1.11	2.1	8.44	25.17
18	0.23	14.6	0.94	2.0	8.15	20.92

Total sugar concentration reached its maximum value for the upper level of the cow manure factor, while regarding the axial point a decrease in concentration was registered. Lower cow manure values reached a high sugar content, which should be evaluated in future experimentations. The substrate was subjected to SSAF for 48 h, and the variation in cow manure, enzymes, and temperature according to the central composite design was analyzed, showing an increase in the concentration of dissolved monosaccharides, as a result of the depolymerization enzymatic activity. Table 5 shows the resulting sugar content analyzed after the complete process.

Maximum sugar concentration was reached, as for the previous stage, with a cow manure content of 100 $gCM·L^{-1}$, together with lower values of the enzyme and temperature factors. When the substrate concentration's extreme point was analyzed after chemical pretreatment, there was a decrease in the reaction mixture's monosaccharide content, which suggests a limitation of the long-chain cleavage and large-molecule survival for a high concentration of suspended solids [31]. After SSAF, this behavior had the same effect on the enzymatic reaction rate. The reason for the sugars remaining after SSAF should be studied, as a way to increase current production ratios.

Various byproducts are created during chemical preparation. There are variations in byproduct production depending on the lignocellulosic source or the pretreatment technique used to decompose the lignocellulose. Furans, organic acids, and phenolics are the three major subgroups of byproducts produced [32]. Acetic acid, furfural, and 5-HMF were measured after pretreatment. Acetic acid was analyzed again after SSAF.

On substrates high in lignocellulosic byproducts, the addition of furfural to precultures of *Bacillus coagulans* may improve growth and lactic acid production [11]. Compared to furfural, 5-HMF inhibits dehydrogenase enzymes to a lesser degree [33]. All strains were inhibited by concentrations of furfural and 5-HMF above 5 $g·L^{-1}$ and 8 $g·L^{-1}$, respectively, while some strains were already noticeably inhibited at concentrations of 1 to 2 $g·L^{-1}$ of furans. At concentrations of 15 $g·L^{-1}$ of acetic acid, growth rates were badly inhibited but productivity was not significantly affected. It should be mentioned that when determining the toxicity of acids, pH is always a crucial consideration. At pH 3.5, 3.5 g·L^{-1} of acetic acid has the same inhibitory impact as 9 g·L^{-1} of acetic acid at pH 5 [34]. When determining the toxicity of lignocellulose substrates as a whole, combined effects should be taken into account. Research carried out with pure compounds provides an indication of the inhibitory effect of lignocellulose byproducts. Ethanol was analyzed and no content was found. The content analysis of inhibitors showed low concentrations, as summarized in Table 6.

Table 6. Inhibitors content.

<table>
<tr><th rowspan="2">Run</th><th>Acetic acid (g·L^{-1})</th><th>Furfural (g·L^{-1})</th><th>5-HMF (g·L^{-1})</th><th>Acetic Acid (g·L^{-1})</th></tr>
<tr><th colspan="3">Chemical Pretreatment</th><th>SSAF</th></tr>
<tr><td>1</td><td>1.44</td><td>0.74</td><td>0.94</td><td>1.80</td></tr>
<tr><td>2</td><td>2.00</td><td>1.26</td><td>5.00</td><td>2.20</td></tr>
<tr><td>3</td><td>2.00</td><td>1.26</td><td>5.00</td><td>1.72</td></tr>
<tr><td>4</td><td>0.96</td><td>1.43</td><td>0.33</td><td>1.08</td></tr>
<tr><td>5</td><td>2.14</td><td>1.14</td><td>1.38</td><td>1.90</td></tr>
<tr><td>6</td><td>0.58</td><td>0.42</td><td>0.48</td><td>0.90</td></tr>
<tr><td>7</td><td>1.44</td><td>0.74</td><td>0.94</td><td>1.80</td></tr>
<tr><td>8</td><td>0.58</td><td>0.42</td><td>0.48</td><td>0.96</td></tr>
<tr><td>9</td><td>0.62</td><td>0.80</td><td>0.38</td><td>1.24</td></tr>
<tr><td>10</td><td>0.81</td><td>1.00</td><td>0.33</td><td>0.81</td></tr>
<tr><td>11</td><td>0.62</td><td>0.80</td><td>0.38</td><td>1.00</td></tr>
<tr><td>12</td><td>2.14</td><td>1.14</td><td>1.38</td><td>1.05</td></tr>
<tr><td>13</td><td>1.36</td><td>0.74</td><td>0.54</td><td>1.38</td></tr>
<tr><td>14</td><td>0.96</td><td>1.43</td><td>0.33</td><td>1.20</td></tr>
<tr><td>15</td><td>0.94</td><td>0.68</td><td>0.61</td><td>1.70</td></tr>
<tr><td>16</td><td rowspan="2">1.04</td><td rowspan="2">0.76</td><td rowspan="2">3.92</td><td>2.00</td></tr>
<tr><td>17</td><td>1.14</td></tr>
<tr><td>18</td><td>1.36</td><td>0.74</td><td>0.54</td><td>1.92</td></tr>
<tr><td>Average</td><td>1.33</td><td>1.04</td><td>1.35</td><td>1.45</td></tr>
</table>

The results of this research, including the dependent (or response) variables lactic acid content, yield, and productivity, are given in Table 7.

The highest values of lactic acid concentration and productivity were reported for a substrate concentration of 80 g/L and a temperature of 50 °C, which is optimal for the growth of *Bacillus coagulans* DSM 2314. In a period of 24 h of SSAF, the highest values of lactic acid content and productivity were reported for substrate concentrations of 60 g·L^{-1} and 80 g·L^{-1}, and temperatures corresponding to the central point (50 °C). The enzyme concentration, with a variable behavior, maximizes both responses for its central point value. If the SSAF time is increased to 48 h, a considerable increase in product formation is also highlighted for cow manure concentrations of 100 g·L^{-1}. On the other hand, the yield increases for substrate concentrations corresponding to the lower limit and the lower axial point, according to the lactic acid content measured. The maximum lactic acid value of 13.65 g·L^{-1} was obtained with 212.5 IU·gCM$_{DM}$$^{-1}$ of enzyme, also achieving the maximum yield (33%) and maximum productivity of 0.57. At 48 h of SSAF, the same conditions showed better results for LA production and yield, as shown in Table 7. The 16th run achieved 15.09 g·L^{-1} of LA at 48 h, while the 10th run achieved a yield of 49.91%. The maximum productivity was achieved at 24 h, obtaining a value of 0.57 g

LA·(g lignocellulosic portion)$^{-1}$. These effects would suggest carrying out an analysis to establish a compromise between the values of the study factors that maximize the responses evaluated.

Table 7. Experimental design results obtained after 24 and 48 h.

Run	24 h			48 h		
	LA ($g{\cdot}L^{-1}$)	Productivity ($g{\cdot}L^{-1}{\cdot}h^{-1}$)	Yield wt.% (LA/Lignocellulosic)	LA ($g{\cdot}L^{-1}$)	Productivity ($g{\cdot}L^{-1}{\cdot}h^{-1}$)	Yield wt.% (LA/ Lignocellulosic)
1	3.21	0.13	6.22	15.06	0.31	29.19
2	2.04	0.09	3.95	2.52	0.05	4.88
3	1.89	0.08	3.66	2.16	0.05	4.19
4	12.45	0.52	30.16	13.2	0.28	31.98
5	2.13	0.09	3.63	2.43	0.05	4.14
6	6.36	0.27	20.54	11.82	0.25	38.18
7	5.34	0.22	10.35	14.34	0.30	27.79
8	4.80	0.20	15.50	7.38	0.15	23.84
9	11.10	0.46	35.85	14.31	0.30	46.22
10	9.18	0.38	38.37	11.94	0.25	49.91
11	11.55	0.48	37.31	12.51	0.26	40.41
12	11.01	0.46	26.67	12.9	0.27	31.25
13	1.83	0.08	4.43	2.01	0.04	4.87
14	11.13	0.46	26.96	12.06	0.25	29.22
15	9.90	0.41	23.98	10.44	0.22	25.29
16	13.65	0.57	33.07	15.09	0.31	36.56
17	11.13	0.46	26.96	12.45	0.26	30.16
18	13.41	0.56	32.49	13.65	0.28	33.07

The reported yield values of the product, taking into account the stoichiometry of the consumed sugars, xylose and glucose, were between 30% and 99%, registered after the use of *Bacillus coagulans* strains, from lignocellulosic raw materials in different operational configurations [35]. The above evidences the potential of cow manure for its use for this purpose. However, the conversion rate of total sugar into lactic acid depends on the conditions, according to the levels established for the experimental factors [36], where elevated xylose and glucose concentrations were reported after SSAF (Table 5). These effects on enzymatic catalysis and fermentation must be evaluated through kinetic studies such as the one developed by [37], which allows the visualization of the growth phases of the microorganism as well as the possible existence of inhibition by substrate or byproduct formation.

3.2. Statistical Evaluation

A regression analysis was performed to fit response functions and experimental data. During the statistical evaluation, the analysis of variance indicated that the adjusted models were significant (p-value < 0.05), while not all the terms had significant results. In addition, determination coefficients (R^2) higher than 80% were verified in all cases. A quadratic model was obtained, excluding those non-significant terms. This procedure was carried out for all models. Two Way Interaction (TWI) for cow manure content against enzyme and temperature versus enzyme was not significant in any case. A summary can be found in Table 8.

Table 8. Model values at 24 h and 48 h for lactic acid, productivity, and yield.

Time	Model	Intercept	Cow Manure Content (CM)	Enzyme E	Temperature (T)	CM:T	CM^2	E^2	T^2	Adjusted R^2	*p*-Value
24 h	LA	-2.2024×10^2	--	--	7.4187	--	-6.6988×10^{-3}	-5.4426 10^{-4}	-7.9398×10^{-2}	0.8537	<0.0001
	Y	-2.5511×10^2	−2.3562	5.7394×10^{-1}	1.4827	3.5200	--	-1.2879 10^{-3}	-1.9009×10^{-1}	0.8647	<0.0001
	P	−5.5552	1.8012×10^{-2}	--	2.4129×10^{-1}	3.6250×10^{-4}	-2.5738×10^{-4}	--	-2.9163×10^{-3}	0.818	<0.0001
48 h	LA	-2.0978×10^2	1.6774	7.1460	--	-2.1375×10^{-2}	-4.4435×10^{-3}	--	-6.1962×10^{-2}	0.8755	<0.0001
	Y	−320.523	0.87509	--	15.3092	−0.02912	--	--	−0.14735	0.9039	<0.0001
	P	−4.3588	3.4893×10^{-2}	--	1.4845×10^{-1}	-4.4500×10^{-4}	-9.2354×10^{-5}	--	-1.2865×10^{-3}	0.8745	<0.0001

Based on the projected regression model, Figure 2 shows the response surfaces to estimate the lactic acid production relative to the independent variables: cow manure, enzyme, and temperature at 24 h of SSAF. Figure 2a shows how LA increases significantly with enzyme and cow manure, until it reaches its maximum at 13.65 $g \cdot L^{-1}$, with cow manure and enzyme values around the central point. In Figure 2b, the maximum is obtained with a lower value of temperature and cow manure. In Figure 2c, the curved graphic shows the maximum lactic acid production with the lower temperature value, and the central point for the enzyme. Figure 2d shows the productivity reaching its maximum of 0.57 $g \cdot L^{-1} \cdot h^{-1}$ around the central point for temperature and cow manure, to decrease again at high values of both variables. Figure 2e shows a linear relationship of the yield against cow manure, while the enzyme maximizes the yield around the central point, decreasing it at high or low values. In Figure 2f, the maximum yield is produced at low values of cow manure and temperature, decreasing when both variables increase. Figure 2g shows the maximum yield at the enzyme's central point and lower temperatures. It decreases when the temperature increases, or if the enzyme is away from the central value.

In Figure 3, the response surfaces estimation at 48 h are presented. Figure 3a shows the lactic acid response against enzymes and cow manure, with a curved relation. The maximum was reached at lower or central values of both cow manure and temperature. In Figure 3b, maximum productivity was also reached at lower or central values of both cow manure and temperature. Figure 3c reaches the maximum yield at lower values of cow manure and temperature.

Based on the model, a numerical optimization according to RSM results was carried out with the software R. The optimal working conditions with a combined desirability bigger than 0.9, based on the three variables, were obtained. Results are summarized in Table 9.

Table 9. Numerical optimization calculated with R.

Process Time (h)	Cow Manure ($g \cdot L^{-1}$)	Enzyme ($IU \cdot gCM_{DM}^{-1}$)	Temperature (°C)	Lactic Acid ($g \cdot L^{-1}$)	Productivity ($g \cdot L^{-1} \cdot h^{-1}$)	Yield (wt. %)	Desirability
24 h	71	220	45	13.4	0.53	36.28	0.952
48 h	86	140	42.5	15.27	0.317	32.76	0.917

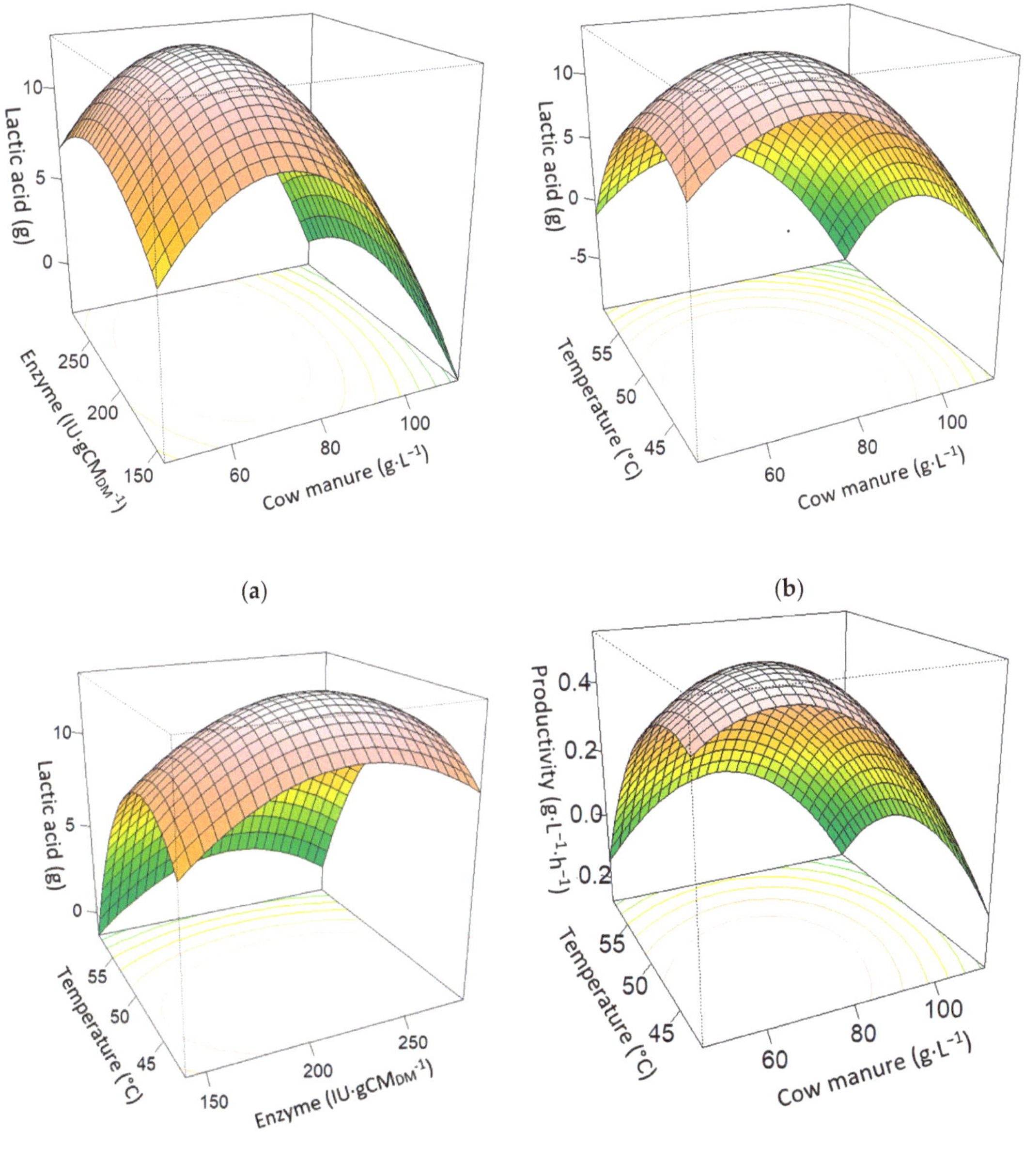

Figure 2. *Cont.*

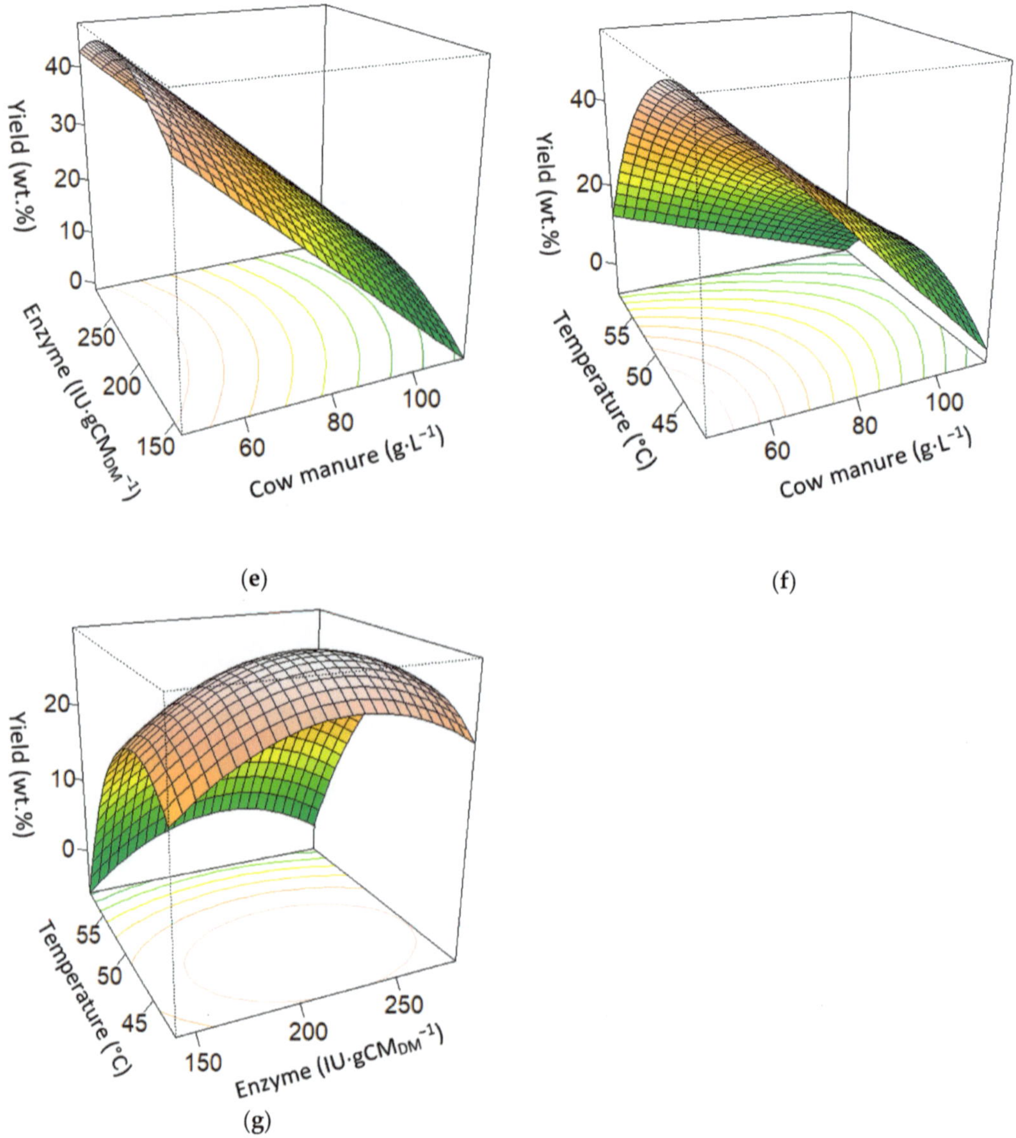

Figure 2. Surface plot effects of enzyme, temperature, and cow manure interactions at 24 h SSAF process. (**a**) Cow manure (g·L^{-1}) and enzyme (IU·gCM$_{DM}$$^{-1}$) on lactic acid (g). (**b**) Cow manure (g·L^{-1}) and temperature (°C) on lactic acid (g). (**c**) Temperature (°C) and enzyme (IU·gCM$_{DM}$$^{-1}$) on lactic acid (g). (**d**) Cow manure (g·L^{-1}) and temperature (°C) on productivity (g·L^{-1}·h^{-1}). (**e**) Cow manure (g·L^{-1}) and enzyme (IU·gCM$_{DM}$$^{-1}$) on yield (wt.%). (**f**) Cow manure (g·L^{-1}) and temperature (°C) on yield (wt.%). (**g**) Enzyme (IU·gCM$_{DM}$$^{-1}$) and temperature (°C) on yield (wt.%).

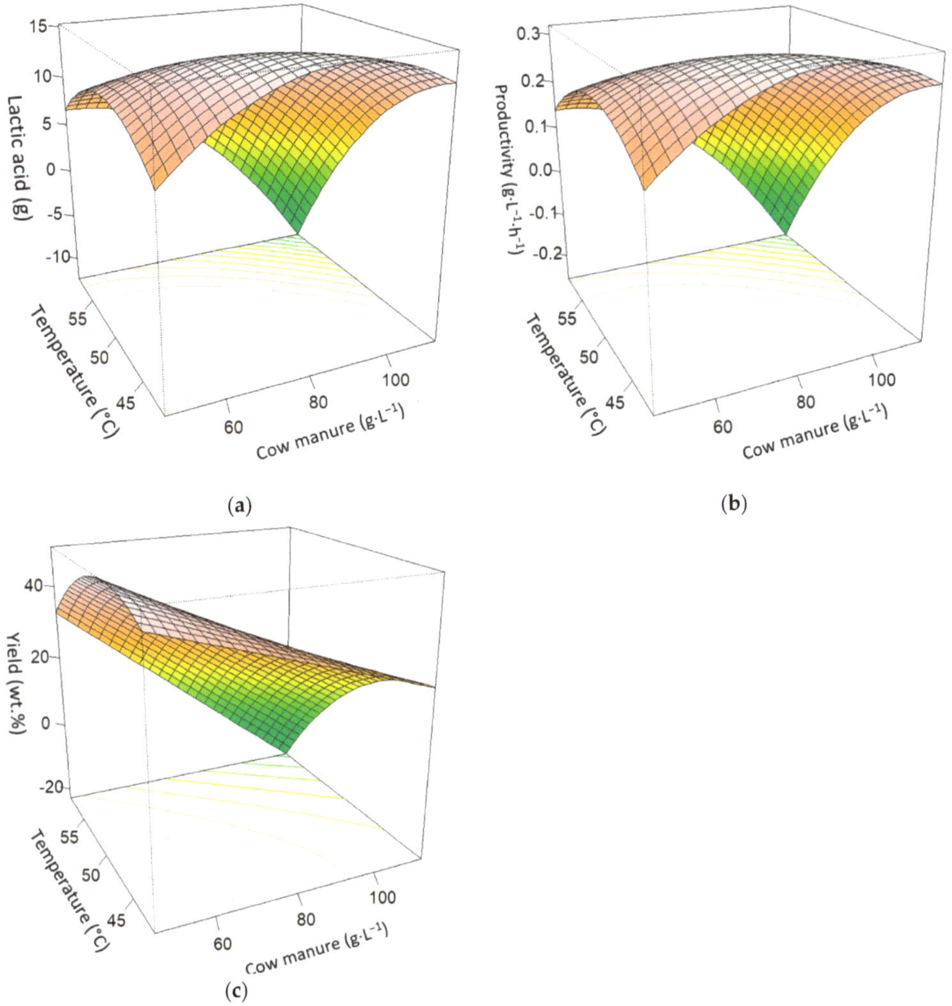

Figure 3. Surface plot effects of enzyme, temperature, and cow manure interactions at 48 h SSAF process. (**a**) Cow manure (g·L^{-1}) and temperature (°C) on lactic acid (g). (**b**) Cow manure (g·L^{-1}) and temperature (°C) on productivity (g·L^{-1}·h^{-1}). (**c**) Cow manure (g·L^{-1}) and temperature (°C) on yield (wt.%).

Comparing 24 with 48 h of SSAF process, there was no considerable increase in lactic acid production. Therefore, yield and productivity decreased. The results for 24 h were lower than those achieved with the use of the same microbial strain on sugarcane bagasse in the same operational strategy [11].

Studies using pretreated bagasse, with same microorganism, and enzyme cocktail GC220, reported a yield of 74%, producing 55.6–59.3 g·L^{-1}. Productivities during SSAF of 0.78–1.14 g·L^{-1} were achieved, which were lower compared to the productivities of 2.5–3 g·L^{-1} reached in research carried out with high-grade sugars [17].

However, the yield was close to values obtained with the same microorganism on wheat straw with basic pretreatment [10], and improved values reported using *Bacillus coagulans* LA-15-2 on rice straw hydrolysate in a batch process [38]. On the other hand, lactic acid values and productivity obtained under these conditions were higher than other studies utilizing different lignocellulosic materials, pretreatments, microorganisms, and operational strategies. Such is the case in the study to obtain this product with *Lactobacillus brevis* ATCC 8287 on spent coffee ground hydrolysate [39] and *Saccharomyces cerevisiae* on spent coffee grounds [40]. In the first case, the reported performance was slightly higher than results obtained in this study, while in the second one it was lower.

4. Conclusions

The production of lactic acid from cow manure was performed using an SSAF process. Lactic acid productivity experimentation showed similar values to fermentations using high-grade sugars or lignocellulose as feedstock. The data collected through the central composite design (24 h and 48 h) fit well to a quadratic model with interactions of all variables. TWI was only significant for cow manure and temperature. The least significant of the three factors analyzed was the enzyme.

The cellulose and hemicellulose composition of cow manure (similar to commonly used lignocellulosic materials, also with low lignin content) makes it a biodegradable material with great potential for obtaining lactic acid.

The maximum lactic acid production conditions were reached at 48 h of fermentation consisting of 15.09 gLA·L^{-1}, productivity of 0.31 gLA·L^{-1} h, and a yield of lactic acid based on cellulose and hemicellulose of 36.56% wt.%. Concerning the 24 h SSAF analysis, higher LA values were obtained around the central point or with low values of temperature and cow manure, achieving values from 11 to 13.5 g·L^{-1}. The enzyme quantity was not so decisive. At high temperatures and/or high cow manure concentrations, LA production was lower. The best yield values were obtained together with higher LA production. With the 48 h SSAF process, moderate increases in the LA product were obtained, reaching 13.5 to 15 g·L^{-1}. The trend continues to reach best values around the central point or for lower temperatures and cow manure. Additionally, the enzyme quantity was not determinative.

Along with statistical evaluation, the adjusted quadratic models turned out to be significant, where the optimal values obtained in the study range showed that if time was increased up to 48 h, LA production increased by 1.87 gLA·L^{-1} and the enzyme amount was reduced, but performance and productivity decreased compared to a 24 h time period. The remaining sugar quantity increased after SSAF, and there is potential to increase LA production in this step.

The experimentally determined composition of byproducts did not impair bacterial activity, so the process does not require additional pretreatment. Nevertheless, the significant presence of sugars in the medium after the SSAF process suggests the possibility of inhibitory effects, which should be further evaluated by means of a kinetic study.

The best operational conditions should be determined based on an economical and environmental function, based on maximizing productivity, lower economic cost, or environmental benefits by using as much cow manure as possible for the process. Possible barriers for product formation should be studied, such as the inhibition of byproducts, to improve the current production, taking into account that not all sugars are consumed after SSAF. Future research on new operational conditions is recommended.

Author Contributions: Conceptualization, V.F. and L.F.C.; methodology, O.P.N. and V.F.; formal analysis, L.F.C. and V.F.; investigation, R.G., O.P.N. and A.A.S.; resources, L.F.C.; data curation, L.F.C.; writing—original draft preparation, R.G., O.P.N. and A.A.S.; writing—review and editing, V.F. and L.F.C.; visualization, R.G.; supervision, V.F. and L.F.C.; project administration, L.F.C.; funding acquisition, L.F.C. All authors have read and agreed to the published version of the manuscript.

Funding: This work received financial support from the Doctorat Industrial grant (2021 DI 22) from the AGAUR through the Secretariat of Universities and Research of the Department of Business and Knowledge of the Generalitat de Catalunya.

Institutional Review Board Statement: Not applicable.

Informed Consent Statement: Not applicable.

Data Availability Statement: Data available on request to the correspondence author.

Acknowledgments: The authors at the University of Lleida would like to thank the Catalan Government for the quality accreditation given to their research group: GREiA (2021 SGR 01615). GREiA is a certified agent TECNIO in the category of technology developers from the Government of Catalonia. This work is partially supported by ICREA under the ICREA Academia programme.

Conflicts of Interest: Author Víctor Falguera was employed by the company AKIS International. The remaining authors declare that the research was conducted in the absence of any commercial or financial relationships that could be construed as a potential conflict of interest.

References

1. Castro-Aguirre, E.; Iñiguez-Franco, F.; Samsudin, H.; Fang, X.; Auras, R. Poly(lactic acid)—Mass production, processing, industrial applications, and end of life. *Adv. Drug Deliv. Rev.* **2016**, *107*, 333–366. [CrossRef] [PubMed]
2. Wee, Y.J.; Kim, J.N.; Ryu, H.W. Biotechnological production of lactic acid and its recent applications. *Food Technol. Biotechnol.* **2006**, *44*, 163–172.
3. Datta, R.; Henry, M. Lactic acid: Recent advances in products, processes and technologies—A review. *J. Chem. Technol. Biotechnol.* **2006**, *81*, 1119–1129. [CrossRef]
4. Datta, R. Technological and economic potential of poly(lactic acid) and lactic acid derivatives. *FEMS Microbiol. Rev.* **1995**, *16*, 221–231. Available online: https://www.academia.edu/25459908/Technological_and_economic_potential_of_poly_lactic_acid_and_lactic_acid_derivatives (accessed on 7 January 2021). [CrossRef]
5. Timbuntam, W.; Sriroth, K.; Tokiwa, Y. Lactic acid production from sugar-cane juice by a newly isolated *Lactobacillus* sp. *Biotechnol. Lett.* **2006**, *28*, 811–814. [CrossRef] [PubMed]
6. Garrido, R.; Cabeza, L.F.; Falguera, V. An Overview of Bioplastic Research on Its Relation to National Policies. *Sustainability* **2021**, *13*, 7848. [CrossRef]
7. Södegard, A.; Stolt, M. Properties of polylactic acid fiber based polymers and their correlation with composition. *Prog. Polym. Sci.* **2002**, *27*, 1123–1163. [CrossRef]
8. Yan, Q.; Liu, X.; Wang, Y.; Li, H.; Li, Z.; Zhou, L.; Qu, Y.; Li, Z.; Bao, X. Cow manure as a lignocellulosic substrate for fungal cellulase expression and bioethanol production. *AMB Express* **2018**, *8*, 190. [CrossRef]
9. Zhao, X.Q.; Zi, L.H.; Bai, F.W.; Lin, H.L.; Hao, X.M.; Yue, G.J.; Ho, N.W. *Bioethanol from Lignocellulosic Biomass*; Bai, F.-W., Liu, C.-G., Huang, H., Tsao, G.T., Eds.; Springer: Berlin/Heidelberg, Germany, 2011; pp. 25–51.
10. Maas, R.H.; Bakker, R.R.; Jansen, M.L.; Visser, D.; De Jong, E.; Eggink, G.; Weusthuis, R.A. Lactic acid production from lime-treated wheat straw by *Bacillus coagulans*: Neutralization of acid by fed-batch addition of alkaline substrate. *Appl. Microbiol. Biotechnol.* **2008**, *78*, 751–758. [CrossRef]
11. van der Pol, E.C.; Eggink, G.; Weusthuis, R.A. Production of l(+)-lactic acid from acid pretreated sugarcane bagasse using *Bacillus coagulans* DSM2314 in a simultaneous saccharification and fermentation strategy. *Biotechnol. Biofuels* **2016**, *9*, 248. [CrossRef]
12. van der Pol, E.C.; Bakker, R.R.; Baets, P.; Eggink, G. By-products resulting from lignocellulose pretreatment and their inhibitory effect on fermentations for (bio)chemicals and fuels. *Appl. Microbiol. Biotechnol.* **2014**, *98*, 9579–9593. [CrossRef]
13. Garrido, R.; Cabeza, L.F.; Falguera, V.; Navarro, O.P. Potential Use of Cow Manure for Poly(Lactic Acid) Production. *Sustainability* **2022**, *14*, 16753. [CrossRef]
14. Castro, Y.P. *Aprovechamiento de Biomasa Lignocelulósica: Algunas Experiencias de Investigación en Colombia*; UTadeo: Bogotá, Colombia, 2014.
15. Miller, C.; Fosmer, A.; Rush, B.; McMullin, T.; Beacom, D.; Suominen, P. *Industrial Production of Lactic Acid*, 2nd ed.; Elsevier B.V.: Amsterdam, The Netherlands, 2011; Volume 3.
16. Singhvi, M.; Gokhale, D. Biomass to biodegradable polymer (PLA). *RSC Adv.* **2013**, 3, 13558. [CrossRef]
17. Van Der Pol, E.C. Development of a Lactic Acid Production Process Using Lignocellulosic Biomass as Feedstock. Ph.D. Thesis, Wageningen University and Research, Wageningen, The Netherlands, 2016.
18. Sluiter, J.S.A.; Hames, B.; Ruiz, R.; Scarlata, C.; Templeton, D. *Determination of Ash in Biomass*, 2005th ed.; NREL/TP-510-42622; NREL: Golden, CO, USA, 2005; Available online: https://www.nrel.gov/docs/gen/fy08/42622.pdf (accessed on 3 January 2021).
19. Sahito, A.R.; Mahar, R.; Siddiqui, Z.; Brohi, K.M. Estimating Calorific Values of Lignocellulosic Biomass from Volatile and Fixed Solids. *Int. J. Biomass Renew.* **2013**, *2*, 1–6.
20. Bezerra, M.A.; Santelli, R.E.; Oliveira, E.P.; Villar, L.S.; Escaleira, L.A. Response surface methodology (RSM) as a tool for optimization in analytical chemistry. *Talanta* **2008**, *76*, 965–977. [CrossRef] [PubMed]
21. Kalavathy, M.H.; Regupathi, I.; Pillai, M.G.; Miranda, L.R. Modelling, analysis and optimization of adsorption parameters for H3PO4 activated rubber wood sawdust using response surface methodology (RSM). *Colloids Surfaces B Biointerfaces* **2009**, *70*, 35–45. [CrossRef] [PubMed]
22. R Core Team. *R: A Language and Environment for Statistical Computing*; R Core Team: Vienna, Austria, 2022; Available online: https://www.r-project.org (accessed on 3 January 2021).
23. Lenth, R.V. Response-Surface Methods in R, Using rsm. *J. Stat. Softw.* **2009**, *32*, 1–17. [CrossRef]

24. Fan, B.; Li, T.; Song, X.; Wu, C.; Qian, C. A rapid, accurate and sensitive method for determination of monosaccharides in different varieties of Osmanthus fragrans Lour by pre-column derivatization with HPLC-MS/MS. *Int. J. Biol. Macromol.* **2019**, *125*, 221–231. [CrossRef]
25. Xia, Y.-G.; Wang, T.-L.; Sun, L.-M.; Liang, J.; Yang, B.-Y.; Kuang, H.-X. A New UPLC-MS/MS Method for the Characterization and Discrimination of Polysaccharides from Genus Ephedra Based on Enzymatic Digestions. *Molecules* **2017**, *22*, 1992. [CrossRef]
26. Sun, X.; Wang, H.; Han, X.; Chen, S.; Zhu, S.; Dai, J. Fingerprint analysis of polysaccharides from different Ganoderma by HPLC combined with chemometrics methods. *Carbohydr. Polym.* **2014**, *114*, 432–439. [CrossRef]
27. Gao, Y.-Y.; Jiang, Y.; Chen, G.-C.; Li, S.-S.; Yang, F.; Ma, Q. A Sensitive and Rapid UPLC-MS/MS Method for Determination of Monosaccharides and Anti-Allergic Effect of the Polysaccharides Extracted from Saposhnikoviae Radix. *Molecules* **2018**, *23*, 1924. [CrossRef]
28. Cheng, Q.; Shi, X.; Liu, Y.; Liu, X.; Dou, S.; Ning, C.; Liu, Z.Q.; Sun, S.; Chen, X.; Ren, X. Production of nisin and lactic acid from corn stover through simultaneous saccharification and fermentation. *Biotechnol. Biotechnol. Equip.* **2018**, *32*, 420–426. [CrossRef]
29. Kumar, A.K.; Sharma, S. Recent updates on different methods of pretreatment of lignocellulosic feedstocks: A review. *Bioresour. Bioprocess.* **2017**, *4*, 7. [CrossRef] [PubMed]
30. van der Pol, E.; Bakker, R.; van Zeeland, A.; Garcia, D.S.; Punt, A.; Eggink, G. Analysis of by-product formation and sugar monomerization in sugarcane bagasse pretreated at pilot plant scale: Differences between autohydrolysis, alkaline and acid pretreatment. *Bioresour. Technol.* **2015**, *181*, 114–123. [CrossRef] [PubMed]
31. Kong, H.; Zou, Y.; Gu, Z.; Li, Z.; Jiang, Z.; Cheng, L.; Hong, Y.; Li, C. Liquefaction concentration impacts the fine structure of maltodextrin. *Ind. Crops Prod.* **2018**, *123*, 687–697. [CrossRef]
32. Palmqvist, E.; Hahn-Hägerdal, B. Fermentation of lignocellulosic hydrolysates. II: Inhibitors and mechanisms of inhibition. *Bioresour. Technol.* **2000**, *74*, 25–33. [CrossRef]
33. Modig, T.; Lidén, G.; Taherzadeh, M.J. Inhibition effects of furfural on alcohol dehydrogenase, aldehyde dehydrogenase and pyruvate dehydrogenase. *Biochem. J.* **2002**, *363*, 769–776. [CrossRef]
34. Taherzadeh, M.J.; Niklasson, C.; Lidén, G. Acetic acid-friend or foe in anaerobic batch conversion of glucose to ethanol by Saccharomyces cerevisiae? *Chem. Eng. Sci.* **1997**, *52*, 2653–2659. [CrossRef]
35. Yankov, D. Fermentative Lactic Acid Production From Lignocellulosic Feedstocks: From Source to Purified Product. *Front. Chem.* **2022**, *10*, 823005. [CrossRef]
36. Farooq, M.; Hussain, M.; Wahid, A.; Siddique, K.H.M.; Aroca, R. *Plant Responses to Drought Stress*; Springer: Berlin/Heidelberg, Germany, 2012.
37. González-Leos, A.; Bustos-Vázquez, M.G.; Rodríguez-Castillejos, G.C.; Rodríguez-Durán, L.V.; Del Ángel-Del Ángel, A. Kinetics of Lactic Acid Fermentation from Sugarcane Bagasse by Lactobacillus Pentosus. *Rev. Mex. Ing. Quim.* **2019**, *19*, 377–386. [CrossRef]
38. Chen, H.; Huo, W.; Wang, B.; Wang, Y.; Wen, H.; Cai, D.; Zhang, C.; Wu, Y.; Qin, P. L-lactic acid production by simultaneous saccharification and fermentation of dilute ethylediamine pre-treated rice straw. *Ind. Crops Prod.* **2019**, *141*, 111749. [CrossRef]
39. Lee, K.H.; Jang, Y.W.; Lee, J.; Kim, S.; Park, C.; Yoo, H.Y. Statistical Optimization of Alkali Pretreatment to Improve Sugars Recovery from Spent Coffee Grounds and Utilization in Lactic Acid Fermentation. *Processes* **2021**, *9*, 494. [CrossRef]
40. Kim, J.; Jang, J.H.; Yeo, H.J.; Seol, J.; Kim, S.R.; Jung, Y.H. Lactic Acid Production from a Whole Slurry of Acid-Pretreated Spent Coffee Grounds by Engineered Saccharomyces cerevisiae. *Appl. Biochem. Biotechnol.* **2019**, *189*, 206–216. [CrossRef] [PubMed]

fermentation

Review

Sugar Beet Pulp as Raw Material for the Production of Bioplastics

Cristina Marzo-Gago *, Ana Belén Díaz and Ana Blandino

Department of Chemical Engineering and Food Technology, Faculty of Sciences, University of Cádiz, Puerto Real, 11510 Cádiz, Spain; anabelen.diaz@uca.es (A.B.D.); ana.blandino@uca.es (A.B.)
* Correspondence: cristina.marzo@uca.es

Abstract: The production of bioplastics from renewable materials has gained interest in recent years, due to the large accumulation of non-degradable plastic produced in the environment. Here, sugar beet pulp (SBP) is evaluated as a potential raw material for the production of bioplastics such as polylactic acid (PLA) and polyhydroxyalkanoates (PHAs). SBP is a by-product obtained in the sugar industry after sugar extraction from sugar beet, and it is mainly used for animal feed. It has a varied composition consisting mainly of cellulose, hemicellulose and pectin. Thus, it has been used to produce different value-added products such as methane, hydrogen, pectin, simple sugars, ethanol, lactic acid and succinic acid. This review focuses on the different bioprocesses involved in the production of lactic acid and PHAs, both precursors of bioplastics, from sugars derived from SBP. The review, therefore, describes the pretreatments applied to SBP, the conditions most frequently used for the enzymatic hydrolysis of SBP as well as the fermentation processes to obtain LA and PHAs.

Keywords: enzymatic hydrolysis; fermentation; bioplastics; lactic acid; polyhydroxyalkanoates; pretreatment; biorefinery; sugar beet pulp; solid-state fermentation

Citation: Marzo-Gago, C.; Díaz, A.B.; Blandino, A. Sugar Beet Pulp as Raw Material for the Production of Bioplastics. *Fermentation* **2023**, *9*, 655. https://doi.org/10.3390/fermentation9070655

Academic Editors: Miguel Ladero and Victoria E. Santos

Received: 21 April 2023
Revised: 4 July 2023
Accepted: 9 July 2023
Published: 12 July 2023

1. Introduction

The use of plastics has increased exponentially in different sectors that affect our daily lives, such as packaging, building, automotive, medical devices, etc. This trend is the result of the good properties of this material, such as its flexibility, lightness, or strength in addition to its simple and productive manufacturing process [1]. However, different characteristics are required for each application and can be changed by switching the component materials. Therefore, the characteristics of plastic can be adjusted for the intended application. The main problem is that plastics are used only for a short period or even just one time, and after that are discarded. If not properly collected, this waste is deposited and accumulates in the environment, contaminating the different ecosystems of all living beings. According to the report data, global plastic production reached 367 million tons in 2020, of which 55 million tons were produced in Europe [2]. On the other hand, only 29 million tons of plastic residue were recovered, corresponding to recycling of only 34.6%. The remainder was used to recover energy (42%) or disposed of in landfill (23.4%) [2]. Consequently, a wide variety of studies are centered on the search for renewable materials with properties similar to plastic but that can be degraded easily after use, thus helping to reduce environmental contamination.

Bioplastics represent an environmentally friendly alternative to plastic. They have the advantage of being biobased, biodegradable or both, which means that they are a product obtained from biomass (biobased) and/or they can be degraded by microorganisms into water and CO_2 (biodegradable) [3]. Thus, bioplastics are classified according to these properties (Figure 1): biobased and non-biodegradable, biobased and biodegradable or fossil-based and biodegradable.

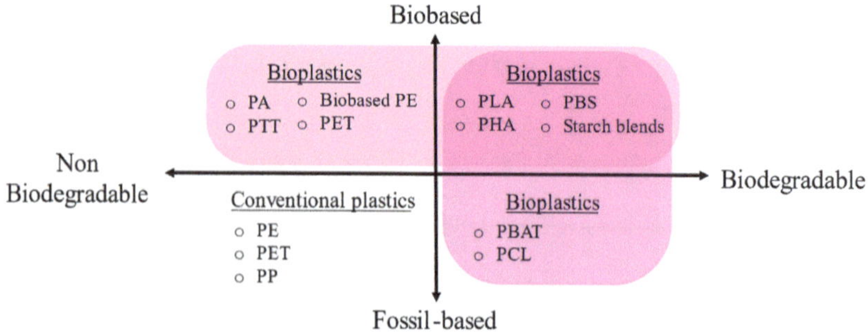

Figure 1. Classification and examples of each type of plastic. PA: polyamide, PTT: polytrimethylene terephthalate; PE: polyethylene; PET: polyethylene terephthalate; PP: polypropylene; PLA: polylactic acid; PHA: polyhydroxyalkanoates; PBS: polybutylene succinate; PBAT: polybutylene adipate terephthalate; PCL: polycaprolactone. Adapted from [4].

The most studied examples are polylactic acid (PLA) and PHAs, as they are biobased and biodegradable bioplastics. PLA is a polymer composed of lactic acid (LA) monomers and qualifies as a linear aliphatic thermoplastic polyester [5]. LA molecules contain chiral carbon, for which there are two isomeric forms: L (+) lactic acid and D (−) lactic acid. The properties of PLA can change according to the proportion of L-LA to D-LA. PLA can be used in different areas of our lives such as in medical devices, 3D printers or food packaging [6–9]. LA can be obtained by chemical synthesis or through fermentation. However, fermentation is usually preferred, as it can produce optical pure isomers of LA instead of a racemic mixture of L-LA and D-LA [5]. Through the fermentation pathway, LA is produced from sugars, mainly glucose, by lactic acid bacteria (LAB) through three different metabolic pathways, which influence the required carbon source and the obtained LA yield.

On the other hand, PHAs are a polymer classified as polyesters consisting of units of (R)-hydroxy fatty acid linked by ester bonds [10]. PHAs are generated by some microorganisms, in which they accumulate when the microorganism is under stress due to a lack of nutrients [11]. Various strategies have been studied to induce the accumulation of PHAs inside cells, for instance via limitation of the nitrogen source or oxygen concentration in fermentation using pure cultures. On the other hand, continuous cycles of fest and famine can be used in mixed cultures to select microorganisms able to accumulate PHAs. To date, more than 150 different monomers have been identified, and they can be combined to modify or improve the properties of the bioplastic. In general, they provide excellent benefits, including 100% biodegradability, biocompatibility, non-toxicity and antioxidant and immunotolerant properties [12]. As in the case of PLA, the bioplastic PHA has multiple uses, such as in packaging, coatings, and pharmaceutical and medical applications [13].

These bioplastics have many benefits over petroleum-based plastics. However, the biggest drawback is their production cost, which is 3 to 4 times higher. Different ways of reducing this cost are currently being studied, one of which is through using cheaper raw materials as alternatives [14]. Bioplastics are usually produced from pure sugars or fatty acids and, in some cases, also from first-generation raw materials such as corn or sugarcane [15]. Although high yields are obtained, the use of edible raw materials for bioplastic production raises concerns about food prices and quantity. As an alternative, second-generation feedstocks, such as agriculture or food industry residues, can also be used. These materials do not compete with human food, are found in abundance and are inexpensive. An additional advantage of these materials is they help to reduce the problems associated with the treatment of these wastes.

The majority of second-generation feedstocks are classified as lignocellulosic biomass (LCB) because they are mainly composed of cellulose, hemicellulose and lignin [16,17].

LCB is considered the largest renewable source on Earth [18]. The composition of this biomass depends on its source, but they are generally composed of 35–50% cellulose, 20–35% hemicellulose and 10–25% lignin [19]. As seen in Figure 2, the typical lignocellulose structure is composed of a matrix of crystalline cellulose surrounded by hemicellulose, pectin and lignin polymers [19]. This structure prevents the degradation of cellulose and confers rigidity to cell wall plants and, thus, resistance against insects and pathogens [20].

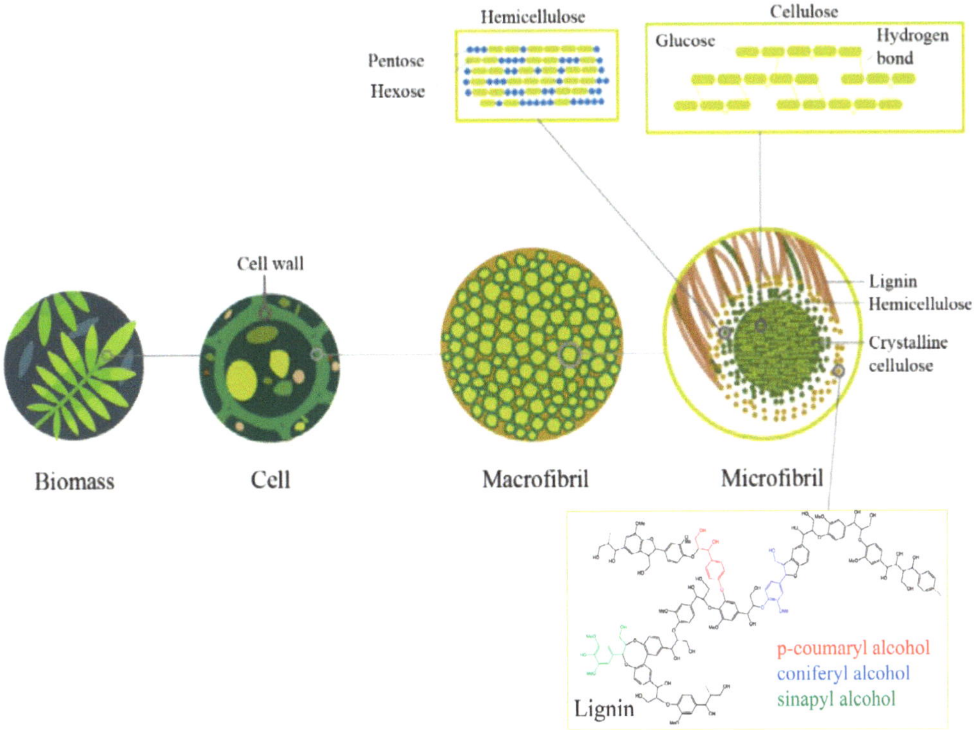

Figure 2. Structure of lignocellulosic biomass. Adapted from [21].

Sugar beet pulp (SBP) is a lignocellulosic by-product of the sugar industry that has traditionally been used for animal feed [22]. However, it has also been used as raw material to produce a wide range of value-added products, such as LA or PHAs, through biotechnological processes. In this review, each of the main stages involved in the production of bioplastics from this by-product are studied in detail (Figure 3). In this way, an evaluation of the different pretreatments applied to SBP to facilitate the subsequent enzymatic hydrolysis has been made. In addition, the influence of the enzymatic cocktail used for the hydrolysis of the biomass is analyzed. Regarding this, and with the aim of reducing the cost of processing the sugar beet pulp, enzymes are sometimes produced directly on the biomass via solid-state fermentation. This novel aspect is also described in detail in the review. Finally, the different strategies used to increase the yield of lactic acid and polyhydroxyalkanoates from SBP hydrolysates will be compared.

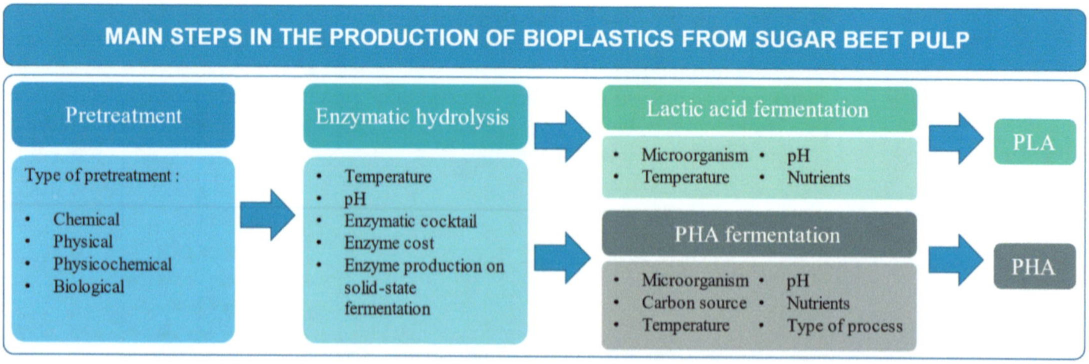

Figure 3. Flowchart showing the main steps in the production of bioplastics from sugar beet pulp, for which the associated issues are addressed in this review.

2. Exhausted Sugar Beet Pulp as a Renewable Feedstock

Sugar beet (*Beta vulgaris* L.) is one of the main sugar crops, along with sugarcane, used to produce sucrose for human consumption. As one of the main sugar crops, sugar beet accounts for about 20% of global sugar production, with sugarcane making up the remaining 80% [22]. This crop thrives in moderate-temperature regions in the northern hemisphere, where the climate is suited to its growth and development. The top sugar beet producers in the world in 2021 were the European Union (112,847.63 tons), the Russian Federation (47,500 tons), and the United States (32,364.15 tons) [23].

The sugar content of sugar beet can vary from 12% to 20%, making it a valuable raw material. The sugar industry can also achieve a financial gain of up to 10% through the use of by-products generated during production. The three main by-products are beet pulp, lime sludge and molasses, which can be used as renewable materials to produce energy or other value-added products [22]. For instance, one ton of sugar beet yields 160 kg of sugar, 500 kg of wet exhausted pulp and 38 kg of molasses [24] (see Figure 4A).

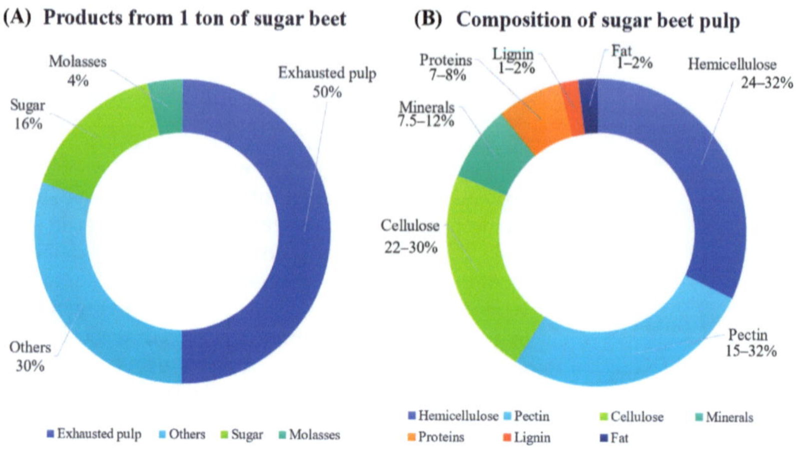

Figure 4. (**A**) Products obtained from 1 ton of sugar beet and (**B**) chemical composition of sugar beet pulp.

2.1. Processing of Sugar Beet

The diagram in Figure 5 shows the steps for producing white sugar from sugar beets. Sugar beetroots, or taproots, are harvested mechanically, separating the leaves from the bulbs. The bulbs are washed, and the washing sludge is decanted to reuse the water [25]. The roots are then cut into thin strips, called cossettes, which are moved to stainless steel tanks, called diffusers, where they are mixed with hot water (55–75 °C) at a solid/liquid ratio (SLR) of 1:2 *w/w* with residence times from 1 to 2 h [22]. The juice obtained, called raw juice, contains mainly sucrose, but other compounds also that have been extracted from the root can hinder the crystallization of sugar. For this reason, the raw juice is purified before crystallization, with calco-carbonic treatment being the most commonly used approach [26,27]. The clean juice obtained after this step, the thin juice, is concentrated with multi-effect evaporators to obtain the final juice (thick juice) [28]. Finally, the thick juice is introduced into the crystallizers, where sugar is produced through crystallization [29,30]. Crystals are later separated from the liquid by centrifugation, dried in continuous rotary dryers and sieved [22]. The residual liquid stream, called molasses, contains 50% *w/w* sugars and other substances, such as oligosaccharides and organic acid salts.

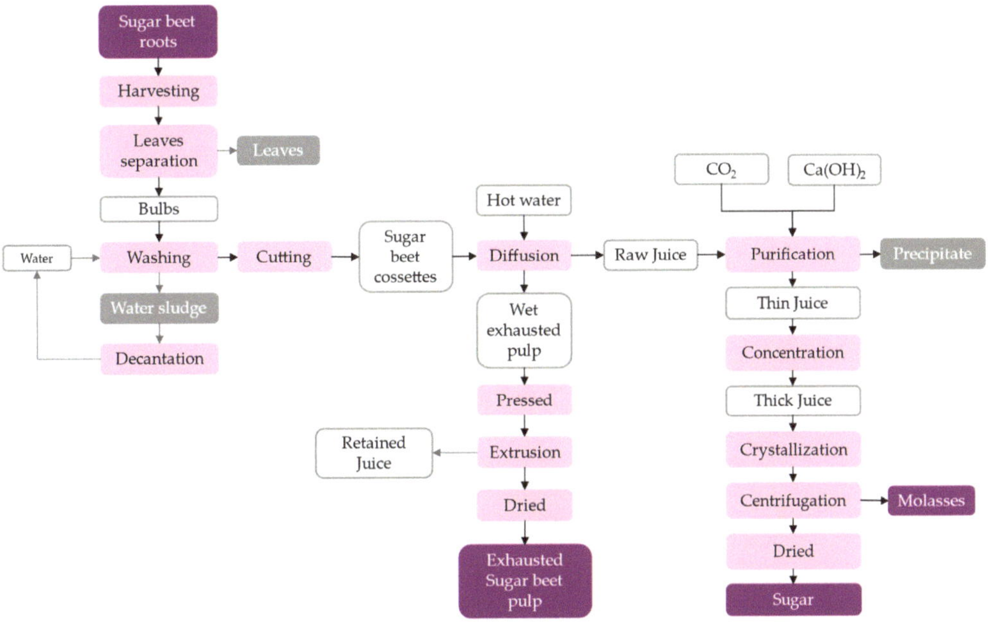

Figure 5. Flowchart of sugar beet processing. Adapted from [22].

The remaining pulp obtained after the diffusion step, named exhausted SBP, is pressed, and the retained juice is recovered. The pulp then contains a humidity of 70% and can be conformed into pellets of approximately 5 cm in length and 0.5 cm in diameter by extrusion [31]. These pellets are preserved by drying to reach a final humidity lower than 10%.

SBP is commonly used as animal feed. However, it has been also used as a raw material for the extraction of pectin and phenolic compounds and for producing several value-added products (VAPs), such as biogas by anaerobic digestion, hydrogen by acidogenic fermentation, or hydrolysates rich in sugars that can be fermented into several bioproducts such as ethanol, lactic acid, succinic acid or polyhydroxyalkanoates, among others. The digestate from the anaerobic digestion of SBB has also been used as an organic fertilizer in agriculture [32–34].

2.2. Chemical Composition of SBP

SBP contains between 87 and 92% of dried matter [35] and is mainly composed of 24–32% of hemicelluloses, 22–30% cellulose, 15–32% pectin, 1–2% lignin, 7–8% protein, 7.5–12% minerals and 1–2% fats [36–38] (see Figure 4B). The pulp composition can vary depending on the process of sugar extraction, the degree of maturation of the root and the sugar beetroot's geographical origin.

Among the sugar industry effluents, beet pulp stands out as a material rich in carbon sources [39]. Hence, component sugars from chemically or enzymatically hydrolyzed polysaccharides from SBP are appropriate raw materials for fermentation. There are several papers reporting on the total sugar composition of SBP after complete hydrolysis as being D-glucose (21–26%), D-galacturonic acid (14–21%) and L-arabinose (21–23%) the majority. The next highest sugar contents were D-galactose (5–6%), L-rhamnose (2.5%), D-xylose (about 2%) and mannose (1%) [40].

In a recent study comparing different agri-food wastes for biopolymer production, the feedstock quality was determined based on having high cellulose and low moisture and lignin content. Low moisture and lignin contents mean lower energy requirements for drying and easier processing, respectively. The residues generated in the harvest and processing stages of the food supply chain of wheat, barley, oats, rapeseed, sugar beet, carrots and onions were analyzed, and the percentage of moisture, cellulose, and lignin ranged from 7–87%, 10–42%, and 2–39%, respectively. Although sugar beet pulp has medium-high cellulose content (around 32%), it showed the lowest lignin (less than 5%). Regarding moisture content, vegetable wastes had a much higher percentage, between 77 and 87%, compared to the cereal residues (less than 20%). However, other factors such as the annual production, seasonality and feedstock cost should be taken into account [41].

3. Production of Bioplastic Precursors from Sugar Beet Pulp

The production of bioplastics, such as PLA and PHAs from SBP, can be performed through a biotechnological process such as the one proposed in Figure 3, which is most applied in the valorization of lignocellulosic biomass. Usually, the biomass is pretreated to enable access to the enzymes used in the saccharification step, where the biomass is hydrolyzed to produce a medium rich in sugars. This medium can be fermented to produce LA or PHAs, which are the precursor molecules to produce PLA and PHAs, respectively.

3.1. Pretreatments Applied to Sugar Beet Pulp

Pretreatment is an important step in the valorization of biomass because it helps to separate lignin and hemicellulose structures from cellulose. In this way, enzymes have easier access to cellulose [42,43]. Depending on the working principle, the pretreatments can be classified as physical, chemical, physicochemical and biological pretreatments [44].

The aim of physical pretreatment is to increase the surface area of the biomass particles. This can be achieved by reducing the particle size with mechanical comminution. Thus, the crystallinity of the cellulose structure is also damaged, which facilitates enzyme accessibility [44,45].

The results achieved with chemical pretreatments will depend on the reagent added, its concentration and the pretreatment duration. Thus, alkalis, such as sodium or potassium hydroxides, are used with the intention to solubilize lignin and hemicellulose [46]. On the other hand, acids, such as sulfuric or hydrochloric acids, are more efficient in degrading only hemicellulose [45]. Moreover, oxidant agents such as alkaline hydrogen peroxide or organic acids are chosen to degrade lignin [42].

Physicochemical pretreatments attempt to solubilize hemicellulose and lignin by changing the temperature, pH and moisture of the biomass [42]. In this way, pretreatments such as steam explosion, ammonia fiber explosion (AFEX), wet oxidation, CO_2 explosion and thermal (liquid hot water) treatment can be performed, where high-temperature exposure is usually combined with high pressure [47].

Biological pretreatments degrade part of the biomass structure by the action of microorganisms [42]. Usually, this is performed using fungi because they can more easily penetrate solids. In this pretreatment, not only might hemicellulose and lignin be degraded but part of the cellulose may also be consumed by the microorganism [45].

Most pretreatments applied to SBP are diluted acid and thermal pretreatment, which produce the highest hydrolysis yields. Other studied SBP pretreatments include size reduction (milling), ultrasound, steam explosion, ensiling and solid-state fermentation, which are all summarized in Table 1. For example, ultrasound pretreatment was applied on SBP using SONOPULS HD 2200 homogenizer in continuous mode with 400 W for 20 min [24]. However, the hydrolysis yield did not increase. On the other hand, for thermal pretreatment of SBP, the hydrolysis yield was increased from 0.70 to 0.85 g of reducing sugars per gram of SBP (g_{RS}/g_{SBP}) [24]. This pretreatment was carried out in an autoclave at 120 °C for 30 or 60 min [24]. Another study performed by Rezic et al. also focused on the effect of these pretreatments on SBP, and similar results were obtained [24,48]. In this case, thermal pretreatment was performed in the autoclave for 20 min, and ultrasound pretreatment was carried out at 200 W for 5 to 45 min [48].

Table 1. Pretreatments applied to SBP. Y_H: hydrolysis yield (g_{RS}/g_{SBP}).

Pretreatment Type		Conditions	Y_H	Reference
Physical	Milling	0.8–1.0 mm	0.71 g/g	[24]
	Milling + Ultrasound	0.8–1.0 mm 50% or 100% amplitude, 20 min, water or 2% *w*/*w* H_2SO_4	0.70–0.76 g/g	[24]
Chemical	Dilute acid	Autoclave, 150 °C, 10 min, 1.1% *w*/*w* H_2SO_4	0.82 g/g	[49]
	Dilute acid	120 °C, 6 min, 0.1 N HCl	0.86 g/g	[50]
	Dilute acid	120 °C, 0.66% H_2SO_4	0.63 g/g	[51]
	Dilute acid	1% H_2SO_4	0.49 g/g	[52]
Physicochemical	Milling + Thermal	Autoclave, 121 °C, 30 or 60 min, 2% *w*/*w* H_2SO_4	0.85 g/g	[24]
	Milling + Thermal	Autoclave, 120 °C, 30 or 60 min, water	0.75 g/g	[24]
	Thermal	Autoclave, 121 °C, 20 min	0.60 g/g	[52]
Biological	Ensiling	Lactobacillus species	0.95 g/g	[53]
	Solid-state fermentation	Aspergillus awamori, 70% moisture, 5 days	0.34 g/g	[52]

Several authors have studied the effect of diluted acid pretreatment on SBP, resulting in the solubilization of pectin and hemicellulose fractions [50]. For instance, SBP was pretreated with sulfuric acid (1.1% *w*/*w*) at 150 °C for 10 min and solid loading of 10% *w*/*w*, and the hydrolysis yield reached 0.82 g_{RS}/g_{SBP} [49]. However, this yield can be improved to 0.86 g_{RS}/g_{SBP} by performing the pretreatment at the optimal conditions found by El-gendy et al. of 120 °C, 0.1 N HCl, 14% *w*/*w* of solid loading and 6 min [50]. Nonetheless, a higher hydrolysis yield (0.92 g_{RS}/g_{SBP}) was achieved when SBP was pretreated with sulfuric acid (0.66%) with a solid loading of 6% at 120 °C [51]. More recently, Marzo et al. performed diluted acid pretreatment on SBP, studying the effect of sulfuric acid concentrations of 0.25, 0.5, or 1% *w*/*v* [52]. The pretreatment was performed by soaking the solid at a solid/liquid ratio of 1:20 (*w*/*v*) with a solution of sulfuric acid and autoclaving the mixture at 120 °C for 20 min. The results showed an increase in the production of glucose as the concentration of acid was increased, reaching a hydrolysis yield of 0.45 g_{RS}/g_{SBP}.

Another lesser-studied pretreatment is the steam explosion. The optimum conditions of pressure and pretreatment time found by Cárdenas–Fernández et al. were 5.3 bar and 24.4 min [54]. With this pretreatment, solubilization of both pectin and the insoluble cellulose fraction was achieved, thereby increasing the ethanol yield due to the increased solubilization of cellulose.

Biological pretreatments such as ensiling or solid-state fermentation have been also studied for SBP. For instance, SBP was pretreated using the ensiling methodology, where several *Lactobacillus* species were involved [53]. This increased the enzymatic digestibility of SBP by 35% [53]. In this case, the microorganism degraded some fraction of the biomass. As this technique is usually employed for long-term storage of biomass, it seems an interesting option to stabilize SBP and pretreat it at the same time. Additionally, SBP was also pretreated via solid-state fermentation [52]. Fermentation was carried out with the fungus *Aspergillus awamori* at 70% moisture content and incubation at 30 °C for 5 days. The results obtained after the enzymatic hydrolysis of the pretreated SBP showed a decrease in the concentration of total reducing sugars produced compared with the non-pretreated solid. However, the same glucose concentration was obtained, showing that only the hemicellulose and pectin content has been removed by the fungus.

3.2. Enzymatic Hydrolysis of SBP

The production of sugars from SBP can be performed through acid or enzymatic hydrolysis. Acid hydrolysis is carried out with concentrated or diluted acids such as hydrochloric acid or sulfuric acid, which degrade the lignocellulose structure [19]. Although a pretreatment step before acid hydrolysis is not necessary, this step has numerous disadvantages compared to enzymatic hydrolysis. For example, a special material is required for the equipment used in this type of hydrolysis to ensure it can resist the corrosion resulting from the acids [19]. However, one of the main issues of this process is the production of compounds that are inhibitory for the microorganisms, such as furfural or its derivates, which are generated after the degradation of sugars due to acidic conditions. On the other hand, enzymatic hydrolysis is a more environmentally friendly process.

Enzymatic hydrolysis is based on the use of hydrolytic enzymes that hydrolyze the lignocellulose structure into simple sugars. There is a wide variety of enzymes that are necessary to hydrolyze the whole biomass, with the most important being cellulases, hemicellulases and pectinases [55]. The proportion of each will depend on the composition of the biomass that is hydrolyzed. One of the main advantages of enzymatic hydrolysis is the use of mild conditions for temperature (45–50 °C) and pH (4.8–5.0) [56]. Moreover, inhibitory compounds are not generated, and the equipment is not damaged by corrosion. On the other hand, the main problem of enzymatic hydrolysis is the high cost of enzymes and the occasional need for prior pretreatment.

Several authors have studied the enzymatic hydrolysis of SBP with the aim of maximizing sugar production from SBP (Table 2). Some of them studied the influence of the enzymatic cocktail. In this sense, the main enzymes used are commercial cocktails composed of a mixture of cellulase, xylanase and pectinase. The optimum conditions of these cocktails are very similar; thus, the hydrolysis of SBP is generally performed at 50 °C, pH 4.8, 150 rpm and 2% *w*/*w* of solid loading (Table 2). One of the main cocktails used is called "Celluclast® 1.5 L", and it is added to correspond to enzyme activity in the range of 4–30 filter paper units per gram of dry matter (FPU/g). Depending on the added enzyme activity, the hydrolysis yield can be improved from 0.27 to 0.49 g_{RS}/g_{SBP} when the cellulase activity is increased from 5 to 30 FPU/g [57]. However, other cellulase cocktails richer in cellobiase, xylanase or pectinase, such as Novozyme 188, Novozyme 431, Viscozyme L, Pectinex®, Celustar XL or Agropect pomace, can be also used. For example, the hydrolysis yield was increased from 0.15 to 0.70 g_{RS}/g_{SBP} when a cocktail combining cellulase and cellobiase was supplemented with hemicellulases and pectinases [58]. Due to the composition of SBP, the addition of these enzymes solubilizes the fraction of hemicellulose and pectin, making cellulose more accessible to enzymes.

Other authors focused their studies on the pretreatment of SBP to improve the enzymatic hydrolysis yield. For example, the hydrolysis yield was increased from 0.15 to 0.5 g_{RS}/g_{SBP} when SBP was pretreated with ammonia (0.5 mL/g of ammonia, 80 °C, 5 min) and the enzymatic hydrolysis was performed with cellulase and cellobiase [58]. However, the hydrolysis yield decreased from 0.7 to 0.61 g_{RS}/g_{SBP} when SBP was pretreated in the same conditions but the hydrolysis was carried out with cellulase, cellobiase, hemicellulase and pectinase [58]. The hydrolysis yield was also increased from 0.2 g_{RS}/g_{SBP}

for non-pretreated SBP to 0.48 and 0.38 g_{RS}/g_{SBP} when SBP was pretreated with thermal and chemical pretreatment with hydrochloric acid, respectively [59]. The same effect was observed in other papers, where the hydrolysis yield was increased from 0.18 g_{RS}/g_{SBP} for non-pretreated SBP to 0.38, 0.49, 0.23 and 0.25 g_{RS}/g_{SBP}, when hydrochloric acid, ammonia, pectinase or ammonium oxalate was added to pretreat SBP [60].

3.3. Enzyme Production by Solid-State Fermentation

Enzymatic hydrolysis is the stage of the process that most influences the overall cost of the process, due to the high cost of the enzymes. To render the process more profitable, various approaches have been explored, such as looking for new sources of enzymes or alternative techniques to submerged fermentation [61]. One such technique is the production of hydrolytic enzymes by solid-state fermentation (SSF) [62,63]. The advantage of this fermentation is that different types of biomass, such as agri-food industry wastes or bioproducts, can be used as raw material, acting as a source of carbon and nutrients and as solid support for fungal growth and enzyme production [64].

SSF is a heterogeneous fermentation that takes place in the absence or near absence of visible water between particles [65]. It is usually carried out by filamentous fungi that grow on the surface of the solid due to their ability to colonize the interparticle spaces of porous materials [66], although bacteria and yeasts can also be used. Most fungal species employed for this technique belong to the genera *Aspergillus*, *Pleurotus* and *Trichoderma* [67].

SSF is a complex process where parameters such as temperature, aeration rate, pH, initial moisture, particle size, agitation, water activity or inoculum concentration should be optimized, and most will depend on the microorganism selected to produce the enzymes [68]. Depending on the fungal species, the optimum temperature for growth can vary from 20 to 55 °C, while the optimum moisture content can change in the range of 40 and 70% [69]. However, the pH is usually adjusted to 5 [70], and aeration is mainly used in pilot-scale reactors [71].

Table 2. Conditions for enzymatic hydrolysis of SBP. T: temperature; SL: solid loading; EA: enzyme activity; YH: hydrolysis yield (g_{RS}/g_{SBP}).

Reference	T	pH	Agitation	SL	Enzyme Type [1]	EA [2]	Pretreatment	Y_H
[58]	40 °C	4.8	-	5% *w/w*	Celluclast® 1.5 L Novozyme 431	4.2 FPU/g d.m. 28.4 CBU/g d.m.	Ammonia 0.5:1, 85 °C, 5 min	0.50 g/g
	40 °C	4.8	-	5% *w/w*	Celluclast® 1.5 L Novozyme 431 Viscozyme L	4.2 FPU/g d.m. 28.4 CBU/g d.m. 0.85 HU/g d.m. 60.2 PGU/g d.m.	Ammonia 0.5:1, 85 °C, 5 min	0.61 g/g
	40 °C	4.8	-	5% *w/w*	Celluclast® 1.5 L Novozyme 431	4.2 FPU/g d.m. 28.4 CBU/g d.m.	Untreated	0.15 g/g
	40 °C	4.8	-	5% *w/w*	Celluclast® 1.5 L Novozyme 431 Viscozyme L	4.2 FPU/g d.m. 28.4 CBU/g d.m. 0.85 HU/g d.m. 60.2 PGU/g d.m.	Untreated	0.70 g/g
[53]	50 °C	4.8	150 rpm	2% *w/w*	Celluclast® 1.5 L Novozyme 188	15 FPU/g d.m.15 CBU/g d.m.	Ensilage 90 days	0.19 g/g
[57]	40 °C	4.8	150 rpm	2.5% *w/w*	Celluclast® 1.5 L	5 FPU/g d.m.	HCl pH 1.5, 85 °C, 4 h	0.27 g/g
	40 °C	4.8	150 rpm	2.5% *w/w*	Celluclast® 1.5 L	30 FPU/g d.m.	HCl pH 1.5, 85 °C, 4 h	0.49 g/g
[72]	50 °C	5	-	10% *w/w*	Celustar XL Agropect	0.75 FPU/g d.m.	Untreated	0.3 g/g
[51]	50 °C	4.8	150 rpm	2% *w/w*	Celluclast® 1.5 L Novozyme 188 Pectinex®	15 FPU/g d.m. 15 CBU/g d.m. 60 PGU/g d.m.	Diluted acid H_2SO_4, 0.66%, 120 °C, 2% solid loading	0.63 g/g

Table 2. *Cont.*

Reference	T	pH	Agitation	SL	Enzyme Type [1]	EA [2]	Pretreatment	Y_H
[59]	45 °C	4.8	-	2% *w/w*	Celluclast® 1.5 L	20 FPU/g d.m.	Untreated	0.20 g/g
	45 °C	4.8	-	2% *w/w*	Celluclast® 1.5 L	20 FPU/g d.m.	HCl pH 1.5, 85 °C, 4 h	0.38 g/g
	45 °C	4.8	-	2% *w/w*	Celluclast® 1.5 L	20 FPU/g d.m.	Autoclave (2.1 bars, 30 min), water 1:20 (*w/v*)	0.44 g/g
[60]	50 °C	4.8	-	5% *w/w*	Cellulase	20 FPU/g d.m.	Untreated	0.18 g/g
	50 °C	4.8	-	5% *w/w*	Cellulase	20 FPU/g d.m.	HCl 1% *w/w*, 80 °C, 6 h	0.38 g/g
	50 °C	4.8	-	5% *w/w*	Cellulase	20 FPU/g d.m.	Ammonia 10% *w/w*, 80 °C, 6 h	0.49 g/g
	50 °C	4.8	-	5% *w/w*	Cellulase	20 FPU/g d.m.	Pectinase 30 U/g, 50 °C, 6 h	0.23 g/g
	50 °C	4.8	-	5% *w/w*	Cellulase	20 FPU/g d.m.	Ammonium oxalate 5% *w/w*, 80 °C, 6 h	0.25 g/g
[73]	50 °C	5.0	150 rpm	10% *w/w*	Celluclast® 1.5 L β-glucosidase xylanase exo-polygalacturonase	2200 FPU/g d.m. 6 CBU/g d.m. 300 HU/g d.m. 110 PGU/g d.m.	Autoclave 120 °C, 20 min	0.71 g/g

[1] Novozyme 188: Cellobiase; Novozyme 431: Cellobiase; Pectinex®: Pectinase; Agropect pomace: Pectinase; Celluclast® 1.5L: Cellulase, xylanase, pectinase, mannanase; Viscozyme L: arabinase, cellulase, β-glucanase, hemicellulase and xylanase; Cellustar XL: xylanase, cellulase, β-glucanase. [2] FPU: filter paper cellulase unit; CBU: cellobiase activity units; PGU: polygalacturonase activity units; HU: hemicellulase activity units; d.m: dry matter.

A noteworthy substrate for this process is lignocellulosic biomass due to its rich composition of cellulose, hemicellulose and pectin. This type of substrate can induce the production of different enzymes by fungi [65,74]. In this sense, SBP seems to be a suitable substrate to produce hydrolytic enzymes. For instance, hydrolytic enzymes, such as xylanase, exo-polygalacturonase and cellulase, were produced via SSF of SBP, reaching 35 U/g of xylanase and 28 U/g of exo-polygalacturonase after 8 days of fermentation with *Aspergillus awamori* [75]. These values were higher than those for production from orange peel waste under the same conditions [75]. Dextranase was also produced by SSF of SBP, reaching peak activity (1057 U/g) after three days of growth of *A. awamori* F-234 [76]. This result was obtained in a study where various residues, such as olive mill solid waste, jojoba mill solid waste and sugar cane bagasse, and different fungi strains, such as *Aspergillus niger* F-93, *Aspergillus fumigatus* F-993, *Penicillium funiculosum* NRC289, *Trichoderma koningii* F-25 and *Aspergillus awamori* F-234, were tested. However, SBP was also studied in combination with other residues such as wheat bran [77]. The results showed that the enzyme production depended on the medium and the fermentation conditions and that polygalacturonase production was induced when wheat bran was mixed with 30% of SBP, reaching 909 U/g with *A. sojae* ATCC 20235 after 8 days at 30 °C [77].

Other studies were performed to increase enzyme production by adding extra nitrogen sources to the raw material. For example, wastewater from monosodium glutamate production was added as a nitrogen and water source to improve pectinase production from the fermentation of SBP with *Aspergillus niger* (CGMCC0455) [78]. Thus, the enzyme activities were increased achieving 15.6×10^{-3} U/g of endopectinase, 3.6×10^{-3} U/g of polygalacturonase and 16×10^{-3} U/g of pectin-lyase [78]. Ammonium sulfate was also tested as a nitrogen source on the production of α-L-arabinofuranosidase by SSF of SBP with *Trichoderma reesei*, reaching 433 U/g of α-L-arabinofuranosidase [79].

To conclude, SSF is a flexible process that can be adapted to produce a wide range of hydrolytic enzymes. However, enzymes are usually extracted and purified to be used in enzymatic hydrolysis, which also raises the overall cost of the process. As an alternative process, some authors have studied the addition of fermented solid directly to the medium to be hydrolyzed, avoiding the enzyme extraction step. Leung et al. were one of the first to propose this process in the SSF of waste bread [80]. They produced protease and glucoamylase by SSF of waste bread and, subsequently, the fermented solid was added to fresh waste bread to produce a hydrolysate, which then was fermented to succinic acid. The same strategy was followed by other authors, such as Pleissner et al., Kwan et al. or Dessie et al., for different wastes, such as food waste, bakery waste, and fruit and vegetable waste [81–83]. More recently, Marzo et al. used the same strategy on SBP, where the enzymes were produced through the SSF of SBP and then added to fresh SBP to produce a hydrolysate rich in sugars [84]. The hydrolysis yield was increased from 0.45 to 0.55 g_{RS}/g_{SBP} when a fed-batch strategy was applied. Thus, 15 g of fermented solid and 13.75 g of fresh solid were mixed at the beginning of the hydrolysis, and then the same amount of fresh solid was added 3 times every 2.5 h [84].

3.4. *Lactic Acid Fermentation*

The medium obtained after enzymatic hydrolysis has a high content of sugars, such as glucose. Therefore, it can be a raw material to produce a wide variety of VAPs through submerged fermentation. LA is a common product obtained through the fermentation of sugars and is commonly used in cosmetic formulations and in the food and pharmaceutical industries. For example, in the food industry, it is used as a flavor-enhancing agent and acidifier [85–87]; in cosmetics formulations, it is added for its emulsifying properties and the moisturizing effects produced on the skin; and in the pharmaceutical industry, it is used for the synthesis of dermatologic products and drugs against osteoporosis [88]. Additionally, LA has recently received special attention for being the precursor of PLA, a bio-degradable and bio-based bioplastic [89]. It is also considered a platform chemical to produce different products, such as acrylic acid, pyruvic acid, ethyl-lactate, 2,3-pentanedione, acetaldehyde and propylene glycol [90].

3.4.1. Metabolic Pathways to Produce LA via Fermentation

LA can be produced by a wide variety of microorganisms such as bacteria, fungi, yeast, cyanobacteria or algae [14]. However, the most studied are LAB, which includes genera such as *Carnobacterium*, *Enterococcus*, *Lactobacillus*, *Lactococcus*, *Leuconostoc*, *Oenococcus*, *Pediococcus*, *Streptococcus*, *Tetragenococcus*, *Vagococcus* and *Weisselll a* [91]. These LAB require the optimum conditions of temperature (30–45 °C) and pH (6–7) to grow and produce LA [92]. They also require the proper amount of nitrogen, vitamins and minerals for optimal growth. These nutrients may be contained in the hydrolysate obtained from agri-food waste, but they are usually supplemented in the medium. For instance, yeast extract is one of the most used nitrogen sources, however, its supplementation in the hydrolysates increases the total cost of the LA production process [93].

LAB are classified as homofermentative, heterofermentative or facultative heterofermentative, according to the metabolic pathways used to produce LA. Thus, the conversion of sugars to LA by LAB is produced through three different pathways (Figure 5): the Embden–Meyerhof–Parnas (EMP) pathway, the pentose phosphate (PP) pathway and the phosphoketolase (PK) pathway [88].

Homofermentative LAB ferment hexoses via the EMP pathway, producing theoretically 2 mol of LA per mol of glucose. In addition, some strains also convert glucose via the PP pathway instead of the EMP pathway, producing 1.67 mol of LA per mol of glucose [88]. On the other hand, heterofermentative LAB can ferment hexoses and pentoses exclusively via the PK pathway, producing LA, acetic acid or ethanol, and CO_2. Thus, the theoretical yield obtained is 1 mol of LA per mol of glucose or pentose [88]. By contrast, facultative heterofermentative LAB can metabolize hexoses via the EMP pathway and pentoses via the PK pathway, obtaining 2 mol of LA per mol of glucose and 1 mol of LA per mol of pentose [88].

3.4.2. Production of LA from SBP Hydrolysates

Different studies have been performed to produce LA from SBP (Table 3). To increase the LA yield, several authors have varied the pH regulation mode, nutrient supplementation and the type of processing, such as separate hydrolysis and fermentation, simultaneous saccharification and fermentation, fed-batch fermentation or continuous fermentation. For instance, Marzo et al. studied the effect of supplementing the SBP hydrolysate with different nitrogen sources, finding that it is necessary to supplement it with yeast extract to increase the production of LA [94]. Additionally, they also studied pH regulation during the lactic fermentation of SBP hydrolysate with *Lactobacillus plantarum*, reaching 30 g/L of LA with a yield of 0.12 g_{LA}/g_{SBP} by adding 27 g/L of $CaCO_3$ [94]. In another report, the same authors evaluated the pretreatment of SBP, testing acid, alkaline and biological pretreatment. They increased the production of LA to 50 g/L with a yield of 0.5 g_{LA}/g_{SBP} when the SBP was pretreated with 1% H_2SO_4 [52]. Díaz et al. studied different simultaneous saccharification and fermentation strategies to produce LA from SBP with *Lactobacillus casei* [95]. They achieved the maximum concentration of 27 g/L of LA with a yield of 0.13 g_{LA}/g_{SBP} using a fed-batch SSF process with pH control (by adding 30 g/L $CaCO_3$) and nutrient supplementation (by adding MRS medium 0.2 mL/mL).

Table 3. Lactic acid production from SBP. YLA: lactic acid yield (g_{LA}/g_{SBP}).

Reference	Strain	Conditions	Concentration	Y_{LA}
[24]	*Lactobacillus plantarum* HII & *Lactobacillus brevis* PCM 488	SSF with co-culture	60 g/L	0.55 g/g
[96]	*L. coryniformis subsp. torquens* DSM 20005 & *L. preudomesenteroides*	SHF with co-culture	22 g/L	0.78 g/g
[94]	*Lactobacillus plantarum*	SHF	30 g/L	0.12 g/g
[52]	*Lactobacillus plantarum*	SHF with pretreated SBP	50 g/L	0.5 g/g
[95]	*Lactobacillus casei*	Fed-fach SSF	27 g/L	0.13 g/g
[97]	*Bacillus coagulans*	Continuous fermentation	35 g/L	0.71 g/g

Due to the composition of SBP, the hydrolysis of SBP produces a medium rich in glucose, but other sugars such as fructose, mannose, arabinose galactose, raffinose, rhamnose, xylose and galacturonic acid are also included [24]. It is difficult for a single microbial species to completely assimilate such a medium. Lactic acid production from arabinose, galactose and xylose derived from SBP is not as efficient as from glucose [98]. For example, *Bacillus coagulans* was used to produce lactic acid from sugar beet pulp hydrolysates [97]. This strain was able to consume glucose and xylose; however, arabinose was not totally consumed at the end of fermentation. Most homofermentative LAB, such as *Lactobacillus delbrueckii* and *Lactobacillus acidophilus*, can produce lactic acid from glucose but not from sugars derived from hemicellulose, such as arabinose and xylose [99]. For better utilization of the substrate, it is necessary to use LAB able to utilize pentoses. However, pentoses are only utilized by some lactobacilli. This is the case of the facultative heterofermentative microorganism *Lactobacillus casei* 2246, which degrades hexoses, mainly glucose, through the EMP pathway, and pentoses by the PK pathway [95]. This strain has been demonstrated to produce lactic acid from glucose and arabinose from simultaneous saccharification and fermentation of SBP. Moreover, lactic acid fermentation can be improved by using a mixed population of LAB with different assimilation profiles [100]. In fact, the effectiveness of LA production can be improved by 10–30% by using mixed cultures. Hence, Berlowska et al. studied the effect of mixed cultures of LAB species through simultaneous saccharification and lactic fermentation [24]. They performed the fermentation in two steps: the first one with a monoculture able to ferment mainly glucose and the second one with another strain able to consume unfermented sugars (arabinose and xylose). Using this procedure, most sugars were consumed by the mixed culture, reaching a LA concentration of 60 g/L with a yield of 0.55 g_{LA}/g_{SBP} with the strain *Lactobacillus plantarum* HII in the first step and *Lactobacillus plantarum* HII and *Lactobacillus brevis* PCM 488 in the second. The same strategy was followed by Alexandri et al. after observing the non-complete consumption of the hydrolyzed sugars from SBP [96]. They produced 22 g/L of LA with a yield of 0.78 g_{LA}/g_{SBP} with the strains *L. coryniformis* subsp. *torquens* DSM 20005 and isolate A250 (most likely *L. preudomesenteroides*), adding the second strain when half of the glucose was consumed [96]. Additionally, a similar procedure was performed by Diaz et al., where SBP hydrolysate was fermented first to ethanol and then to LA with the strain *Lactobacillus plantarum* [73]. In that study, the complete use of sugars was achieved by producing ethanol and LA sequentially.

An interesting biorefinery process was presented by Oliveira et al. [97]. This process can be implemented inside the processing of sugar beet to produce sugar. They produce LA from SBP via continuous fermentation with *Bacillus coagulans*, reaching 2781.01 g of LA from 3916.91 g of sugars with a maximum productivity of 18.06 g/L/h.

3.5. *Polyhydroxyalkanoates Fermentation*

The first identification of poly-3-hydroxybutyrate (P(3HB)) was performed in the strain *Bacillus megaterium* by Lemoigne in 1926 [101]. Afterward, more than 300 bacterial strains have been identified as PHA producers. Some examples include *Cupriavidus necator* (also named *Ralstonia eutropha*), *Pseudomonas aeruginosa*, *Pseudomonas oleovarans* and *Pseudomonas stutzeri* [102]. These bacteria are able to accumulate PHA inside the cell cytoplasm as a reserve of carbon and energy. Therefore, they are produced when the microorganisms cannot grow normally due to a deficiency of nutrients and an excess of carbon.

3.5.1. Metabolic Pathway to Produce PHAs via Fermentation

The production of PHAs is induced by the limitation of nutrients in the medium used for microorganism growth. This limitation activates the pathways of metabolism that bacteria use to produce PHAs [11]. The metabolic pathways involved in the production of PHA depend on the carbon source and the microorganism used. These bacteria can use different carbon sources to produce PHAs, such as carbohydrates or volatile fatty acids (VFAs) [103].

In pure culture fermentation, carbohydrates are metabolized to pyruvate via the Enter–Doudoroff pathway. Afterward, pyruvate is converted to acetyl-CoA and, under growth-limiting conditions, acetyl-CoA is transformed into P(3HB) by the action of three enzymes: 3-ketothiolase (PhaA), acetoacetyl-CoA reductase (PhaB) and PHA synthase (PhaC) [103]. Through this pathway, the obtained yield will depend on the carbon source. Thus, glucose produces the highest yield (0.40 g PHA/g); however, other sugars such as xylose or arabinose can also be used to produce PHAs, although at lower yields (0.17–0.19 g PHA/g) [104].

The production of PHAs from VFAs follows the same route for metabolism once they are converted to the corresponding acyl-CoA, being the precursors for different hydroxyalkanoate monomers. In this way, P(3HB) is produced from acetate, and 3HV, 3-hydroxy-2-methylvalerate (3H2MV) or 3-hydroxy-2-methylbutyrate (3H2MB) are produced from propionate [11].

PHAs can be produced from pure cultures following a two-step process: growth phase and accumulation phase [11]. Firstly, the strain is cultivated in a medium with a proper ratio of carbon, nitrogen and nutrients. Secondly, one of the essential nutrients (nitrogen, phosphorus or oxygen) is restricted to induce the accumulation of PHA inside the cell. PHAs can also be produced by mixed cultures through the fermentation of VFAs and the process requires two stages [11]. First of all, microorganisms with high PHA storage capacity are selected by imposing cycles where an essential nutrient is restricted. Afterward, the selected microorganisms are cultured to produce PHAs by restricting the same essential nutrient restricted in the previous step. The advantage of using mixed culture is the reduction in production cost due to sterilization is not required and the culture can adapt to various complex feedstocks [105,106].

3.5.2. Production of PHAs from SBP

As previously mentioned, SBP hydrolysate is a complete medium with a great variety of sugars, including glucose, rhamnose, arabinose, galactose, xylose and galacturonic acid. However, there are no published papers on the production of PHA from SBP-derived sugars different from glucose.

Kurt-Kızıldoğan et al. produced poly(3-hydroxybutyrate) (PHB), one of the main PHAs, with *Haloarcula* sp. TG1 from glucose derived from different agricultural wastes [107]. Among them, the highest PHA content (45.6% of the biomass) was achieved using SBP treated with rCKT3eng, a recombinant endoglucanase of *Haloarcula* sp. CKT3 expressed in *E. coli* BL21 (DE3) cells (pH 7.35, 72 h, 37 °C).

The most common pathway in bacteria, yeast and fungi to convert glucose to PHAs is the EMP [108,109]. However, the model organism for PHA production, *C. necator*, lacks the gene for the key enzyme of this pathway, utilizing the Entner–Doudoroff (ED) pathway instead [110]. There are only a few strains able to convert C5 sugars and produce PHAs, but when they do, the conversion efficiencies are very low [111]. It has been demonstrated that some microorganisms are able to perform this conversion from hemicellulose hydrolysates derived from rice straw, sugarcane bagasse, sugar maple wood chips, etc., such as *Bacillus firmus*, *Cupriavidus necator*, *Burkholderia cepacia*, etc. [109]. However, no papers have been found describing PHA production from C5 sugars derived from SBP. Numerous metabolic engineering strategies have been explored for metabolizing pentose sugars available in lignocellulosic hydrolysates. For example, a strain of *C. necator* has been modified to metabolize arabinose through heterologous expression of a set of *E. coli* genes for L-arabinose uptake and metabolism [112].

As stated above, VFAs are another carbon source to produce PHAs. SBP has also been studied as raw material to produce VFAs, with acetic, lactic, caproic and butyric being the main acids produced [113]. These acids are commonly produced through acidogenic fermentation by mixed microbial cultures. This fermentation involves two steps of the anaerobic digestion process, hydrolysis and acidogenesis, and the inhibition of methanogenesis [114]. In this process, the organic matter is hydrolyzed into sugars, amino acids and fatty acids. Afterward, these compounds are converted into VFAs by acidogenic bacteria.

4. Conclusions and Future Directions

Bioplastic production from renewable sources is a topic that has gained interest in recent years. In this direction, SBP is a versatile by-product from the food industry that can be used as raw material to produce value-added products, such as LA and PHAs, through biotechnological processes. Although several studies have been performed on pretreatments and enzymatic hydrolysis of SBP, only a few focused on the production of LA or PHAs, which can be used as precursors of bioplastics. Only one study was found where PHAs are obtained from SBP through a sequential process of hydrolysis and fermentation.

Focusing on another alternative process to produce PHAs, that is from VFAs, there is also a lack of studies where VFAs produced from SBP are used to obtain PHAs. However, there are several studies based on the production of hydrogen and VFAs through the anaerobic digestion of SBP. Hence, it seems that it is a feasible process that has not yet been studied in depth.

Additionally, SBP is composed of similar percentages of cellulose, hemicellulose and pectin. This characteristic makes it an interesting raw material to use in the framework of a biorefinery. Only a few authors have proposed a biorefinery process applied to this material, where pectin, phenolic compounds and sugar-rich hydrolysates can be fermented to ethanol, LA or succinic acid, among others, facilitating the preparation of valuable products. Further research should be carried out in this direction to implement a cost-effective process in the sugar industry.

Author Contributions: Conceptualization, C.M.-G., A.B. and A.B.D.; methodology, C.M.-G.; investigation, C.M.-G.; writing—original draft preparation, C.M.-G.; writing—review and editing, A.B. and A.B.D.; visualization, C.M.-G.; project administration, A.B.; funding acquisition, A.B. All authors have read and agreed to the published version of the manuscript.

Funding: This research was funded by "Ministerio de Economía, Industria y Competitividad", "Agencia Estatal de Investigación (AEI)" and "Fondo Europeo de Desarrollo Regional (FEDER)" (CTM2016-79071-R), and by "Ministerio de Universidades" and "European Union-Next GenerationEU" through the Margarita Salas Post-doctoral fellowship (2021-067/PN/MS-RECUAL/CD).

Institutional Review Board Statement: Not applicable.

Informed Consent Statement: Not applicable.

Data Availability Statement: Not applicable.

Acknowledgments: The authors thank "Ministerio de Economía, Industria y Competitividad", "Agencia Estatal de Investigación (AEI) and "Fondo Europeo de Desarrollo Regional (FEDER)" for the financial support of this study. C.M.-G thanks "Ministerio de Universidades del Gobierno de España" for the allocated Budget from the NextGenerationEU program to public universities for the requalification of the Spanish university system, which funds her postdoctoral position at the University of Cádiz in the form of a Margarita Salas fellowship (ref. 2021-067/PN/MS-RECUAL/CD).

Conflicts of Interest: The authors declare no conflict of interest.

Abbreviations

CBU	Cellobiase activity units
PGU	Polygalacturonase activity units
COD	Chemical oxygen demand
d.m.	Dry matter
FPU	Filter paper unit
HU	Hemicellulase activity units
LA	Lactic acid
LAB	Lactic acid bacteria
LCB	Lignocellulosic biomass
MRS	Man, Rogosa and Sharpe broth
PA	Polyamide
PBAT	Polybutylene adipate terephthalate
PBS	Polybutylene succinate
PCL	Polycaprolactone
PE	Polyethene
PET	Polyethene terephthalate
PGU	Polygalacturonase activity units
PHA	Polyhydroxyalkanoate
PLA	Poly-lactic acid
POS	Pectin-derived oligosaccharides
PP	Polypropylene
PTT	Polytrimethylene terephthalate
RS	Reducing sugars
SBP	Sugar beet pulp
SLR	Solid-liquid ratio
SSF	Solid-state fermentation
TS	Total solids
U	Unit of enzyme
VAP	Value-added product
VFAs	Volatile fatty acids
VS	Volatile solid
Y_H	Hydrolysis yield

References

1. Narancic, T.; O'Connor, K.E. Plastic Waste as a Global Challenge: Are Biodegradable Plastics the Answer to the Plastic Waste Problem? *Microbiology* **2019**, *165*, 129–137. [CrossRef]
2. Plasctics Europe Plastics-the Facts 2021 An Analysis of European Plastics Production, Demand and Waste Data. Available online: https://plasticseurope.org/knowledge-hub/plastics-the-facts-2021/ (accessed on 24 March 2022).
3. Rivero, C.P.; Hu, Y.; Kwan, T.H.; Webb, C.; Theodoropoulos, C.; Daoud, W.; Lin, C.S.K. Bioplastics From Solid Waste. In *Current Developments in Biotechnology and Bioengineering: Solid Waste Management*; Elsevier: Amsterdam, The Netherlands, 2016; pp. 1–26, ISBN 9780444636751.
4. European Bioplastic European Bioplastic. Available online: http://www.european-bioplastics.org (accessed on 3 February 2023).
5. Martin, O.; Avérous, L. Poly(Lactic Acid): Plasticization and Properties of Biodegradable Multiphase Systems. *Polymer* **2001**, *42*, 6209–6219. [CrossRef]
6. Bonartsev, A.P.; Bonartseva, G.A.; Reshetov, I.V.; Kirpichnikov, M.P.; Shaitan, K.V. Application of Polyhydroxyalkanoates in Medicine and the Biological Activity of Natural Poly(3-Hydroxybutyrate). *Acta Nat.* **2019**, *11*, 4–16. [CrossRef]
7. Lasprilla, A.J.R.; Martinez, G.A.R.; Lunelli, B.H.; Jardini, A.L.; Filho, R.M. Poly-Lactic Acid Synthesis for Application in Biomedical Devices—A Review. *Biotechnol. Adv.* **2012**, *30*, 321–328. [CrossRef]
8. Saniei, H.; Mousavi, S. Surface Modification of PLA 3D-Printed Implants by Electrospinning with Enhanced Bioactivity and Cell Affinity. *Polymer* **2020**, *196*, 122467. [CrossRef]
9. Samrot, A.V.; Sean, T.C.; Kudaiyappan, T.; Bisyarah, U.; Mirarmandi, A.; Faradjeva, E.; Abubakar, A.; Ali, H.; Lavanya, J.; Angalene, A.; et al. Production, Characterization and Application of Nanocarriers Made of Polysaccharides, Proteins, Bio-Polyesters and Other Biopolymers: A Review. *Int. J. Biol. Macromol.* **2020**, *165*, 3088–3105. [CrossRef]
10. Sharma, V.; Sehgal, R.; Gupta, R. Polyhydroxyalkanoate (PHA): Properties and Modifications. *Polymer* **2020**, *212*, 123161. [CrossRef]
11. Reis, M.; Albuquerque, M.; Villano, M.; Majone, M. Mixed Culture Processes for Polyhydroxyalkanoate Production from Agro-Industrial Surplus/Wastes as Feedstocks. In *Comprehensive Biotechnology*, 2nd ed.; Elsevier: Amsterdam, The Netherlands, 2011; Volume 6, pp. 669–683, ISBN 9780080885049.
12. Sen, K.Y.; Baidurah, S. Renewable Biomass Feedstocks for Production of Sustainable Biodegradable Polymer. *Curr. Opin. Green Sustain. Chem.* **2021**, *27*, 100412. [CrossRef]
13. Tarrahi, R.; Fathi, Z.; Seydibeyoğlu, M.Ö.; Doustkhah, E.; Khataee, A. Polyhydroxyalkanoates (PHA): From Production to Nanoarchitecture. *Int. J. Biol. Macromol.* **2020**, *146*, 596–619. [CrossRef] [PubMed]
14. Abdel-Rahman, M.A.; Tashiro, Y.; Sonomoto, K. Recent Advances in Lactic Acid Production by Microbial Fermentation Processes. *Biotechnol. Adv.* **2013**, *31*, 877–902. [CrossRef] [PubMed]
15. Wellenreuther, C.; Wolf, A.; Zander, N. Cost Competitiveness of Sustainable Bioplastic Feedstocks – A Monte Carlo Analysis for Polylactic Acid. *Clean. Eng. Technol.* **2022**, *6*, 100411. [CrossRef]
16. Yousuf, A.; Pirozzi, D.; Sannino, F. Fundamentals of Lignocellulosic Biomass. In *Lignocellulosic Biomass to Liquid Biofuels*; Elsevier: Amsterdam, The Netherlands, 2019; pp. 1–15, ISBN 9780128159361.
17. Sani, R.K. *Biorefining of Biomass to Biofuels: Opportunities and Perception*; Springer: Berlin/Heidelberg, Germany, 2018; Volume 4, ISBN 978-3-319-67678-4.
18. Obruca, S.; Benesova, P.; Marsalek, L.; Marova, I. Use of Lignocellulosic Materials for PHA Production. *Chem. Biochem. Eng. Q.* **2015**, *29*, 135–144. [CrossRef]
19. Al-Battashi, H.S.; Annamalai, N.; Sivakumar, N.; Al-Bahry, S.; Tripathi, B.N.; Nguyen, Q.D.; Gupta, V.K. Lignocellulosic Biomass (LCB): A Potential Alternative Biorefinery Feedstock for Polyhydroxyalkanoates Production. *Rev. Environ. Sci. Biotechnol.* **2019**, *18*, 183–205. [CrossRef]
20. Isikgor, F.H.; Becer, C.R. Lignocellulosic Biomass: A Sustainable Platform for the Production of Bio-Based Chemicals and Polymers. *Polym. Chem.* **2015**, *6*, 4497–4559. [CrossRef]
21. Streffer, F. Lignocellulose to Biogas and Other Products. *JSM Biotechnol. Biomed. Eng.* **2014**, *2*, 1023.
22. Marzo, C.; Díaz, A.B.; Caro, I.; Blandino, A. Status and Perspectives in Bioethanol Production From Sugar Beet. In *Bioethanol Production from Food Crops*; Elsevier: Amsterdam, The Netherlands, 2019; pp. 61–79.
23. OECD; FAO. *OECD-FAO Agricultural Outlook 2022-2031*; OECD: Paris, France, 2022.
24. Berlowska, J.; Pielech-przybylska, K.; Balcerek, M.; Dziekonska-kubczak, U.; Patelski, P.; Dziugan, P.; Kregiel, D. Simultaneous Saccharification and Fermentation of Sugar Beet Pulp for Efficient Bioethanol Production. *Biomed Res. Int.* **2016**, *2016*, 10. [CrossRef] [PubMed]
25. Panella, L. Sugar Beet as an Energy Crop. *Sugar Tech.* **2010**, *12*, 288–293. [CrossRef]
26. Schonhoff, A.; Ihling, N.; Schreiber, A.; Zapp, P. Environmental Impacts of Biosurfactant Production Based on Substrates from the Sugar Industry. *Sustain. Chem. Eng.* **2022**, *10*, 9345–9358. [CrossRef]
27. Pezzi, G. Considerations of the Technological Quality Assessment of Sugarbeet. *Sugar Ind.* **2011**, *136*, 85–89. [CrossRef]
28. Cubero, M.T.G.; Sanz, M.C.; Benito, G.G.; Puerta, M.G.D. Distribution of Colorants in the Evaporation and First Crystallization Stages. *Sugar Ind.* **2004**, *129*, 660–666.
29. Vaccari, G.; Wawro, P.; Tamburini, E.; Sgualdino, G.; Bernardi, T. Cooling Crystallization of Raw Juice: Laboratory Investigations of Sucrose Crystal Growth Kinetics. *Sugar Ind.* **1996**, *121*, 111–117.

30. Vaccari, G.; Wawro, P.; Tamburini, E.; Sgualdino, G.; Bernardi, T. Cooling Crystallization of Microfiltered Raw Juice and of Traditional Thick Juice: A Comparison. *Sugar Ind.* **2002**, *127*, 22–28.
31. Merkes, R.; Kröhl, M.; Mugele, H.; Sauer, M. Sugarbeet Production Technology in the Year 2000: Cost Reduction, Environmental Protection, Sustainability. *Sugar Ind.* **2001**, *126*, 804–811.
32. Baryga, A.; Połeć, B.; Klasa, A.; Olejnik, T.P. Application of Sugar Beet Pulp Digestate as a Soil Amendment in the Production of Energy Maize. *Processes* **2021**, *9*, 765. [CrossRef]
33. Baryga, A.; Połeć, B.; Klasa, A. Quality of Sugar Beets under the Effects of Digestate Application to the Soil. *Processes* **2020**, *8*, 1402. [CrossRef]
34. Baryga, A.; Polec, B.; Skibniewska, K.A.; Seciu, E.; Grabara, J. Utilisation of Residual Waste From Sugar Beet Pulp Fermentation As Fertiliser in Sustainable Agriculture. *J. Environ. Prot. Ecol.* **2016**, *17*, 1048–1057.
35. Rana, A.K.; Kumar Gupta, V.; Newbold, J.; Roberts, D.; Rees, R.M.; Krishnamurthy, S.; Kumar Thakur, V. Sugar Beet Pulp: Resurgence and Trailblazing Journey towards a Circular Bioeconomy. *Fuel* **2022**, *312*, 122953. [CrossRef]
36. Dinand, E.; Chanzy, H.; Vignon, M.R. Parenchymal Cell Cellulose from Sugar Beet Pulp: Preparation and Properties. *Cellulose* **1996**, *3*, 183–188. [CrossRef]
37. Dinand, E.; Chanzy, H.; Vignon, R.M. Suspensions of Cellulose Microfibrils from Sugar Beet Pulp. *Food Hydrocoll.* **1999**, *13*, 275–283. [CrossRef]
38. Michel, F.; Thibault, J.-F.; Barry, J.-L.; de Baynast, R. Preparation and Characterisation of Dietary Fibre from Sugar Beet Pulp. *J. Sci. Food Agric.* **1988**, *42*, 77–85. [CrossRef]
39. Grahovac, J.; Rončević, Z. Environmental Impacts of the Confectionary Industry. In *Environmental Impact of Agro-Food Industry and Food Consumption*; Elsevier: Amsterdam, The Netherlands, 2021; pp. 189–216, ISBN 9780128213636.
40. Ward, D.P.; Cárdenas-Fernández, M.; Hewitson, P.; Ignatova, S.; Lye, G.J. Centrifugal Partition Chromatography in a Biorefinery Context: Separation of Monosaccharides from Hydrolysed Sugar Beet Pulp. *J. Chromatogr. A* **2015**, *1411*, 84–91. [CrossRef] [PubMed]
41. Bolaji, I.; Nejad, B.; Billham, M.; Mehta, N.; Smyth, B.; Cunningham, E. Multi-Criteria Decision Analysis of Agri-Food Waste as a Feedstock for Biopolymer Production. *Resour. Conserv. Recycl.* **2021**, *172*, 105671. [CrossRef]
42. Abo, B.O.; Gao, M.; Wang, Y.; Wu, C.; Ma, H.; Wang, Q. Lignocellulosic Biomass for Bioethanol: An Overview on Pretreatment, Hydrolysis and Fermentation Processes. *Rev. Environ. Health* **2019**, *34*, 57–68. [CrossRef]
43. Solarte-Toro, J.C.; Romero-García, J.M.; Martínez-Patiño, J.C.; Ruiz-Ramos, E.; Castro-Galiano, E.; Cardona-Alzate, C.A. Acid Pretreatment of Lignocellulosic Biomass for Energy Vectors Production: A Review Focused on Operational Conditions and Techno-Economic Assessment for Bioethanol Production. *Renew. Sustain. Energy Rev.* **2019**, *107*, 587–601. [CrossRef]
44. Haldar, D.; Purkait, M.K. A Review on the Environment-Friendly Emerging Techniques for Pretreatment of Lignocellulosic Biomass: Mechanistic Insight and Advancements. *Chemosphere* **2021**, *264*, 128523. [CrossRef]
45. Alvira, P.; Tomás-Pejó, E.; Ballesteros, M.; Negro, M.J. Pretreatment Technologies for an Efficient Bioethanol Production Process Based on Enzymatic Hydrolysis: A Review. *Bioresour. Technol.* **2010**, *101*, 4851–4861. [CrossRef]
46. Xu, J.K.; Sun, R.C. Recent Advances in Alkaline Pretreatment of Lignocellulosic Biomass. In *Biomass Fractionation Technologies for a Lignocellulosic Feedstock Based Biorefinery*; Elsevier: Amsterdam, The Netherlands, 2016; pp. 431–459, ISBN 9780128025611.
47. Rahmati, S.; Doherty, W.; Dubal, D.; Atanda, L.; Moghaddam, L.; Sonar, P.; Hessel, V.; Ostrikov, K.K. Pretreatment and Fermentation of Lignocellulosic Biomass: Reaction Mechanisms and Process Engineering. *React. Chem. Eng.* **2020**, *5*, 2017–2047. [CrossRef]
48. Rezicó, T.; Oros, D.; Markovicó, I.; Kracher, D.; Ludwig, R.; Šantek, B.; Rezic, T.; Oros, D.; Markovicó, I.; Kracher, D.; et al. Integrated Hydrolyzation and Fermentation of Sugar Beet Pulp to Bioethanol. *J. Microbiol. Biotechnol.* **2013**, *23*, 1244–1252. [CrossRef] [PubMed]
49. Donkoh, E.; Degenstein, J.; Tucker, M.; Ji, Y. Optimization of Enzymatic Hydrolysis of Dilute Acid Pretreated Sugar Beet Pulp Using Response Surface Design. *J. Sugarbeet Res.* **2012**, *49*, 26–38. [CrossRef]
50. El-gendy, N.S.; Madian, H.R.; Nassar, H.N. Response Surface Optimization of the Thermal Acid Pretreatment of Sugar Beet Pulp for Bioethanol Production Using Trichoderma Viride and Saccharomyces Cerevisiae. *Recent Pat. Biotechnol.* **2015**, *9*, 50–62. [CrossRef]
51. Zheng, Y.; Lee, C.; Yu, C.; Cheng, Y.S.; Zhang, R.; Jenkins, B.M.; VanderGheynst, J.S. Dilute Acid Pretreatment and Fermentation of Sugar Beet Pulp to Ethanol. *Appl. Energy* **2013**, *105*, 1–7. [CrossRef]
52. Marzo, C.; Díaz, A.B.; Caro, I.; Blandino, A. Effect of Several Pretreatments on the Lactic Acid Production from Exhausted Sugar Beet Pulp. *Foods* **2021**, *10*, 2414. [CrossRef] [PubMed]
53. Zheng, Y.; Yu, C.; Cheng, Y.S.; Zhang, R.; Jenkins, B.; VanderGheynst, J.S. Effects of Ensilage on Storage and Enzymatic Degradability of Sugar Beet Pulp. *Bioresour. Technol.* **2011**, *102*, 1489–1495. [CrossRef] [PubMed]
54. Cárdenas-Fernández, M.; Bawn, M.; Hamley-Bennett, C.; Bharat, P.K.V.; Subrizi, F.; Suhaili, N.; Ward, D.P.; Bourdin, S.; Dalby, P.A.; Hailes, H.C.; et al. An Integrated Biorefinery Concept for Conversion of Sugar Beet Pulp into Value-Added Chemicals and Pharmaceutical Intermediates. *Faraday Discuss.* **2017**, *202*, 415–431. [CrossRef]
55. Valdés, G.; Mendonça, R.T.; Aggelis, G. Lignocellulosic Biomass as a Substrate for Oleaginous Microorganisms: A Review. *Appl. Sci.* **2020**, *10*, 7698. [CrossRef]

56. Canilha, L.; Chandel, A.K.; dos Santos Milessi, T.S.; Fernandes Antunes, F.A.; Luiz Da Costa Freitas, W.; Almeida Felipe, M.D.G.; Silva, S.S. Bioconversion of Sugarcane Biomass into Ethanol: An Overview about Composition, Pretreatment Methods, Detoxification of Hydrolysates, Enzymatic Saccharification, and Ethanol Fermentation. *J. Biomed. Biotechnol.* **2012**, *2012*, 15. [CrossRef]
57. Ivetic, D.Z.; Marina, B.S.; Antov, M.G.; Ivetić, D.Ž.; Šćiban, M.B.; Antov, M.G. Enzymatic Hydrolysis of Pretreated Sugar Beet Shreds: Statistical Modeling of the Experimental Results. *Biomass Bioenergy* **2012**, *7*, 387–394. [CrossRef]
58. Foster, B.L.; Dale, B.E.; Doran-Peterson, J.B. Enzymatic Hydrolysis of Ammonia-Treated Sugar Beet Pulp. *Appl. Biochem. Biotechnol.* **2001**, *91*, 91–93. [CrossRef]
59. Ivetic, D.T.; Antov, M.G. The Impact of Pretreatments on Cellulose from Sugar Beet Shreds and Its Susceptibility to Enzymatic Hydrolysis. *Cellul. Chem. Technol.* **2016**, *50*, 139–146.
60. Li, G.; Sun, Y.; Guo, W.; Yuan, L. Comparison of Various Pretreatment Strategies and Their Effect on Chemistry and Structure of Sugar Beet Pulp. *J. Clean. Prod.* **2018**, *181*, 217–223. [CrossRef]
61. Maitan-Alfenas, G.P.; Visser, E.M.; Guimarães, V.M. Enzymatic Hydrolysis of Lignocellulosic Biomass: Converting Food Waste in Valuable Products. *Curr. Opin. Food Sci.* **2015**, *1*, 44–49. [CrossRef]
62. Anand, G.; Yadav, S.; Yadav, D. Production, Purification and Biochemical Characterization of an Exo-Polygalacturonase from Aspergillus Niger MTCC 478 Suitable for Clarification of Orange Juice. *3 Biotech* **2017**, *7*, 122. [CrossRef]
63. Díaz, A.B.; De Ory, I.; Caro, I.; Blandino, A. Enhance Hydrolytic Enzymes Production by Aspergillus Awamori on Supplemented Grape Pomace. *Food Bioprod. Process.* **2012**, *90*, 72–78. [CrossRef]
64. Brijwani, K.; Oberoi, S.; Vadlani, P.V.; Oberoi, H.S.; Vadlani, P.V. Production of a Cellulolytic Enzyme System in Mixed-Culture Solid-State Fermentation of Soybean Hulls Supplemented with Wheat Bran. *Process Biochem.* **2010**, *45*, 120–128. [CrossRef]
65. Diaz, A.B.; Blandino, A.; Webb, C.; Caro, I. Modelling of Different Enzyme Productions by Solid-State Fermentation on Several Agro-Industrial Residues. *Appl. Microbiol. Biotechnol.* **2016**, *100*, 9555–9566. [CrossRef] [PubMed]
66. Verduzco-Oliva, R.; Gutierrez-Uribe, J.A. Beyond Enzyme Production: Solid State Fermentation (SSF) as an Alternative Approach to Produce Antioxidant Polysaccharides. *Sustainability* **2020**, *12*, 495. [CrossRef]
67. Mansour, A.A.; Arnaud, T.; Lu-Chau, T.A.; Fdz-Polanco, M.; Moreira, M.T.; Rivero, J.A.C. Review of Solid State Fermentation for Lignocellulolytic Enzyme Production: Challenges for Environmental Applications. *Rev. Environ. Sci. Biotechnol.* **2016**, *15*, 31–46. [CrossRef]
68. Yazid, N.A.; Barrena, R.; Komilis, D.; Sánchez, A. Solid-State Fermentation as a Novel Paradigm for Organic Waste Valorization: A Review. *Sustainability* **2017**, *9*, 224. [CrossRef]
69. Bhargav, S.; Panda, B.P.; Ali, M.; Javed, S. Solid-State Fermentation - An Overview. *Chem. Biochem. Eng* **2008**, *22*, 49–70.
70. Marques, N.P.; de Cassia Pereira, J.; Gomes, E.; da Silva, R.; Araújo, A.R.; Ferreira, H.; Rodrigues, A.; Dussán, K.J.; Bocchini, D.A. Cellulases and Xylanases Production by Endophytic Fungi by Solid State Fermentation Using Lignocellulosic Substrates and Enzymatic Saccharification of Pretreated Sugarcane Bagasse. *Ind. Crops Prod.* **2018**, *122*, 66–75. [CrossRef]
71. Cerda, A.; Gea, T.; Vargas-García, M.C.; Sánchez, A. Towards a Competitive Solid State Fermentation: Cellulases Production from Coffee Husk by Sequential Batch Operation and Role of Microbial Diversity. *Sci. Total Environ.* **2017**, *589*, 56–65. [CrossRef]
72. Zieminski, K.; Romanowska, I.; Kowalska, M.; Ziemiński, K.; Romanowska, I.; Kowalska, M. Enzymatic Pretreatment of Lignocellulosic Wastes to Improve Biogas Production. *Waste Manag.* **2012**, *32*, 1131–1137. [CrossRef]
73. Díaz, A.B.; Marzo, C.; Caro, I.; de Ory, I.; Blandino, A. Valorization of Exhausted Sugar Beet Cossettes by Successive Hydrolysis and Two Fermentations for the Production of Bio-Products. *Bioresour. Technol.* **2017**, *225*, 225–233. [CrossRef]
74. Rodriguez Couto, S. Exploitation of Biological Wastes for the Production of Value-Added Products Uncler Solid-State Fermentation Conditions. *Biotechnol. J.* **2008**, *3*, 859–870. [CrossRef] [PubMed]
75. Marzo, C.; Díaz, A.B.; Caro, I.; Blandino, A. Valorization of Agro-Industrial Wastes to Produce Hydrolytic Enzymes by Fungal Solid-State Fermentation. *Waste Manag. Res.* **2018**, *32*, 1–8. [CrossRef] [PubMed]
76. Mouafi, F.E.; Karam, E.A.; Hassan, H.M. Production of Dextranase from Agro-Industrial Wastes by Aspergillus Awamori F-234 under Solid State Fermentation. *Res. J. Pharm. Biol. Chem. Sci.* **2016**, *7*, 1451–1459.
77. Heerd, D.; Diercks-Horn, S.; Fernández-Lahore, M. Efficient Polygalacturonase Production from Agricultural and Agro-Industrial Residues by Solid-State Culture of Aspergillus Sojae under Optimized Conditions. *Springerplus* **2014**, *3*, 1–14. [CrossRef] [PubMed]
78. Bai, Z.H.; Zhang, H.X.; Qi, H.Y.; Peng, X.W.; Li, B.J. Pectinase Production by Aspergillus Niger Using Wastewater in Solid State Fermentation for Eliciting Plant Disease Resistance. *Bioresour. Technol.* **2004**, *95*, 49–52. [CrossRef] [PubMed]
79. Roche, N.; Berna, P.; Desgranges, C.; Durand, A. Substrate Use and Production of α-l-Arabinofuranosidase during Solid-State Culture of Trichoderma Reesei on Sugar Beet Pulp. *Enzyme Microb. Technol.* **1995**, *17*, 935–941. [CrossRef]
80. Leung, C.C.J.; Cheung, A.S.Y.; Zhang, A.Y.Z.; Lam, K.F.; Lin, C.S.K.; Chark Joe Leung, C.; Siu Yeung Cheung, A.; Yan-Zhu Zhang, A.; Fung Lam, K.; Sze Ki Lin, C. Utilisation of Waste Bread for Fermentative Succinic Acid Production. *Biochem. Eng. J.* **2012**, *65*, 10–15. [CrossRef]
81. Pleissner, D.; Kwan, T.H.; Lin, C.S.K. Fungal Hydrolysis in Submerged Fermentation for Food Waste Treatment and Fermentation Feedstock Preparation. *Bioresour. Technol.* **2014**, *158*, 48–54. [CrossRef]
82. Kwan, T.H.; Hu, Y.; Lin, C.S.K. Valorisation of Food Waste via Fungal Hydrolysis and Lactic Acid Fermentation with Lactobacillus Casei Shirota. *Bioresour. Technol.* **2016**, *217*, 129–136. [CrossRef] [PubMed]

83. Dessie, W.; Zhang, W.; Xin, F.; Dong, W.; Zhang, M.; Ma, J.; Jiang, M. Succinic Acid Production from Fruit and Vegetable Wastes Hydrolyzed by On-Site Enzyme Mixtures through Solid State Fermentation. *Bioresour. Technol.* **2018**, *247*, 1177–1180. [CrossRef] [PubMed]
84. Marzo, C.; Díaz, A.B.; Caro, I.; Blandino, A. Conversion of Exhausted Sugar Beet Pulp into Fermentable Sugars from a Biorefinery Approach. *Foods* **2020**, *9*, 1351. [CrossRef]
85. García, C.; Bautista, L.; Rendueles, M.; Díaz, M. A New Synbiotic Dairy Food Containing Lactobionic Acid and Lactobacillus Casei. *Int. J. Dairy Technol.* **2019**, *72*, 47–56. [CrossRef]
86. Shafi, A.; Naeem Raja, H.; Farooq, U.; Akram, K.; Hayat, Z.; Naz, A.; Nadeem, H.R. Antimicrobial and Antidiabetic Potential of Synbiotic Fermented Milk: A Functional Dairy Product. *Int. J. Dairy Technol.* **2019**, *72*, 15–22. [CrossRef]
87. Silva, H.L.A.; Balthazar, C.F.; Silva, R.; Vieira, A.H.; Costa, R.G.B.; Esmerino, E.A.; Freitas, M.Q.; Cruz, A.G. Sodium Reduction and Flavor Enhancer Addition in Probiotic Prato Cheese: Contributions of Quantitative Descriptive Analysis and Temporal Dominance of Sensations for Sensory Profiling. *J. Dairy Sci.* **2018**, *101*, 8837–8846. [CrossRef]
88. Cubas-Cano, E.; González-Fernández, C.; Ballesteros, M.; Tomás-Pejó, E. Biotechnological Advances in Lactic Acid Production by Lactic Acid Bacteria: Lignocellulose as Novel Subtrate. *Biofuels Bioprod. Biorefining* **2018**, *12*, 290–303. [CrossRef]
89. Ahmed, T.; Shahid, M.; Azeem, F.; Rasul, I.; Shah, A.A.; Noman, M.; Hameed, A.; Manzoor, N.; Manzoor, I.; Muhammad, S. Biodegradation of Plastics: Current Scenario and Future Prospects for Environmental Safety. *Environ. Sci. Pollut. Res.* **2018**, *25*, 7287–7298. [CrossRef]
90. López-Gómez, J.P.; Pérez-Rivero, C.; Venus, J. Valorisation of Solid Biowastes: The Lactic Acid Alternative. *Process Biochem.* **2020**, *99*, 222–235. [CrossRef]
91. Masutani, K.; Kimura, Y. *Synthesis, Structure and Properties of Poly (Lactic Acid)*; Springer: Berlin/Heidelberg, Germany, 2018; Volume 279, ISBN 978-3-319-64229-1.
92. Tan, J.; Abdel-Rahman, M.A.; Sonomoto, K. Biorefinery-Based Lactic Acid Fermentation: Microbial Production of Pure Monomer Product. In *Advances in Polymer Science*; Springer: New York, NY, USA, 2018; Volume 279, pp. 27–66.
93. Nancib, A.; Nancib, N.; Meziane-Cherif, D.; Boubendir, A.; Fick, M.; Boudrant, J. Joint Effect of Nitrogen Sources and B Vitamin Supplementation of Date Juice on Lactic Acid Production by Lactobacillus Casei Subsp. Rhamnosus. *Bioresour. Technol.* **2005**, *96*, 63–67. [CrossRef]
94. Marzo, C.; Díaz, A.B.; Caro, I.; Blandino, A. Valorisation of Fungal Hydrolysates of Exhausted Sugar Beet Pulp for Lactic Acid Production. *J. Sci. Food Agric.* **2021**, *101*, 4108–4117. [CrossRef] [PubMed]
95. Díaz, A.B.; González, C.; Marzo, C.; Caro, I.; Blandino, A. Feasibility of Exhausted Sugar Beet Pulp as Raw Material for Lactic Acid Production. *J. Sci. Food Agric.* **2020**, *100*, 3036–3045. [CrossRef]
96. Alexandri, M.; Hübner, D.; Schneider, R.; Fröhling, A.; Venus, J. Towards Efficient Production of Highly Optically Pure D-Lactic Acid from Lignocellulosic Hydrolysates Using Newly Isolated Lactic Acid Bacteria. *New Biotechnol.* **2022**, *72*, 1–10. [CrossRef]
97. Oliveira, R.A.; Schneider, R.; Lunelli, B.H.; Rossell, C.E.V.; Filho, R.M.; Venus, J. A Simple Biorefinery Concept to Produce 2G-Lactic Acid from Sugar Beet Pulp (SBP): A High-Value Target Approach to Valorize a Waste Stream. *Molecules* **2020**, *25*, 2113. [CrossRef]
98. Santosh, I.; Ashtavinayak, P.; Amol, D.; Sanjay, P. Enhanced Bioethanol Production from Different Sugarcane Bagasse Cultivars Using Co-Culture of Saccharomyces Cerevisiae and Scheffersomyces (Pichia) Stipitis. *J. Environ. Chem. Eng.* **2017**, *5*, 2861–2868. [CrossRef]
99. Cui, F.; Li, Y.; Wan, C. Lactic Acid Production from Corn Stover Using Mixed Cultures of Lactobacillus Rhamnosus and Lactobacillus Brevis. *Bioresour. Technol.* **2011**, *102*, 1831–1836. [CrossRef]
100. Berlowska, J.; Cieciura, W.; Borowski, S.; Dudkiewicz, M.; Binczarski, M.; Witonska, I.; Otlewska, A.; Kregiel, D. Simultaneous Saccharification and Fermentation of Sugar Beet Pulp with Mixed Bacterial Cultures for Lactic Acid and Propylene Glycol Production. *Molecules* **2016**, *21*, 1380. [CrossRef]
101. Lemoigne, M. Produits de Deshydration et de Polymerisation de l'acide B= Oxybutyrique. *Bull. Soc. Chim. Biol.* **1926**, *8*, 770–782.
102. Muneer, F.; Rasul, I.; Azeem, F.; Siddique, M.H.; Zubair, M.; Nadeem, H. Microbial Polyhydroxyalkanoates (PHAs): Efficient Replacement of Synthetic Polymers. *J. Polym. Environ.* **2020**, *28*, 2301–2323. [CrossRef]
103. Jiang, G.; Hill, D.J.; Kowalczuk, M.; Johnston, B.; Adamus, G.; Irorere, V.; Radecka, I. Carbon Sources for Polyhydroxyalkanoates and an Integrated Biorefinery. *Int. J. Mol. Sci.* **2016**, *17*, 1157. [CrossRef]
104. Bertrand, J.L.; Ramsay, B.A.; Chavarie, C. Biosynthesis of Poly-β-Hydroxyalkanoates from Pentoses by Pseudomonas Pseudoflava. *Appl. Environ. Microbiol.* **1990**, *56*, 3133–3138. [CrossRef] [PubMed]
105. Laycock, B.; Halley, P.; Pratt, S.; Werker, A.; Lant, P. The Chemomechanical Properties of Microbial Polyhydroxyalkanoates. *Prog. Polym. Sci.* **2013**, *38*, 536–583. [CrossRef]
106. Pakalapati, H.; Chang, C.K.; Show, P.L.; Arumugasamy, S.K.; Lan, J.C.W. Development of Polyhydroxyalkanoates Production from Waste Feedstocks and Applications. *J. Biosci. Bioeng.* **2018**, *126*, 282–292. [CrossRef]
107. Kurt-Kızıldoğan, A.; Türe, E.; Okay, S.; Otur, Ç. Improved Production of Poly(3-Hydroxybutyrate) by Extremely Halophilic Archaeon Haloarcula Sp. TG1 by Utilization of RCKT3eng-Treated Sugar Beet Pulp. *Biomass Convers. Biorefinery* **2021**. [CrossRef]
108. Jeffries, T.W. Utilization of Xylose by Bacteria, Yeasts, and Fungi. *Adv. Biochem. Eng. Biotechnol.* **1983**, *27*, 1–32. [CrossRef]
109. Dietrich, K.; Dumont, M.J.; Del Rio, L.F.; Orsat, V. Sustainable PHA Production in Integrated Lignocellulose Biorefineries. *New Biotechnol.* **2019**, *49*, 161–168. [CrossRef]

110. Pohlmann, A.; Fricke, W.F.; Reinecke, F.; Kusian, B.; Liesegang, H.; Cramm, R.; Eitinger, T.; Ewering, C.; Pötter, M.; Schwartz, E.; et al. Genome Sequence of the Bioplastic-Producing "Knallgas" Bacterium Ralstonia Eutropha H16. *Nat. Biotechnol.* **2006**, *24*, 1257–1262. [CrossRef]
111. Cesário, M.T.F.; de Almeida, M.C.M.D. Lignocellulosic Hydrolysates for the Production of Polyhydroxyalkanoates. In *Microorganisms in Biorefineries*; Springer: Berlin/Heidelberg, Germany, 2015; pp. 79–104.
112. Lu, X.; Liu, G.; Wang, Y.; Ding, J.; Weng, W. Engineering of an L-Arabinose Metabolic Pathway in Ralstonia Eutropha W50. *Wei Sheng Wu Xue Bao* **2013**, *53*, 1267–1275.
113. Cieciura-Włoch, W.; Borowski, S.; Domański, J. Dark Fermentative Hydrogen Production from Hydrolyzed Sugar Beet Pulp Improved by Nitrogen and Phosphorus Supplementation. *Bioresour. Technol.* **2021**, *340*, 125622. [CrossRef]
114. Carvalheira, M.; Marreiros, B.C.; Reis, M. Acids (VFAs) and Bioplastic (PHA) Recovery. In *Clean Energy and Resource Recovery*; Elsevier: Amsterdam, The Netherlands, 2022; pp. 245–254, ISBN 9780323901789.

fermentation

Article

Study on the Operational Modes Using Both Growing and Resting Cells for Succinic Acid Production from Xylose Kinetic Modelling

Itziar A. Escanciano [1], Vanessa Ripoll [2], Miguel Ladero [1,*] and Victoria E. Santos [1,*]

1 FQPIMA Group, Department of Chemical and Materials Engineering, Faculty of Chemistry, Complutense University of Madrid, 28040 Madrid, Spain; itziaria@ucm.es
2 Facultad de Ciencias Experimentales, Universidad Francisco de Vitoria (UFV), Ctra. Pozuelo Majadahonda km 1.800, Pozuelo de Alarcón, 28223 Madrid, Spain; vanessa.ripoll@ufv.es
* Correspondence: mladerog@ucm.es (M.L.); vesantos@ucm.es (V.E.S.)

Abstract: Succinic acid (SA) is one of the most prominent C4 biomass-based platform chemicals that can be biologically obtained. This article verifies, for the first time, the possibility of producing succinic acid with fed-batch or repeated batch operations with *Actinobacillus succinogenes* in a resting state, that is, in the absence of a nitrogen source. In this work it is possible to optimise separately the stages of cell growth and production in the fed-batch or repeated batch modes, minimising the costs associated with the nitrogen source and facilitating the subsequent purification of SA. These experiments were carried out with xylose, the most abundant monosaccharide in hemicelluloses, with the results subsequently being compared to those obtained in equivalent operations carried out with cells in a state of growth. First, a cost-effective synthetic growth medium was proposed and successfully employed for SA production. Biocatalysts' reutilisation showed that the bioprocess can be carried out successfully in repeated batch and fed-batch modes. The best mode for growing cells is repeated batch, achieving a maximum productivity of 0.77 $g \cdot L^{-1} \cdot h^{-1}$, a selectivity of 53% and a yield of 51% with respect to xylose consumed. In contrast, the fed-batch mode was found to be the most convenient mode with resting cell biocatalyst, reaching a maximum productivity of 0.83 $g \cdot L^{-1} \cdot h^{-1}$, a selectivity of 0.78 $g \cdot g^{-1}$ and a yield of 68% with respect to the xylose consumed. In addition, by-product formation is significantly reduced when employing resting cells. An unstructured non-segregated kinetic model was developed for both biocatalysts, capable of simulating cell growth, xylose consumption, SA production and by-product generation, with successful estimation of kinetic parameters supported by statistical criteria.

Keywords: succinic acid; xylose; *Actinobacillus succinogenes*; resting cells; kinetic model

Citation: Escanciano, I.A.; Ripoll, V.; Ladero, M.; Santos, V.E. Study on the Operational Modes Using Both Growing and Resting Cells for Succinic Acid Production from Xylose Kinetic Modelling. *Fermentation* **2023**, *9*, 663. https://doi.org/10.3390/fermentation9070663

Academic Editor: Massimiliano Fabbricino

Received: 19 June 2023
Revised: 9 July 2023
Accepted: 10 July 2023
Published: 14 July 2023

1. Introduction

Currently, one of the main challenges in the fight against climate change is the search for renewable raw materials that replace fossil resources to generate energy, chemicals and materials. As a consequence, there is an urgent need for platform chemicals obtained through sustainable bioprocesses as alternatives to processes and products obtained via petroleum-based conventional refineries [1,2]. The majority of the processes and products deemed promising are being developed in second-generation biorefineries based on lignocellulosic biomass (LCB). LCB is especially attractive, as it is the most abundant and sustainable group of raw materials available worldwide and does not interfere in any food chain [3]. LCB is mainly composed of cellulose (40–50%), hemicellulose (25–30%) and lignin (15–20%). Depolymerised hemicellulose is composed of 90% xylose, being this monosaccharide the second most abundant sugar available in LCB after glucose, a constituent of cellulose [4]. However, the number of wild-type microorganisms that can metabolise xylose as a carbon source is very low, meaning that the hemicellulosic part of the waste is usually disposed of instead of being used as a carbon source.

One of the most promising microorganisms in the conversion of xylose into SA is a facultative anaerobic bacterium isolated from the rumen of cattle, *Actinobacillus succinogenes*. This bacterium is capable of naturally generating C4 dicarboxylic acids during the pregastric digestion of polysaccharides. In other words, it can produce high amounts of succinic acid from a wide variety of carbon sources. In addition, its use does not generate excessive difficulties in fermentation or in the separation and purification processes, which occurs with certain fungi. This microorganism transforms xylose into glyceraldehyde 3-phosphate through the pentose phosphate pathway. Subsequently, the bacterium incorporates the final product of glycolysis, phosphoenolpyruvate, into the tricarboxylic acid cycle, as shown in Figure 1. This occurs under anaerobic conditions and is facilitated by CO_2 insufflation, resulting in the production of a substantial amount of succinic acid (SA) [5,6]. According to the US Department of Energy (US DOE), SA is one of the main chemical platforms due to its potential to produce a great diversity of chemicals traditionally produced from fossil sources [7]. This organic acid is widely used in the production of polybutylene succinate, polyester, polyols and in the food and pharmaceutical industries as well as in resins, coatings, pigments and biodegradable polymers [8].

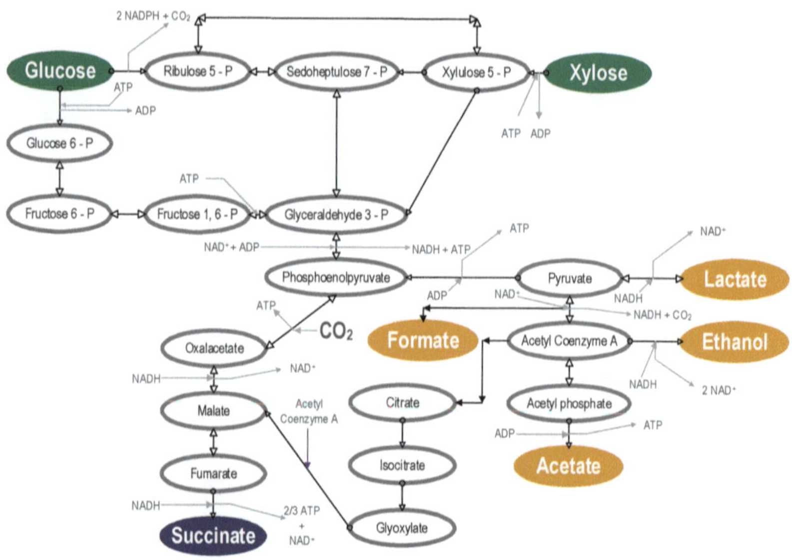

Figure 1. Metabolic pathways of *A. succinogenes* for SA production from glucose and xylose.

In recent years, the lack of sustainability of traditional petrochemical SA production has led to a considerable increase in its production through fermentation processes. In 2015, the market price of this acid produced by the biological route was 2.86 USD·kg^{-1}, while that obtained by the traditional route was 2.50 USD·kg^{-1} [9]. However, due to the great interest in biotechnological processes involving SA, currently, some economically competitive processes have already been developed and even implemented at an industrial level, producing SA with a market price between 2.00 and 2.50 USD·kg^{-1} [10]. With a 20% compound annual growth rate (CAGR), the bio-derived SA market is expected to reach 900 USD million by 2026, much higher than that of 2017 (175.7 USD million) [11]. The three main factors to consider in the implementation of a successful SA bioprocess at an industrial scale are the following: (i) availability of raw material, (ii) adequate SA productivity and titre and (iii) economically viable isolation and purification steps [12].

Ferone et al. (2017) [13] studied the production of succinic acid in batch operational mode by *A. succinogenes* employing a synthetic mixture of sugars representative of a lignocellulosic hydrolysis, obtaining a concentration of SA (27 $g \cdot L^{-1}$) higher than the ones obtained by a combination of fermentations of each single sugar, as well as a better selectivity for SA. Using the same microorganism, Bukhari et al. (2020) [14] used oil palm trunk as carbon source after its hydrolysis with oxalic, formic and acetic acid. Carrying out the bioprocess to obtain SA with hydrolysed waste, they obtained a yield of 0.47 $g \cdot g^{-1}$ in batch bottle fermentation and a maximum concentration of succinic acid of 10.62 $g \cdot L^{-1}$. In order to enhance yield and productivity, some researchers have recently focused on optimising the operational mode. Bradfield et al. (2016) [15] reached productivities of 3.4 $g \cdot L^{-1} \cdot h^{-1}$ and a maximum SA titre of 10.9 $g \cdot L^{-1}$ from xylose operating in continuous mode in a biofilm reactor. Jokodola et al. (2022) [16] opted for a fed-batch operational mode, producing 33.6 and 28.7 $g \cdot L^{-1}$ of succinate from hydrolysates of olive pits and sugarcane bagasse, respectively, with same conversion yield (0.27 $g \cdot g^{-1}$). Considering that downstream operations represent approximately 60% of the overall operating costs in the SA production process, it is essential to reduce the by-product generation during fermentation in order to design a competitive process on an industrial level. To do this, this research group carried out a previous study on the viability of succinic acid production employing a biocatalyst composed of *A. succinogenes* cells in a resting state [17]. Operating under nitrogen-limited conditions, and formulating a medium composed exclusively of a carbon source and a buffer solution to maintain osmotic pressure, resting cells are metabolically active even though cell growth is impeded. In this way, the metabolic pathways are active towards SA production and cellular maintenance, but there is no bacterial growth. Concurrently, the number of by-products is also reduced dramatically [18–20], this being one of the main advantages associated with this type of operation. In fact, Escanciano et al. (2022) [17] managed to reduce the quantity of by-products generated by 27.5% in the production of SA from xylose. In order to obtain cells for subsequent use in the production stage in resting state, it is necessary to carry out a conventional fermentation, although it requires the optimisation of conditions so that the cells have the appropriate metabolic state for their subsequent use in resting cells. A notable advantage of separating the stages of growth (or obtaining the biocatalyst) and production (with the cells in the resting cell state) is the culture medium needed in the stages. Generally, to carry out production with growing cells, culture media with components such as yeast extract are used, which, although they do not have a high concentration in the medium, if the total mass is calculated to carry out the process in industrial bioreactors, would entail a high price. When the biomass is produced, it is done on a much smaller scale, so the need for such culture media is considerably reduced. However, the medium used for resting cell production is very simple and cost-effective, devoid of nitrogen sources and usually expensive. Separating growth and production into two different steps allows for a better optimisation of each step and of the overall process. Considering the scale-up, the proposed bioprocess would lead to higher productivity linked to the reduction of dead times as well as to avoid substrate inhibition that was observed when using growing cells as biocatalyst [18,20,21].

In this present study, SA production using *A. succinogenes* in both growth and resting states was researched, exploring different forms of operation through biocatalyst reuse by means of fed-batch and repeated batch operations. This is the first work in which the operation with resting cells is studied in different types of operation, allowing us to optimise the stages of cell growth and SA production separately and reduce the operating costs associated with the nitrogen source and SA purification. Additionally, the effect on production with cells in growth and resting state of the use of a cheaper culture medium than the one used in previous studies of the research group was compared [17]. Finally, kinetic modelling of the bioprocess was developed to predict the evolution of biomass, xylose, succinic acid and the by-product concentration throughout the fermentation time.

2. Materials and Methods

2.1. Bacterial Strain, Adaptation, Preinoculum and Inoculum Stages

The strain *Actinobacillus succinogenes* DSM 22,257 was provided by the Leibniz Institute DSMZ-German Collection of Microorganisms and Cultures GmbH (https://www.dsmz.de/accessed on 15 February 2023). The bacterial strain was reactivated on one agar plate with Brain Heart Infusion (BHI) for two days. A single bacterial colony was inoculated in a bottle containing 60 mL of sterile Tryptic Soy Broth (TSB) [22–24]. The bottle was incubated for 1 day at 37 °C. For the long-term preservation, the culture was mixed with glycerol (1:1 $v \cdot v^{-1}$) and stored at −80 °C.

Following the procedure described in a previous study [17], thawed bacteria were incubated at 37 °C for 24 h in anaerobic bottles with TSB medium. Subsequently, a two-step procedure was developed to adapt the biocatalyst to the carbon source. Broth from anaerobic bottles was employed as inoculum in a 5% ($v \cdot v^{-1}$) ratio by inoculating it in a second bottle containing the production medium (PM) at 37 °C [25]. This production medium contained (in $g \cdot L^{-1}$): yeast extract, 10; K_2HPO_4, 3; $MgCl_2.6H_2O$, 0.427; $CaCl_2$, 0.2; NaCl, 1; $NaHCO_3$, 10; and commercial xylose, 10. Using an inoculum of this second bottle, another stage of adaptation in a third bottle was carried out, increasing the concentration of xylose and $NaHCO_3$ to 20 $g \cdot L^{-1}$ and the biomass initial inoculation volume from the previous step to 10% ($v \cdot v^{-1}$).

2.2. Succinic Acid Production Employing Growing Cells as the Biocatalyst

Runs were carried out in duplicate in a 2 L stirred tank bioreactor (STBR) BIOSTAT B-Plus (Sartorius AG, Göttingen, Germany) with a working volume of 1 L. Batch experiments were carried out with PM medium. The initial amount of xylose was 20 $g \cdot L^{-1}$ in all runs. The pH was maintained at 6.8 (NaOH 5 M) at 37 °C and CO_2 was sparged at 0.1 vvm with a stirring speed of 300 rpm. The experiments were carried out starting from 0.05 $g \cdot L^{-1}$ of biomass in the exponential phase of growth, obtained in the final adaptation step in bottles.

2.3. Succinic Acid Production Employing Resting Cells as the Biocatalyst

To successfully perform a fermentation process employing cells in a resting state as the biocatalyst, it is necessary to previously carry out a batch operational mode with growing cells as was described in previously, in order to obtain a high amount of initial biomass. After 15 h of fermentation, suspended biomass was separated from the broth by centrifugation (9000 rpm, 5 min), washed with a K_2HPO_4 solution and transferred to a 1 L STBR with 0.5 L of working volume at 37 °C and 300 rpm. The production media consisted of a buffered solution (K_2HPO_4 50 mM) of xylose 20 $g \cdot L^{-1}$. The pH was controlled at 6.8 (NaOH 5 M), and carbon dioxide was supplied at 0.1 vvm. Runs were performed in duplicate.

2.4. Succinic Acid Production Operating Fed-Batch and Repeated Batch Fermentations

Fed-batch and repeated batch runs were carried out in duplicate with PM medium both with growing cells and resting cells biocatalysts, In the case of the fed-batch operation, three stages were carried out, feeding a concentrated solution of xylose at the beginning of each one of them. In repeated batch fermentations, at the end of the first and second stages, the suspended biomass was separated from the liquid broth by centrifugation and subsequently inoculated in the reactor of the next stage. The pH was maintained at 6.8 (NaOH 5 M), temperature at 37 °C and the stirring speed at 300 rpm, and CO_2 was sparged at 0.1 vvm.

2.5. Analytical Methods

Biomass concentration was determined by measuring the optical density of broth samples at 600 nm using a spectrophotometer (Shimadzu UV-Vis spectrophotometer UV-1603, Kyoto, Japan).

Xylose and fermentation products (ethanol and succinic, lactic, acetic and formic acids) were analysed through an Agilent Technologies 100 series equipment by high-performance liquid chromatography (HPLC) equipped with a refractive index detector (RID) at 55 °C and a REZEX ROA-Monosaccharide H^+ (8%) column (300 × 7.8 mm, Phenomenex, Torrance, CA, USA) at 80 °C. Acid water (H_2SO_4 5 mM) was eluted at a flow of 0.5 mL·min^{-1}.

2.6. Mathematical Methods

The estimation of the model kinetic parameters was carried out by means of the computer software Aspen Custom Modeler v11 (AspenTech, Bedford, MA, USA), using an implicit Euler method to integrate the ODEs of the kinetic model coupled to a non-linear least-squares solver algorithm (NL2SOL) to obtain optimal values of the kinetic parameters.

To assess the adequacy of the fit, it is necessary to ensure that Fisher's F-value (F)—as seen in Equation (1)—has a higher value than the critical value at a 95% confidence level. Additionally, the sum of squared residuals (SSR) and the residual mean-squared error (RMSE) should be minimised, approaching zero as closely as possible—as seen in Equation (2)—and the variation explained (VE) should be close to or equal to 100%—as seen in Equation (3).

$$F = \frac{\sum_{i=1}^{N}\left(\frac{y_{i,\,calc}}{K}\right)^2}{\sum_{i=1}^{N}\left(\frac{SSR}{N-K}\right)} \tag{1}$$

$$RMSE = \sqrt{\frac{SSR}{N-K}} \tag{2}$$

$$VE(\%) = 100\left(1 - \frac{\sum_{l=1}^{L} SSQ_l}{\sum_{l=1}^{L} SSQ_{mean_l}}\right) \tag{3}$$

where K is the number of parameters, SSR is the squared sum of residues, N is the total number of experimental data, SSQ_l is the sum of the quadratic residues and SSQ_{meanl} is the squared sum of deviations between the experimental and the mean score with respect to the calculated values.

In order to determine the influence of initial substrate concentration, consumed substrate, product distribution and cell metabolism on succinic acid production, the following parameters were defined:

$$\text{Succinic acid yield}: \eta_{SA} = [SA]^{max}/[Xyl]_0 \tag{4}$$

$$\text{Succinic acid macroscopic yield}: \; Y_{SA/Xyl.cons} = [SA]^{max}/[Xyl]_{cons} \tag{5}$$

$$\text{Specific succinic acid yield}: \; Y_{SA/X} = [SA]^{max}/[X]^{max} \tag{6}$$

$$\text{Selectivity on succinic acid}: \; S_{SA} = [SA]^{max}/\left([SA]^{max} + [BP]^{max}\right) \tag{7}$$

$$\text{Succinic acid productivity}: \; P_{SA} = [SA]^{max}/time \tag{8}$$

$$\text{Specific succinic acid productivity}: \; P_{SA/X} = [SA]^{max}/\left([X]^{max}\cdot time\right) \tag{9}$$

where $[SA]^{max}$ is the maximum concentration of succinic acid (SA) (g_{SA}·L^{-1}), $[Xyl]_0$ is the initial concentration of xylose ($g_{Xyl\cdot 0}$·L^{-1}), $[Xyl]_{cons}$ is the concentration of consumed xylose ($g_{Xlycons}$·L^{-1}), $[X]^{max}$ is the maximum concentration of biomass (g_X·L^{-1}) and $[BP]^{max}$ is the maximum concentration of by-products $[BP]$ (g_{BP}·L^{-1}).

3. Results and Discussion

3.1. Influence of Medium Composition on Succinic Acid Production Employing Growing Cells and Resting Cells as the Biocatalyst

Growth medium composition as well as type and availability of carbon source play an important role in SA titre and productivity. Furthermore, the development of an effective biocatalyst is also a key factor that determines the viability of the bioprocess. Recently, *A. succinogenes* in the resting state was proven to be a promising biocatalyst, revealing a higher selectivity to the target acid when compared to growing cells [17]. In that previous study [17], a complex growth medium, TSB, had been selected due to its richness in nutrients, considered in the bibliography as a preferred medium to promote cell growth [22–24]. However, with the aim of reducing the cost of the process, the present study verifies the viability of carrying out cell growth, prior to the operation with resting cells, with a lower-cost synthetic medium, PM.

Table 1 summarises the obtained results in batch runs, in terms of the maximum product concentration achieved; its selectivity, yields and productivities in each evaluated medium culture (PM and TSB); and the biocatalyst state (growing and resting cells). Regarding product concentration, it is noticed that the synthetic medium promotes SA production, whose final concentration in broth reaches the highest values when the PM medium is employed (11.7 $g \cdot L^{-1}$ and 12.5 $g \cdot L^{-1}$ using growing and resting cells, respectively). As a consequence, in these conditions, higher values of SA yield with respect to initial carbon source are achieved (0.55 $g_{SA} \cdot g_{Xyl \cdot 0}{}^{-1}$ for growing state and 0.65 $g_{SA} \cdot g_{Xyl \cdot 0}{}^{-1}$ for resting state) than in runs in which the TSB medium was used. However, in fermentations with PM medium culture, an increase of by-product production (acetic acid and formic acid) is observed, a fact that is reflected in a slight reduction of selectivity values.

Table 1. Concentrations, selectivity, yield and productivity of SA production with growing and resting cells and two different complex growth mediums in batch runs employing 20 $g \cdot L^{-1}$ of carbon source.

Biocatalyst	Growing Cells		Resting Cells	
Growth medium	TSB	MP	TSB	MP
C_{SA} ($g_{SA} \cdot L^{-1}$)	8.72	11.7	8.51	12.5
S_{SA} ($g_{SA} \cdot g_{SA+SP}{}^{-1}$)	0.64	0.55	0.75	0.68
η_{SA} ($g_{SA} \cdot g_{Xyl \cdot 0}{}^{-1}$)	0.44	0.55	0.43	0.65
$Y_{SA/Xyl \cdot cons}$ ($g_{SA} \cdot g_{Xylcons}{}^{-1}$)	0.44	0.55	0.81	0.65
$Y_{SA/X}$ ($g_{SA} \cdot g_X{}^{-1}$)	2.80	2.74	2.03	2.91
P_{SA} ($g_{SA} \cdot L^{-1} \cdot h^{-1}$)	0.36	0.25	0.18	0.43
$P_{SA/X}$ ($g_{SA} \cdot g_X{}^{-1} \cdot h^{-1}$)	0.12	0.06	0.04	0.10
Reference	[17]	This work	[17]	This work

Medium composition barely affects SA yield respect to final biocatalyst concentration employing growing cells. However, a biocatalyst composed of resting cells cultivated in synthetic broth presents cell metabolism pathways that more actively produce SA. This is reflected in the increase of this acid yield respect to biocatalyst concentration (2.91 $g_{SA} \cdot g_X{}^{-1}$) in comparison to the value when TSB is employed (2.03 $g_{SA} \cdot g_X{}^{-1}$). A similar tendency is observed for productivity (P_{SA}), whose meaning is closely linked to a macroscopic SA production rate. This acid productivity is boosted by the PM medium employing the resting cells as a biocatalyst. However, this is not improved in growing cell. As the results showed, acid production from resting cells takes place 2.4 times faster due to the change of the medium composition (with synthetic medium culture: 0.43 $g_{SA} \cdot L^{-1} \cdot h^{-1}$; when using complex medium culture: 0.18 $g_{SA} \cdot L^{-1} \cdot h^{-1}$).

Based on these results, it can be concluded that the employment of the synthetic growth medium PM using *A. succinogenes* in growing and resting states enhances SA production in terms of final titre and yield with respect to initial xylose concentration. In addition, productivity and yield with respect to cell concentration are also improved for the biocatalyst in resting cells. In short, an expensive complex growth medium such as TSB is not necessary when producing with resting cells; an economical and synthetic medium such as PM is a promising alternative for the development of the bioprocess on an industrial scale.

Figure 2 shows the evolution of xylose (Xyl), product (SA), by-product (BP) and biomass (X) concentrations in batch bioreactor employing the PM medium. Substrate exhaustion is reached at around 30 h for both biocatalysts. However, the experimental tendencies are different: whereas lag phase of growth leads to slower substrate consumption and product production rates at the beginning of the fermentation (Figure 2A), a quick decrease on xylose concentration and, therefore, SA production is observed when a resting cells biocatalyst is employed (Figure 2B). Even if the achieved target product concentration is similar in both cases, the global by-product concentration is reduced employing resting cells biocatalyst. This behaviour is also reflected in the parameters showed in Table 1. When synthetic medium is used, all the evaluated parameters are improved by resting cells in comparison to growing biocatalyst. This means that cells under nitrogen-limited conditions reveal a metabolism more specific and selective towards the production of SA from xylose as sole carbon source.

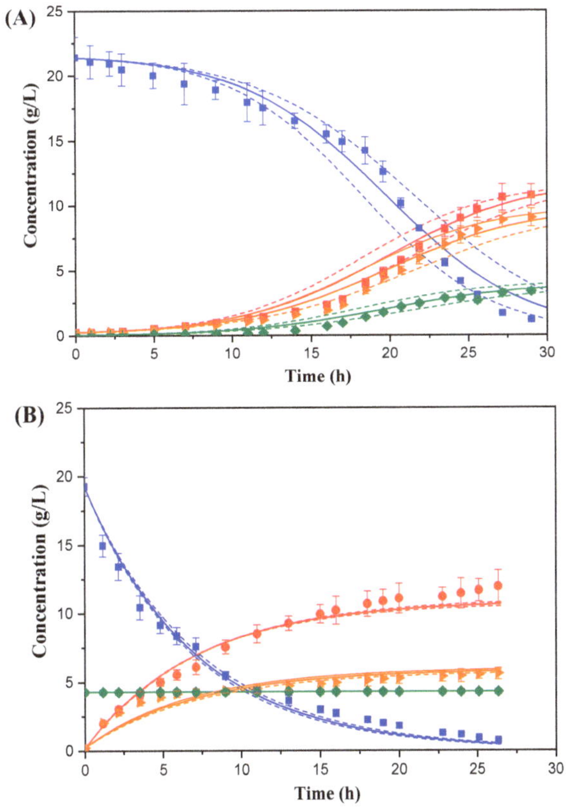

Figure 2. Evolution of substrate, product, by-product and biomass concentrations in batch bioreactor employing (**A**) growing cells as the biocatalyst and (**B**) resting cells as the biocatalyst. *Key:* SA (●), Xyl (■), BP (►), X (♦); dots: experimental data; solid lines: model prediction; dashed lines: error model prediction.

3.2. Succinic Acid Production in a Repeated Batch Bioreactor

As previously mentioned, some authors had managed to significantly increase their yields and/or productivities by carrying out fermentations with growing cells in other modes of operation [15,16,26]. In addition, in previous studies this research group demonstrated in previous works that successive growth stages prior to the inoculum stage improve the yield and productivity of the process. Therefore, successive repeat batch fermentations could have this same effect [17]. For this reason, with the aim of improving the production process of succinic acid with resting cells, as well as further reducing the costs associated with the separation stages, it was decided to carry out fermentations through a repeated batch operation.

In the present section, the possibility of reusing the biocatalyst in three consecutive cycles was evaluated. Essentially, repeated batch experiments explored the viability of separating the biocatalyst at the end of the fermentation and, afterwards, harvested cells were employed as inoculum in the next cycle, as was explained in Section 2.4. Whereas the PM synthetic medium is used for the maintenance of the growing biocatalyst in repeated batch runs, only the carbon source and a buffer solution are employed as the medium for resting cells.

Each batch run was conducted employing approximately 20 $g \cdot L^{-1}$ of initial xylose concentration, which corresponds to the concentration range of carbon source found in raw vegetable materials [27]. When carbon substrate concentration reached 5 $g \cdot L^{-1}$, the experiment was stopped in order to avoid a metabolic stress to the bacterial cells. Subsequently, the biocatalyst was separated from the broth by means of centrifugation and the recovered pellet was employed as a biocatalyst in the next batch run. The main advantage of repeated batch in comparison to fed-batch operation is that repeated batch avoids inhibition due to the accumulation of cytotoxic compounds in the broth, as may occur with the fed-batch operation [28].

Figure 3 shows the evolution of xylose, biomass, SA and by-product concentrations in repeated batch cycles evaluating the influence of the state of the biocatalyst: growing cells (Figure 3A) and resting cells (Figure 3B). While three consecutive cycles employing growing cells took less than 60 h, the use of the resting cells biocatalyst extended the fermentation time by 40 h. However, the first cycle using a biocatalyst composed of resting cells was much faster than the analogous experiment for growing cells, which is probably due to the greater biocatalyst concentration at the beginning and throughout the run. In the following steps, a deceleration of substrate consumption rate is observed when resting cells are reused. Resting cells concentration decreases progressively, due to the experimental cell recovery procedure between cycles. It was observed that, although in the second cycle the cells were still considerably active, in the third cycle, the process suffered a significant slowdown, probably caused by the mechanical stress of the biocatalyst separation process and the scarcity of nutrients during the consecutive fermentation cycles.

Regarding repeated batch working with the growing biocatalyst, it is to be noted that productivity is enhanced throughout the cycles, in terms of both substrate consumption and acid production, as is shown in Table 2. Although the initial biomass concentration at the beginning of each step is increased, the importance of the bacteria adaptation to carbon source also plays an important role in the bioprocess, as has previously been demonstrated by Escanciano et al. [17]. In fact, with growing cells, the yield with respect to the xylose consumed and the productivity in the third stage are 72% and 56% higher than those corresponding to the first stage, respectively, reaching an average yield with respect to xylose consumed of 0.51 $g\ g^{-1}$ and a productivity of 0.53 $g \cdot L^{-1} \cdot h^{-1}$ (Table 2). Regarding specific SA productivity, cells in growth state obtain higher values than the resting cells, maintaining practically constant productivities throughout the three stages, while in the runs with resting cells, the specific productivity is already reduced by 70% from the first to the second stage.

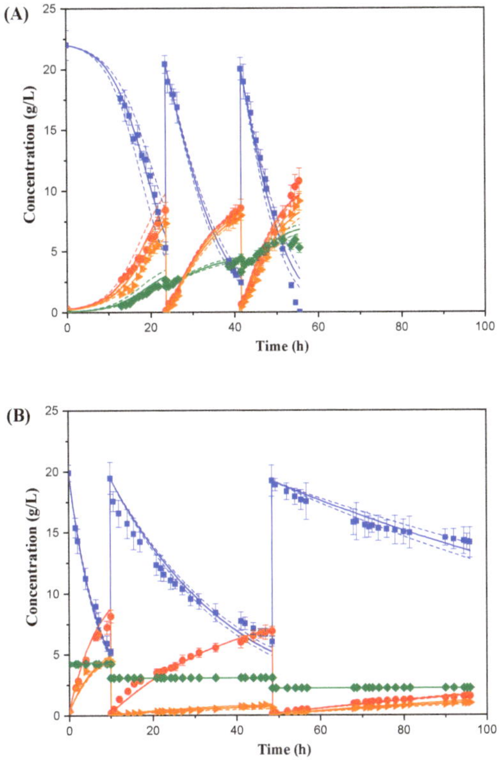

Figure 3. Evolution of substrate, product, by-product, and biomass concentrations in repeated batch bioreactor employing (**A**) growing cells as the biocatalyst and (**B**) resting cells as the biocatalyst. *Key:* SA (●), Xyl (■), BP (►), X (♦) dots: experimental data; solid lines: model prediction; dashed lines: error model prediction.

Table 2. Concentrations, selectivity, yields and productivity of SA production with growing and resting cells operating in repeated batch runs.

Biocatalyst State	**Growing Cells**				**Resting Cells**			
Cycle	**1**	**2**	**3**	**Average**	**1**	**2**	**3**	**Average**
C_{SA} ($g_{SA}{\cdot}L^{-1}$)	8.39	8.56	10.8	9.25	8.13	6.90	1.55	5.53
S_{SA} ($g_{SA}{\cdot}g_{SA+BP}{}^{-1}$)	0.53	0.52	0.54	0.53	0.65	0.92	0.60	0.72
η_{SA} ($g_{SA}{\cdot}g_{Xyl{\cdot}0}{}^{-1}$)	0.38	0.42	0.54	0.45	0.41	0.35	0.08	0.28
$Y_{SA/Xyl{\cdot}cons}$ ($g_{SA}{\cdot}g_{Xyl{\cdot}cons}{}^{-1}$)	0.50	0.48	0.54	0.51	0.55	0.51	0.30	0.45
$Y_{SA/X}$ ($g_{SA}{\cdot}g_X{}^{-1}$)	3.21	1.95	1.83	2.33	1.94	2.25	0.71	1.63
P_{SA} ($g_{SA}{\cdot}L^{-1}{\cdot}h^{-1}$)	0.35	0.48	0.77	0.53	0.83	0.18	0.03	0.35
$P_{SA/X}$ ($g_{SA}{\cdot}g_X{}^{-1}{\cdot}h^{-1}$)	0.13	0.11	0.13	0.12	0.20	0.06	0.01	0.09

In conclusion, although selectivity was favoured by production with resting cells, the reuse of growing cells led to better overall SA production, in terms of concentration, yield, and productivity, which results improve from batch to the next one.

3.3. Succinic Acid Production in a Fed-Batch Bioreactor

To avoid cell damage during separation processes between cycles, fed-batch operation was explored as an alternative to extend operational time and, therefore, increase final SA titre and productivity. Three feeding pulses were carried out (at the beginning and when substrate concentration was around 5 g·L^{-1}) to obtain similar conditions as the repeated batch operation, which allows for the comparison of the results.

Figure 4 shows the evolution of xylose, biomass, SA and by-product concentrations in the fed-batch operation employing different biocatalysts: growing cells (Figure 4A) and resting cells (Figure 4B). Table 3 summarises the obtained results in terms of final concentrations, selectivity, yields and productivity after the consumption of each feeding pulse as well as overall parameters.

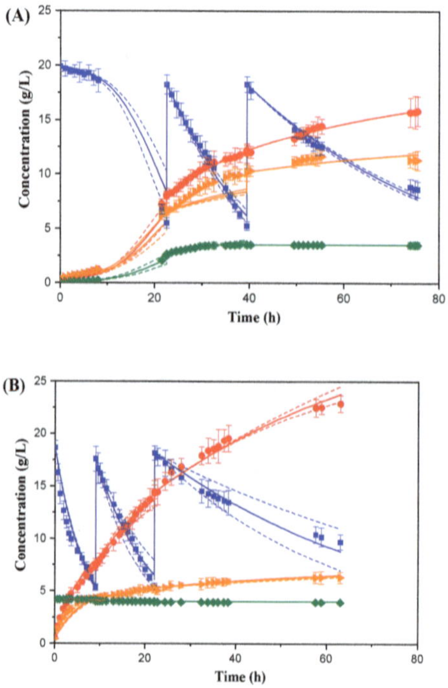

Figure 4. Evolution of substrate, product, by-product, and biomass concentrations in fed-batch bioreactor employing (**A**) growing cells as the biocatalyst and (**B**) resting cells as the biocatalyst. *Key:* SA (•), Xyl (■), BP (▶), X (♦); dots: experimental data; solid lines: model prediction; dashed lines: error model prediction.

As can be observed, the duration of the first two stages of the fed-batch operation with resting cells was considerably shorter than the corresponding ones with growing cells. In both cases, the total substrate consumption was not reached in the last step. In fact, in the operation with resting cells, after 30 h of fermentation, only approximately half of the xylose fed in the third stage was consumed.

The slowdown of the processes in the last part of the experiments can be caused by various factors: the depletion of essential nutrients that are not added in the pulse feeding, the aging of the cells, the accumulation of acids (both the target product as well as by-products) or toxins with an inhibitory effect [29]. This reduction in productivity at long fed-batch fermentation times has also been observed by other authors, such as Jiang et al. [25], who, with fermentation from glucose, noted a drop in the growth rate of *A. succinogenes* and of the succinic acid generated from the third production cycle (14 h of fermentation), which became especially pronounced in the fourth cycle.

Table 3. Concentrations, selectivity, yields and productivity of SA production with growing and resting cells operating in fed-batch runs.

Biocatalyst	**Growing Cells**				**Resting Cells**			
Cycle	1	2	3	**Overall**	1	2	3	**Overall**
C_{SA} ($g_{SA}\cdot L^{-1}$)	8.02	3.85	3.64	15.9	7.50	6.99	8.45	22.9
S_{SA} ($g_{SA}\cdot g_{SA+BP}{}^{-1}$)	0.54	0.52	0.76	0.57	0.63	0.87	0.89	0.78
η_{SA} ($g_{SA}\cdot g_{Xyl\cdot 0}{}^{-1}$)	0.40	0.21	0.20	0.35	0.40	0.40	0.47	0.53
$Y_{SA/Xyl\cdot cons}$ ($g_{SA}\cdot g_{Xyl\cdot cons}{}^{-1}$)	0.56	0.30	0.38	0.43	0.56	0.57	1.00	0.68
$Y_{SA/X}$ ($g_{SA}\cdot g_X{}^{-1}$)	3.15	1.09	1.03	4.48	1.79	1.74	2.15	5.46
P_{SA} ($g_{SA}\cdot L^{-1}\cdot h^{-1}$)	0.36	0.23	0.10	0.21	0.83	0.54	0.21	0.36
$P_{SA/X}$ ($g_{SA}\cdot g_X{}^{-1}\ h^{-1}$)	0.14	0.07	0.03	0.06	0.20	0.13	0.05	0.09

It should also be noted that, when using cells in resting state, the selectivity of SA is fully enhanced, reaching a final concentration of SA 3.6 times higher than the global concentration of by-products. However, in the case of using growing cells, only a final concentration of SA 1.4 times higher than the total concentration of by-products is reached. Regarding SA yields, they had a decreasing trend between cycles in the growing cell experiment (cycle 1: 0.40 $g\cdot g^{-1}$, cycle 2: 0.21 $g\cdot g^{-1}$, cycle 3: 0.20 $g\cdot g^{-1}$), while the opposite behaviour was observed with resting cells (cycle 1: 0.40 $g\cdot g^{-1}$, cycle 2: 0.21 $g\cdot g^{-1}$, cycle 3: 0.20 $g\cdot g^{-1}$). The concentration of the final product was 1.5 times higher using resting cells (22.9 $g\cdot L^{-1}$). In both cases, productivity was drastically reduced throughout the cycles, with a higher average value in the case of operating with resting cells (0.36 $g\cdot L^{-1}\cdot h^{-1}$). These results showed that the metabolism of the biocatalyst slowed down throughout the cycles but was more selective. In conclusion, the fed-batch operation with cells in the resting state leads to higher yield, productivity and selectivity than with growing cells.

Based on these results, the possibility of a reusing biocatalyst by means of repeated batch and fed-batch operational modes was successfully proven. Specifically, employing growing cells under repeated batch conditions and resting cells in fed-batch mode led to better SA production yields (0.51 $g\cdot g^{-1}$ and 0.68 $g\cdot g^{-1}$, respectively). In terms of productivity, production by growing cells in repeated batch mode took place faster (0.12 $g\cdot L^{-1}\cdot h^{-1}$ and 0.09 $g\cdot L^{-1}\cdot h^{-1}$, respectively). Moreover, working with growing cells in repeated batch mode, the third cycle results did not show inhibition or aging cell phenomena. However, in fermentation with resting cells in fed-batch mode, it was possible to increase the selectivity by 32% with respect to the operation in repeated batch with growing cells. This is of great importance if we consider that one of the main operating costs in the production of succinic acid is found in the purification process [30].

Table 4 summarises published results regarding the production of SA mainly from xylose, using *A. succinogenes* as s biocatalyst. Up to now, most of the studies have been undertaken with conventional submerged growth culture. The authors who carried out batch-type operations from xylose with growing cells did not achieve yields greater than 55% [13] nor SA productivities higher than 0.36 $g\cdot L^{-1}\cdot h^{-1}$ [17]. However, substantially higher yields have been achieved with other types of operation, as in the case of Patsalou et al. [31], who managed to produce SA using citrus peel waste hydrolysate as a renewable carbon source with fed-batch type fermentation, reaching a yield of 0.73 $g\cdot g^{-1}$. To date, the fermentations carried out operating in repeated batch mode have mostly been studied with glucose as carbon source (or hydrolysates rich in this sugar) and with immobilised cells, which is why this study has focused in part on the production of SA through this operation mode from xylose with free cells, achieving a yield of 51% and an average productivity of 0.12 $g\cdot L^{-1}\cdot h^{-1}$.

Productions with *A. succinogenes* in a resting state were evaluated for the first time in a previous study of this research group [17], achieving a batch yield of 43% while the productivity was 0.18 $g\cdot L^{-1}\cdot h^{-1}$. In this present work, it was possible to double the productivity thanks to the operation in fed-batch conditions. Therefore, these results are considered to pave the way for development novel biocatalyst and give new insight towards operational modes in SA production.

Table 4. Summary of published studies aimed to SA production under different conditions (biocatalyst state, operational mode and carbon and nitrogen sources).

Biocatalyst State		Type of Operation	Substrate	Nitrogen Source	η_{SA} ($g_{SA} \cdot g_{substrate}^{-1}$)	P_{SA} ($g_{SA} \cdot L^{-1} \cdot h^{-1}$)	C_{SA} ($g_{SA} \cdot L^{-1}$)	Reference
Growing/Resting	Free/Immobilised							
Growing	Free	**Batch**	Xylose	Yeast extract	0.42	0.15	3.94	[32]
			Glucose, mannose, arabinose, xylose	Yeast extract	0.55	0.22	27.0	[13]
			Xylose	Yeast extract	0.44	0.36	8.7	[17]
			Xylose	**Yeast extract**	**0.55**	**0.25**	**11.7**	**This study**
		Fed-batch	Xylose	Yeast extract	0.27	0.51	36.7	[16]
			Xylose	**Yeast extract**	**0.35**	**0.21**	**15.9**	**This study**
			Citrus peel waste	Yeast extract	0.73	0.45	22.4	[31]
		Repeated batch	**Xylose**	**Yeast extract**	**0.45 ***	**0.12 ***	**9.25 ***	**This study**
	Immobilised	Repeated batch	Tequilana agave bagasse	Yeast extract	0.39 *	1.10 *	~6.72 *	[33]
			Glucose	Yeast extract	0.49 *	~0.38 *	~30*	[34]
Resting	Free	**Batch**	Xylose	None	0.43	0.18	8.51	[17]
			Xylose	**None**	**0.65**	**0.43**	**12.5**	**This study**
		Fed-batch	**Xylose**	**None**	**0.53**	**0.36**	**22.9**	**This study**
		Repeated batch	**Xylose**	**None**	**0.28 ***	**0.09 ***	**5.53 ***	**This study**

* average value.

3.4. Kinetic Modelling of the Bioprocess

Kinetic modelling enables the simulation of bio/chemical processes and facilitates their implementation and operational control. To obtain a mathematical model able to fit the kinetic data in this work, a very simple unstructured and non-segregated kinetic model was proposed, which describes the observed experimental tendencies regarding xylose, biomass, SA and by-product concentrations. The following considerations were taken into account:

- A laboratory-scale bioreactor was considered as a completely mixed tank bioreactor.
- The only by-products detected in the culture medium were acetic and formic acids, which are formed almost directly from pyruvate, as is reflected in Figure 1. Moreover, the evolution of their concentrations exhibited remarkable similarity across all experiments. Based on that, acetic and formic acids concentrations were lumped or summed to obtain a global by-product concentration in the kinetic model.
- Growth, SA production and BP generation depend proportionally on the availability of carbon source and the biocatalyst concentration.

Considering this information, a sole kinetic model (Equations (10)–(15)) was developed to describe the behaviour of growing cells and resting cells as biocatalysts for all operational modes. Employing resting cells as a biocatalyst, the biomass concentration remains constant, resulting in zero yield of biomass from product and biomass growth rate.

- Reaction network

$$Y_{S/P}\, S \xrightarrow{r} Y_{X/P}\, X + P + Y_{BP/P}\, BP \tag{10}$$

- Reaction rate

$$r = k_P \cdot [X] \cdot [S] \tag{11}$$

- Production and consumption rates

  Substrate consumption rate

$$R_S = \frac{d[S]}{dt} = -Y_{S/P} \cdot r \tag{12}$$

  Product production rate

$$R_P = \frac{d[P]}{dt} = r \tag{13}$$

  Biomass growth rate

$$R_X = \frac{d[X]}{dt} = Y_{X/P} \cdot r \tag{14}$$

  By-product production rate

$$R_{BP} = \frac{d[BP]}{dt} = Y_{BP/P} \cdot r \tag{15}$$

where $Y_{i/j}$ is the macroscopic yield of compound "i" with respect to compound "j", and k_P is the second-order kinetic constant in the SA production.

The fitting of the proposed kinetic equation to experimental data (substrate, biomass, product and by-products) for each cycle employing both biocatalysts was carried out to estimate the value of each kinetic parameter involved in the kinetic model. The model prediction fits very reasonably to all relevant data, as is reflected in Figures 3 and 4 as solid lines. Consequently, the hypotheses assumed in the model are valid in the studied conditions, where no substrate or product inhibition phenomena were considered.

Tables 5 and 6 show the estimated kinetic parameters, as well as the statistical parameters that provide the information on the goodness of fit. Fitting of experimental data regarding batch run and first cycles of repeated batch and fed-batch runs were carried out together. Goodness-of-fit statistical parameters indicate a high value for Fisher's F, considerably above the limiting value, and a low value for the RMSE and the SSR. Moreover, the experimental tendencies and those predicted from the model are quite similar, as the higher than 90% VE percentages showed.

Table 5. Kinetic and statistical parameter values calculated by fitting the kinetic model to experimental data of growing cells runs.

Cycle	Kinetic Parameter	Operational Mode		
		Batch	Repeated Batch	Fed-Batch
1	k_P ($L \cdot g^{-1} \cdot h^{-1}$)		$(3.02 \pm 0.11) \cdot 10^{-2}$	
	$Y_{X/P}$ ($g \cdot g^{-1}$)		$(3.41 \pm 0.15) \cdot 10^{-1}$	
	$Y_{S/P}$ ($g \cdot g^{-1}$)		1.83 ± 0.03	
	$Y_{BP/P}$ ($g \cdot g^{-1}$)		$(8.21 \pm 0.20) \cdot 10^{-1}$	
2	k_P ($L \cdot g^{-1} \cdot h^{-1}$)	-	$(1.51 \pm 0.04) \cdot 10^{-2}$	$(6.50 \pm 0.19) \cdot 10^{-3}$
	$Y_{X/P}$ ($g \cdot g^{-1}$)	-	$(2.65 \pm 0.16) \cdot 10^{-1}$	$(2.88 \pm 0.25) \cdot 10^{-1}$
	$Y_{S/P}$ ($g \cdot g^{-1}$)	-	2.14 ± 0.04	2.94 ± 0.08
	$Y_{BP/P}$ ($g \cdot g^{-1}$)	-	$(9.46 \pm 0.22) \cdot 10^{-1}$	$(9.39 \pm 0.03) \cdot 10^{-1}$
3	k_P ($L \cdot g^{-1} \cdot h^{-1}$)	-	$(1.38 \pm 0.07) \cdot 10^{-2}$	$(2.33 \pm 0.01) \cdot 10^{-3}$
	$Y_{X/P}$ ($g \cdot g^{-1}$)	-	$(3.89 \pm 0.42) \cdot 10^{-1}$	0.00
	$Y_{S/P}$ ($g \cdot g^{-1}$)	-	1.92 ± 0.08	2.77 ± 0.12
	$Y_{BP/P}$ ($g \cdot g^{-1}$)	-	$(8.64 \pm 0.54) \cdot 10^{-1}$	$(4.45 \pm 0.05) \cdot 10^{-1}$
Cycle	**Statistical Parameter**	**Operational Mode**		
		Batch	**Repeated Batch**	**Fed-Batch**
1	F		4715	
	RMSE		0.59	
	SSR		32.1	
	VE (%)		96.4	
2	F	-	8723	21,562
	RMSE	-	0.24	0.25
	SSR	-	2.15	3.88
	VE (%)	-	99.7	98.8
3	F	-	1129	18,932
	RMSE	-	0.80	0.32
	SSR	-	33.4	4.89
	VE (%)	-	96.0	96.1

Table 6. Kinetic and statistical parameter values calculated by fitting the kinetic model to experimental data of resting cells runs.

Cycle	Kinetic Parameter	Operational Mode		
		Batch	Repeated Batch	Fed-Batch
1	k_P ($L \cdot g^{-1} \cdot h^{-1}$)		$(1.84 \pm 0.03) \cdot 10^{-2}$	
	$Y_{X/P}$ ($g \cdot g^{-1}$)		-	
	$Y_{S/P}$ ($g \cdot g^{-1}$)		1.80 ± 0.03	
	$Y_{BP/P}$ ($g \cdot g^{-1}$)		$(5.30 \pm 0.16) \cdot 10^{-1}$	
2	k_P ($L \cdot g^{-1} \cdot h^{-1}$)	-	$(5.36 \pm 0.14) \cdot 10^{-3}$	$(1.10 \pm 0.02) \cdot 10^{-2}$
	$Y_{X/P}$ ($g \cdot g^{-1}$)	-	-	-
	$Y_{S/P}$ ($g \cdot g^{-1}$)	-	2.08 ± 0.06	1.74 ± 0.04
	$Y_{BP/P}$ ($g \cdot g^{-1}$)	-	$(9.17 \pm 2.49) \cdot 10^{-2}$	$(1.64 \pm 0.02) \cdot 10^{-1}$
3	k_P ($L \cdot g^{-1} \cdot h^{-1}$)	-	$(9.11 \pm 0.57) \cdot 10^{-4}$	$(4.43 \pm 0.10) \cdot 10^{-3}$
	$Y_{X/P}$ ($g \cdot g^{-1}$)	-	-	-
	$Y_{S/P}$ ($g \cdot g^{-1}$)	-	3.80 ± 0.25	$(9.99 \pm 0.29) \cdot 10^{-1}$
	$Y_{BP/P}$ ($g \cdot g^{-1}$)	-	$(6.39 \pm 0.74) \cdot 10^{-1}$	$(1.24 \pm 0.21) \cdot 10^{-1}$
Cycle	**Statistical Parameter**	**Operational Mode**		
		Batch	**Repeated Batch**	**Fed-Batch**
1	F		3512	
	RMSE		0.61	
	SSR		57.1	
	VE (%)		96.8	
2	F	-	5090	19,278
	RMSE	-	0.49	0.28
	SSR	-	20.4	4.54
	VE (%)	-	94.8	98.5
3	F	-	25,099	20,931
	RMSE	-	0.27	0.36
	SSR	-	6.09	7.15
	VE (%)	-	90.7	96.5

The influence of biocatalyst state and reutilisation cycles on the kinetic parameters values is shown in Figure 5. The second-order kinetic parameter k_P decreases along the cycles of reutilisation, independently of biocatalyst state and operational mode. These results point at a biocatalyst deactivation throughout the fermentation cycles. However, the deactivation of growing cells under repeated batch takes place slower than it does under other conditions. Regarding yields (substrate from product, by-product from product and biomass from product), smaller values indicate an effective SA production bioprocess, as it requires less substrate and biomass for product generation and minimises side products generation. As it is reflected on yield tendencies, biocatalyst composed of resting cells is more specific for SA production, especially in the second and third uses.

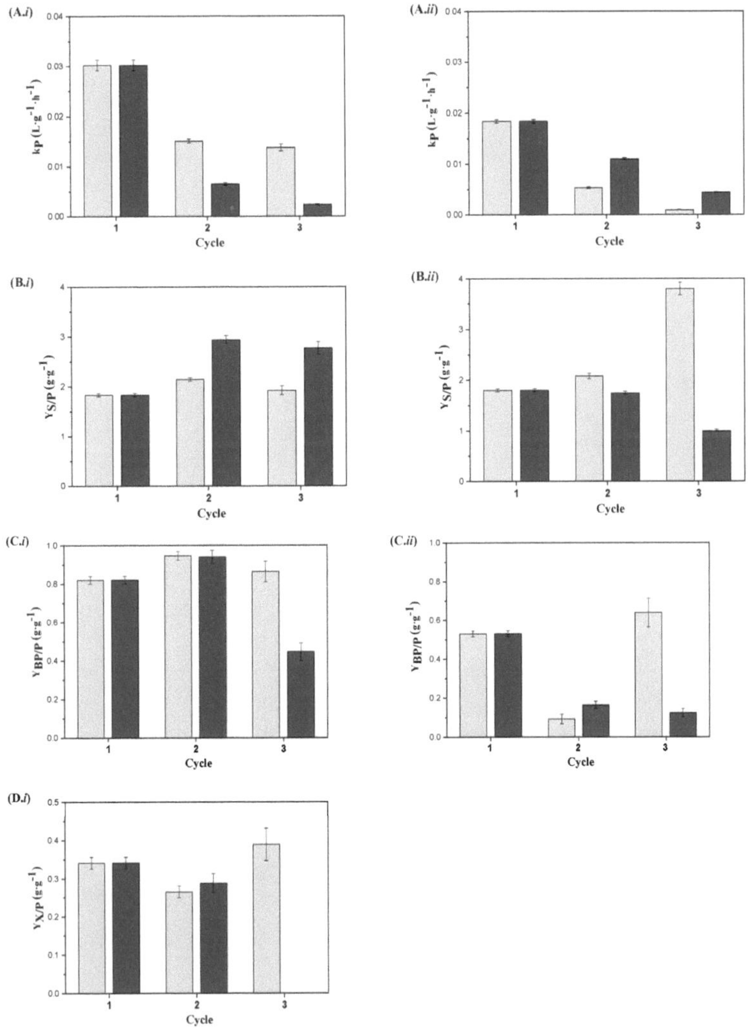

Figure 5. Influence of reusing (*i*) growing cells and (*ii*) resting cells as the biocatalyst under different operational mode on (**A**) second-order kinetic parameter of product formation, (**B**) pseudo-empirical stoichiometric coefficient substrate from product, (**C**) pseudo-empirical stoichiometric coefficient by-product from product and (**D**) pseudo-empirical stoichiometric coefficient biomass from product. *Key:* light grey: repeated batch, dark grey: fed batch.

4. Conclusions

In the present study, a bioprocess for SA production using a synthetic growth medium and xylose as the carbon source was analysed, which proved to offer economic advantages over a medium based on complex carbon and nitrogen sources. Furthermore, for the first time, the reutilisation of the biocatalyst composed by *A. succinogenes* cells under growing and resting states was explored under diverse operation modes. Interestingly, whereas the highest average SA yield and productivity in repeated batch mode were observed with growing cells, the best results in fed-batch mode were reached with resting cells, which had limited nutrients availability. In fact, in fed-batch mode, the yield of SA with respect to xylose consumed was 66% higher with resting cells than with growing cells. Furthermore, the selectivity was 37% higher. This means that, for the first time, it was possible to produce succinic acid during three cycles of xylose feeding in the absence of a nitrogen source from a single stage of cell growth, also achieving yields and selectivities superior to the equivalent operation with growing cells, in other words, in a culture medium with nitrogen source.

Carrying out the bioprocess with resting cells presents several biological and technical advantages: (i) metabolic pathways towards the target product are promoted, as indicated by the reduction in by-product generation; (ii) the culture medium only contains phosphate and sugar, eliminating the use of expensive nitrogen sources; and (iii) SA is accumulated in the broth during the cycles, resulting in a higher final concentration compared to using a growing cells biocatalyst. All these factors contribute to cost reduction in subsequent purification and isolation operations. Thus, the possibility of producing SA with *A. succinogenes* in resting state becomes an innovative option with great potential and opens up new opportunities for bioprocess intensification (e.g., biocatalyst immobilisation).

Finally, a simple non-structured non-segregated kinetic model was successfully developed to describe cell growth, SA production and by-product generation. The estimated kinetic parameters are supported by the observed experimental trends and provide valuable information for a later bioprocess scale-up. In short, this research affords key progresses towards the incorporation of SA in biorefineries on an industrial scale.

Author Contributions: Conceptualisation, V.E.S. and M.L.; methodology, I.A.E.; software, I.A.E. and V.R.; validation, V.E.S. and M.L.; formal analysis, I.A.E., V.R., V.E.S. and M.L.; investigation, I.A.E.; resources, V.E.S. and M.L.; data curation, I.A.E. and V.R.; writing—original draft preparation, I.A.E. and V.R.; writing—review and editing, V.E.S. and M.L.; supervision, V.E.S. and M.L.; project administration, V.E.S. and M.L.; funding acquisition, V.E.S. and M.L. All authors have read and agreed to the published version of the manuscript.

Funding: This work was kindly supported by the Spanish Science and Innovation Ministry though three research projects: CTQ-2013-45970-C2-1-R, CTQ2017-84963-C2-1-R and PID2020-114365RB-C21, funding that is gratefully acknowledged.

Institutional Review Board Statement: Not applicable.

Informed Consent Statement: Not applicable.

Data Availability Statement: All data will be provided to any interested part upon its requesting.

Conflicts of Interest: The authors declare no conflict of interest.

References

1. Singh, N.; Singhania, R.R.; Nigam, P.S.; Dong, C.D.; Patel, A.K.; Puri, M. Global Status of Lignocellulosic Biorefinery: Challenges and Perspectives. *Bioresour. Technol.* **2022**, *344*, 126415. [CrossRef] [PubMed]
2. Patel, A.; Shah, A.R. Integrated Lignocellulosic Biorefinery: Gateway for Production of Second Generation Ethanol and Value Added Products. *J. Bioresour. Bioprod.* **2021**, *6*, 108–128. [CrossRef]
3. Usmani, Z.; Sharma, M.; Awasthi, A.K.; Lukk, T.; Tuohy, M.G.; Gong, L.; Nguyen-Tri, P.; Goddard, A.D.; Bill, R.M.; Nayak, S.C.; et al. Lignocellulosic Biorefineries: The Current State of Challenges and Strategies for Efficient Commercialization. *Renew. Sustain. Energy Rev.* **2021**, *148*, 111258. [CrossRef]

4. Narisetty, V.; Cox, R.; Bommareddy, R.; Agrawal, D.; Ahmad, E.; Pant, K.K.; Chandel, A.K.; Bhatia, S.K.; Kumar, D.; Binod, P.; et al. Valorisation of Xylose to Renewable Fuels and Chemicals, an Essential Step in Augmenting the Commercial Viability of Lignocellulosic Biorefineries. *Sustain. Energy Fuels* **2022**, *6*, 29–65. [CrossRef] [PubMed]
5. Yang, Q.; Wu, M.; Dai, Z.; Xin, F.; Zhou, J.; Dong, W.; Ma, J.; Jiang, M.; Zhang, W. Comprehensive Investigation of Succinic Acid Production by *Actinobacillus succinogenes*: A Promising Native Succinic Acid Producer. Biofuels, Bioprod. *Biorefining* **2020**, *14*, 950–964. [CrossRef]
6. Dessie, W.; Wang, Z.; Luo, X.; Wang, M.; Qin, Z. Insights on the Advancements of in Silico Metabolic Studies of Succinic Acid Producing Microorganisms: A Review with Emphasis on *Actinobacillus Succinogenes*. *Fermentation* **2021**, *7*, 220. [CrossRef]
7. Bikiaris, R.D.; Ainali, N.M.; Christodoulou, E.; Nikolaidis, N.; Lambropoulou, D.A.; Papageorgiou, G.Z. Thermal Stability and Decomposition Mechanism of Poly(Alkylene Succinate)S. *Macromol* **2022**, *2*, 58–77. [CrossRef]
8. Dai, Z.; Guo, F.; Zhang, S.; Zhang, W.; Yang, Q.; Dong, W.; Jiang, M.; Ma, J.; Xin, F. Bio-based Succinic Acid: An Overview of Strain Development, Substrate Utilization, and Downstream Purification. Biofuels, Bioprod. *Biorefining* **2019**, *14*, 965–985. [CrossRef]
9. Dienst, S.; Onderzoek, L. Strategic Thinking in Sustainable Energy From the Sugar Platform to Biofuels and Biochemicals. Final Report for the European Commission Directorate-General Energy. Consorzio per La Ricerca e La Dimostrazione Sulle Energie Rinnovabili (RE-CORD); Report N° ENER/C2/423-2012/SI2.673791. April 2015. Available online: https://www.researchgate.net/publication/279188465_From_the_Sugar_Platform_to_biofuels_and_biochemicals (accessed on 5 July 2023).
10. Mancini, E.; Dickson, R.; Fabbri, S.; Udugama, I.A.; Ullah, H.I.; Vishwanath, S.; Gernaey, K.V.; Luo, J.; Pinelo, M.; Mansouri, S.S. Economic and Environmental Analysis of Bio-Succinic Acid Production: From Established Processes to a New Continuous Fermentation Approach with in-Situ Electrolytic Extraction. *Chem. Eng. Res. Des.* **2022**, *179*, 401–414. [CrossRef]
11. Jegatheesan, J.; Shu, L.; Lens, P.; Chiemchaisri, C. (Eds.) *Applied Environmental Science and Engineering for a Sustainable Future*; Springer: Berlin/Heidelberg, Germany, 2020; ISSN 2570-2173.
12. Dickson, R.; Mancini, E.; Garg, N.; Woodley, J.M.; Gernaey, K.V.; Pinelo, M.; Liu, J.; Mansouri, S.S. Sustainable Bio-Succinic Acid Production: Superstructure Optimization, Techno-Economic, and Lifecycle Assessment. *Energy Environ. Sci.* **2021**, *14*, 3542–3558. [CrossRef]
13. Ferone, M.; Raganati, F.; Olivieri, G.; Salatino, P.; Marzocchella, A. Biosuccinic Acid from Lignocellulosic-Based Hexoses and Pentoses by *Actinobacillus succinogenes*: Characterization of the Conversion Process. *Appl. Biochem. Biotechnol.* **2017**, *183*, 1465–1477. [CrossRef] [PubMed]
14. Bukhari, N.A.; Jahim, J.M.; Loh, S.K.; Nasrin, A.B.; Harun, S.; Abdul, P.M. Organic Acid Pretreatment of Oil Palm Trunk Biomass for Succinic Acid Production. *Waste Biomass Valorization* **2020**, *11*, 5549–5559. [CrossRef]
15. Bradfield, M.F.A.; Nicol, W. Continuous Succinic Acid Production from Xylose by *Actinobacillus succinogenes*. *Bioprocess Biosyst. Eng.* **2016**, *39*, 233–244. [CrossRef]
16. Oreoluwa Jokodola, E.; Narisetty, V.; Castro, E.; Durgapal, S.; Coulon, F.; Sindhu, R.; Binod, P.; Rajesh Banu, J.; Kumar, G.; Kumar, V. Process Optimisation for Production and Recovery of Succinic Acid Using Xylose-Rich Hydrolysates by *Actinobacillus succinogenes*. *Bioresour. Technol.* **2022**, *344*, 126224. [CrossRef]
17. Escanciano, I.A.; Ladero, M.; Santos, V.E. On the Succinic Acid Production from Xylose by Growing and Resting Cells of *Actinobacillus succinogenes:* A Comparison. *Biomass Conv. Bioref.* **2022**. [CrossRef]
18. Ripoll, V.; Ladero, M.; Santos, V.E. Kinetic Modelling of 2,3-Butanediol Production by *Raoultella terrigena* CECT 4519 Resting Cells: Effect of Fluid Dynamics Conditions and Initial Glycerol Concentration. *Biochem. Eng. J.* **2021**, *176*, 108185. [CrossRef]
19. de la Torre, I.; Ladero, M.; Santos, V.E. D-Lactic Acid Production from Orange Waste Enzymatic Hydrolysates with *L. delbrueckii* Cells in Growing and Resting State. *Ind. Crops Prod.* **2020**, *146*, 112176. [CrossRef]
20. de la Morena, S.; Wojtusik, M.; Santos, V.E.; Garcia-Ochoa, F. Kinetic Modeling of Dihydroxyacetone Production from Glycerol by *Gluconobacter oxydans* ATCC 621 Resting Cells: Effect of Fluid Dynamics Conditions. *Catalysts* **2020**, *10*, 101. [CrossRef]
21. de la Torre, I.; Acedos, M.G.; Ladero, M.; Santos, V.E. On the Use of Resting *L. delbrueckii* Spp. *delbrueckii* Cells for D-Lactic Acid Production from Orange Peel Wastes Hydrolysates. *Biochem. Eng. J.* **2019**, *145*, 162–169. [CrossRef]
22. Bumyut, A.; Champreda, V.; Singhakant, C.; Kanchanasuta, S. Effects of Immobilization of *Actinobacillus succinogenes* on Efficiency of Bio-Succinic Acid Production from Glycerol. *Biomass Convers. Bioref* **2020**, *12*, 643–654. [CrossRef]
23. Louasté, B.; Eloutassi, N. Succinic Acid Production from Whey and Lactose by *Actinobacillus succinogenes* 130Z in Batch Fermentation. *Biotechnol. Rep.* **2020**, *27*, 23–27. [CrossRef]
24. Kim, S.Y.; Park, S.O.; Yeon, J.Y.; Chun, G.T. Development of a Cell-Recycled Continuous Fermentation Process for Enhanced Production of Succinic Acid by High-Yielding Mutants of *Actinobacillus succinogenes*. *Biotechnol. Bioprocess Eng.* **2021**, *26*, 125–136. [CrossRef]
25. Jiang, M.; Dai, W.; Xi, Y.; Wu, M.; Kong, X.; Ma, J.; Zhang, M.; Chen, K.; Wei, P. Succinic Acid Production from Sucrose by *Actinobacillus succinogenes* NJ113. *Bioresour. Technol.* **2014**, *153*, 327–332. [CrossRef] [PubMed]
26. Ferone, M.; Raganati, F.; Olivieri, G.; Marzocchella, A. Bioreactors for Succinic Acid Production Processes. *Crit. Rev. Biotechnol.* **2019**, *39*, 571–586. [CrossRef] [PubMed]
27. Djukić-Vuković, A.; Mladenović, D.; Radosavljević, M.; Kocić-Tanackov, S.; Pejin, J.; Mojović, L. Wastes from Bioethanol and Beer Productions as Substrates for l(+) Lactic Acid Production—A Comparative Study. *Waste Manag.* **2016**, *48*, 478–482. [CrossRef] [PubMed]

28. Ramírez, N.; Ubilla, C.; Campos, J.; Valencia, F.; Aburto, C.; Vera, C.; Illanes, A.; Guerrero, C. Enzymatic Production of Lactulose by Fed-Batch and Repeated Fed-Batch Reactor. *Bioresour. Technol.* **2021**, *341*, 125769. [CrossRef]
29. Adnan, A.; Nair, G.; Lay, M.; Swan, J. Bacterial Cellulose Synthesis by *Gluconacetobacter xylinus*: Enhancement via Fed-Batch Fermentation Strategies in Glycerol Media. *Trends Sci.* **2021**, *18*, 18–25. [CrossRef]
30. Salma, A.; Djelal, H.; Abdallah, R.; Fourcade, F.; Amrane, A. Platform Molecule from Sustainable Raw Materials; Case Study Succinic Acid. *Braz. J. Chem. Eng.* **2021**, *38*, 215–239. [CrossRef]
31. Patsalou, M.; Chrysargyris, A.; Tzortzakis, N.; Koutinas, M. A Biorefinery for Conversion of Citrus Peel Waste into Essential Oils, Pectin, Fertilizer and Succinic Acid via Different Fermentation Strategies. *Waste Manag.* **2020**, *113*, 469–477. [CrossRef]
32. Almqvist, H.; Pateraki, C.; Alexandri, M.; Koutinas, A.; Lidén, G. Succinic Acid Production by *Actinobacillus succinogenes* from Batch Fermentation of Mixed Sugars. *J. Ind. Microbiol. Biotechnol.* **2016**, *43*, 1117–1130. [CrossRef]
33. Corona-González, R.I.; Varela-Almanza, K.M.; Arriola-Guevara, E.; de Jesús Martínez-Gómez, Á.; Pelayo-Ortiz, C.; Toriz, G. Bagasse Hydrolyzates from Agave Tequilana as Substrates for Succinic Acid Production by *Actinobacillus succinogenes* in Batch and Repeated Batch Reactor. *Bioresour. Technol.* **2016**, *205*, 15–23. [CrossRef] [PubMed]
34. Ercole, A.; Raganati, F.; Salatino, P.; Marzocchella, A. Continuous Succinic Acid Production by Immobilized Cells of *Actinobacillus succinogenes* in a Fluidized Bed Reactor: Entrapment in Alginate Beads. *Biochem. Eng. J.* **2021**, *169*, 107968. [CrossRef]

fermentation

MDPI

Article

Lactic Acid Production from Steam-Exploded Sugarcane Bagasse Using *Bacillus coagulans* DSM2314

William Rodrigues Alves [1], Thiago Alessandre da Silva [2], Arion Zandoná Filho [2] and Luiz Pereira Ramos [1,2,*]

1 Graduate Program in Chemistry, Federal University of Paraná, Curitiba 81531-980, Brazil; william.alves@ufpr.br

2 Graduate Program in Chemical Engineering, Federal University of Paraná, Curitiba 81531-980, Brazil; thiago.alessandre@ufpr.br (T.A.d.S.); a.zandona@ufpr.br (A.Z.F.)

* Correspondence: luiz.ramos@ufpr.br

Abstract: This work aimed at producing lactic acid (LA) from sugarcane bagasse after steam explosion at 195 °C for 7.5 and 15 min. Enzymatic hydrolysis was carried out with Cellic CTec3 and Cellic HTec3 (Novozymes), whereas fermentation was performed with *Bacillus coagulans* DSM2314. Water washing of pretreated solids before enzymatic hydrolysis improved both hydrolysis and fermentation yields. The presence of xylo-oligosaccharides (XOS) in substrate hydrolysates reduced hydrolysis efficiency, but their effect on fermentation was negligible. The presence of fermentation inhibitors in C5 streams was circumvented by adsorption on activated carbon powder with no detectable sugar losses. High carbohydrates-to-LA conversions ($Y_{p/s}$) of 0.88 $g{\cdot}g^{-1}$ were obtained from enzymatic hydrolysates of water-washed steam-exploded materials that were produced at 195 °C, in 7.5 min, and the use of centrifuged-but-never-washed pretreated solids decreased $Y_{p/s}$ by 16%. However, when the detoxified C5 stream was added at a 10% ratio, $Y_{p/s}$ was raised to 0.93 $g{\cdot}g^{-1}$ for an LA productivity of 2.55 $g{\cdot}L^{-1}{\cdot}h^{-1}$. Doubling the pretreatment time caused a decrease in $Y_{p/s}$ to 0.78 $g{\cdot}g^{-1}$, but LA productivity was the highest (3.20 $g{\cdot}L^{-1}{\cdot}h^{-1}$). For pretreatment at 195 °C for 7.5 min, the elimination of water washing seemed feasible, but the use of longer pretreatment times made it mandatory to eliminate fermentation inhibitors.

Keywords: sugarcane bagasse; steam explosion; enzymatic hydrolysis; fermentation; lactic acid

Citation: Alves, W.R.; da Silva, T.A.; Zandoná Filho, A.; Pereira Ramos, L. Lactic Acid Production from Steam-Exploded Sugarcane Bagasse Using *Bacillus coagulans* DSM2314. *Fermentation* **2023**, *9*, 789. https://doi.org/10.3390/fermentation9090789

Academic Editors: Miguel Ladero and Victoria E. Santos

Received: 21 July 2023
Revised: 21 August 2023
Accepted: 24 August 2023
Published: 26 August 2023

1. Introduction

Lactic acid is a valuable chemical platform with applications in different industrial sectors such as food, cosmetics, textiles, pharmaceuticals, and chemical synthesis [1]. The global lactic acid market increased from 1220 kilotons in 2016 to 1960 kilotons in 2025. This represents a revenue of USD 11.51 billion globally [2,3]. The production of polylactic acid (PLA), a biodegradable polymer, accounts for about 50% of the lactic acid demand [4]. The other half is mainly used as an acidulant, preservative, flavoring, emulsifier, and pH regulator in the food industry [5].

Lactic acid has two types of enantiomers (L or D). The pure enantiomers have greater value than the racemic mixture because they are used for special industrial applications. For instance, the L isomer is preferable for food, beverages, and pharmaceuticals because it is metabolized more rapidly by the human body than the D isomer. L-Lactic acid is used to synthesize poly-L-lactic acid (PLLA), while D-lactic acid is used to produce poly-D-lactic acid (PDLA). Both are semi-crystalline bioplastics, while PDLLA, made with the racemic mixture, is amorphous and relatively easy to break down, ideal for developing drug delivery systems. However, due to its biocompatibility and high mechanical strength, the L isomer predominates in biomedical applications, including bone fixation supports and biodegradable sutures [6]. Food packaging, injection molding, and additive manufacturing (resins for 3D printing) are other applications in which PLA has been widely used [7].

Optically pure lactic acid can be produced by fermentation, while chemical synthesis leads to racemic mixtures [8]. Fermentation is also advantageous because it uses renewable resources and mild process conditions [9]. Several bacteria, fungi, and yeasts can produce high optical purity lactic acid in high yields [10]. However, lactic acid bacteria (LAB) such as *Lactobacillus delbruckii*, *Lb. rhamnosus*, *Lb. casei*, and *Lb. plantarum*, and bioengineered bacteria, such as *Escherichia coli* and *Corynebacterium* spp., are the most widely used for lactic acid fermentation [2,11]. The process is performed by submerged fermentation, and the substrate accounts for almost 70% of the production cost. Starch is the predominant raw material for industrial manufacturing, with about 90% of globally traded lactic acid being produced from corn [12]. Thus, identifying cheaper and widely available substrates is pivotal to reduce process costs [13].

Agro-industrial residues such as sugarcane bagasse (SCB) are lignocellulosic materials with a high potential to reduce industrial costs. However, before fermentation, plant cell wall polysaccharides such as cellulose and hemicelluloses must be converted to simple sugars (primarily glucose and xylose). Pretreatment techniques such as hydrothermolysis, steam explosion, acid-catalyzed organosolv, and dilute acid hydrolysis can provide high yields of fermentable sugars in the form of C5 and C6 streams [14]. C5 sugars are obtained in pretreatment acid hydrolysates, while C6 sugars are derived from enzymatic hydrolysis of pretreated cellulosic materials. However, the use of high pretreatment temperatures (>200 °C) or long residence times result in partial dehydration of pentoses and hexoses, causing the release of fermentation inhibitors such as furfural and 5-(hydroxymethyl)furfural (5-HMF) that reduce process yields [15,16]. Such aromatic aldehydes are known to inhibit key enzymes of microbial carbon metabolism [17]. Other inhibitors such as low molar mass phenolic compounds may also be released from lignin. Also, mild pretreatment severities are enough to release inhibitory acetic acid (pKa = 4.76) from hemicellulose *O*-acetyl groups [16]. Weak acids diffuse through the cell membrane and lower the intracellular pH, affecting cell growth due to their effect on the proton transport activity of the plasma membrane [18,19].

Steam explosion uses saturated steam at high pressures to produce pretreated cellulosic materials with high accessibility to enzymatic hydrolysis [20]. While enzymatic hydrolysates of water-washed steam-exploded materials are easy to ferment, the C5 fraction typically contains inhibitory concentrations of organic acids (primarily acetic) and furan compounds (mostly furfural) [21]. Oligosaccharides released from hemicelluloses can also act as inhibitors for enzymatic hydrolysis [22], and in both situations, the release of inhibitory compounds will largely depend on pretreatment conditions and feedstock composition. High pretreatment severities will release more fermentation inhibitors by carbohydrate dehydration and lignin hydrolysis. At the same time, oligosaccharides will prevail at low severities, particularly when pretreatment is carried out without an exogenous acid catalyst [23].

Several detoxification techniques have been used to reduce the inhibitory effect of biomass acid hydrolysates. Furans can be removed by physical adsorption [24], liquid–liquid extraction [25], evaporation [26], freeze-drying [27], enzymatic treatments using laccases and other oxidative enzymes [28], or overliming [29] while reducing acetic acid to non-inhibitory concentrations may be more problematic. The most widely used detoxification techniques are adsorption on activated carbon powder and overliming. However, their efficiency depends on the type and concentration of fermentation inhibitors released in pretreatment liquors [30]. Physical adsorption was chosen in this work for simplicity, efficiency, and selectivity toward fermentation inhibitors. Furthermore, physical adsorption would not dilute the sugar stream while bringing these chemicals to non-inhibitory levels.

Bacillus sp. strains are more tolerant to inhibitory compounds. In a hydrolysate broth containing 4.01 $g{\cdot}L^{-1}$ acetic acid, 0.08 $g{\cdot}L^{-1}$ formic acid, 0.05 $g{\cdot}L^{-1}$ furfural, and 0.08 $g{\cdot}L^{-1}$ 5-HMF, *B. coagulans* IPE22 converted 96% of sugars into LA [31]. Also, *B. coagulans* JI12 could tolerate up to 20 $g{\cdot}L^{-1}$ acetic acid and 4 $g{\cdot}L^{-1}$ furfural by metabolizing it to 2-furoic acid [32], while *Bacillus* sp. P38 was tolerant to 10 $g{\cdot}L^{-1}$ furfural and 6 $g{\cdot}L^{-1}$ vanillin or acetic acid [33]. This indicates that *Bacillus* spp. may be promising organisms to produce L-LA from biomass hydrolysate without a robust detoxification step. No information was found in the literature about the tolerance and inhibitory levels of *B. coagulans* DSM2314 to the organic acids and furan compounds listed above.

Second-generation lactic acid can be produced by C5 plus C6 fermentation or by co-fermentation of C5/C6 mixtures. For acid pretreatments such as steam explosion, C6 sugars are mainly produced from enzymatic hydrolysis of water-washed pretreated materials. By contrast, C5 sugars are recovered in pretreatment liquors (C5 streams) that must be detoxified before fermentation. *B. coagulans* has become one of the most popular organisms due to its capacity to metabolize C5 sugars via the pentoses phosphate (PP) pathway and produce optically pure L-LA with high yields [34]. Enzymatic hydrolysis and fermentation can be performed separately or simultaneously. Based on this, different bioprocessing strategies have been designed to produce biobased materials such as separate hydrolysis and fermentation (SHF), separate hydrolysis and co-fermentation (SHCF), simultaneous saccharification and fermentation (SSF), and simultaneous saccharification and co-fermentation (SSCF). SHF and SHCF involve enzymatic hydrolysis of the polysaccharides and subsequent fermentation of the sugars released. By contrast, SSF and SSCF are one-pot methods where enzymatic hydrolysis and microbial fermentation occur simultaneously. Combining these operations results in lower capital cost and higher productivity since enzymes perform better due to lower levels of end-product inhibition and sugars are released and readily consumed [33–40].

Michelson et al. [41] compared the performance and yield of two LA producers, *Lb. delbrueckii* ssp. *lactis* DSM 20,073 and *B. coagulans* SIM-7. The latter strain achieved final LA concentrations of 91.5 $g{\cdot}L^{-1}$ and 91.6 $g{\cdot}L^{-1}$ in batch and fed-batch cultivations for 23 and 21 h, respectively. The LA concentration in 10 h was already 56 $g{\cdot}L^{-1}$, whereas comparable results (52 $g{\cdot}L^{-1}$) were achieved only in 24 h by DSM 20073. The maximal production rates of SIM-7 and DSM 20,073 strains were 9.9 and 5.6 $g{\cdot}L^{-1}{\cdot}h^{-1}$, respectively.

Different enzymatic hydrolysis and fermentation conditions were used in this work to produce LA from steam-exploded SCB. Hydrolysis was performed with Cellic CTec3 (Novozymes, Bagsværd, Denmark) cellulases in the absence and presence of Cellic CTec3 hemicellulases at relatively high total solids (TS), using water-washed and centrifuged-but-never-washed steam-exploded materials. Fermentation inhibitors were removed from C5 streams using physical adsorption on activated carbon powder, while fermentation was carried out with *B. coagulans* DSM2314 using both SHF and SHCF protocols.

2. Materials and Methods

The overview of the experimental setup is given in Figure 1, in which the complexity and interrelationship of the main activities (chemical characterization, pretreatment, hydrolysis, and fermentation) are observed. Also, the sequence in which the experiments are performed is inferred by step connectors, while red and blue circles indicate processes of intermediates and final products, respectively. Further details about the experimental setup are given below.

2.1. Sugarcane Bagasse (SCB) Pretreatment and Characterization

Fresh SCB was kindly donated by Raízen (Piracicaba, SP, Brazil). Pretreatment was carried out by steam explosion at 195 °C for 7.5 and 15 min using a 10 L stainless-steel high-pressure steam reactor and SCB with a moisture content of 50 wt% [27]. Pretreatment slurries (20–25 wt% total solids) were centrifuged inside a cotton fiber bag to remove water-soluble hemicellulose sugars and low molar mass lignin components (C5 stream). Half of the unwashed centrifuged fiber cake (SEB7.5-UW and SEB15-UW) was reserved for enzymatic hydrolysis. The other half was water washed at 5 wt% TS for 1 h at room temperature (~25 °C) under constant mechanical stirring, followed by centrifugation to recover the water-washed fiber cake (SEB7.5-WW and SEB15-WW) for its subsequent characterization and enzymatic hydrolysis. Both SEB-UW and SEB-WW substrates, with

35–40 wt% total solids after centrifugation, were stored in vacuum-sealed plastic bags at 4 °C before chemical characterization and enzymatic hydrolysis.

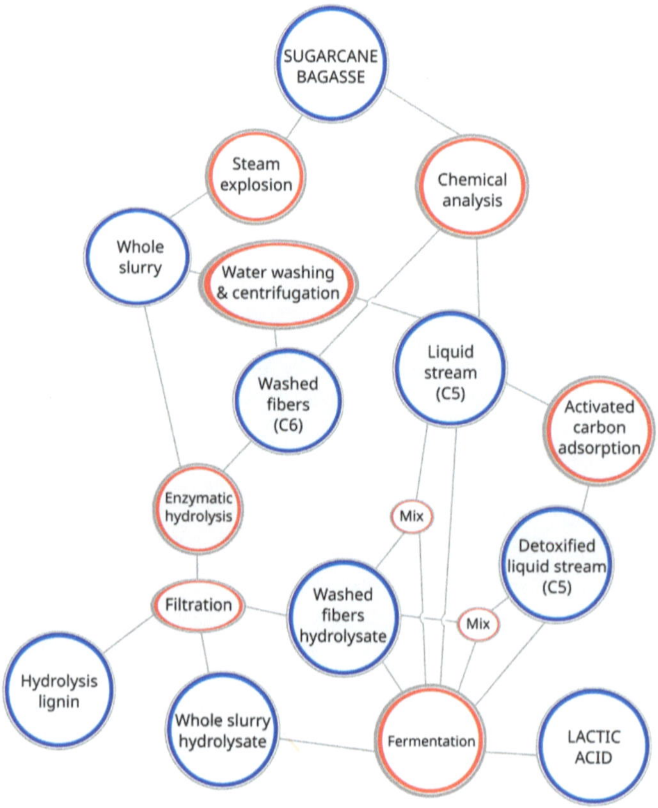

Figure 1. Experimental setup for lactic acid production from sugarcane bagasse.

The composition of untreated and pretreated materials was characterized following National Renewable Energy Laboratory (NREL, Golden, CO, USA) protocols for total moisture (drying at 105 °C until constant mass) [42], ash (calcination at 575 °C) [43], total extractives (exhaustive Soxhlet extractions with water and ethanol 95%) [44], and carbohydrates plus total lignin content (acid-soluble lignin and acid-insoluble lignin) following a two-stage sulfuric acid hydrolysis [45]. All reagents and solvents were obtained in analytical grade from Labsynth (Diadema, Brazil) and used as received. Chromatographic standards (>98% purity) were purchased from Sigma-Aldrich (St. Louis, MO, USA). Mass balances and process yields were calculated according to the above-mentioned standard procedures. To this end, theoretical conversion factors were considered to express recovery yields concerning each raw SCB macromolecular component.

The pretreatment liquor (C5 fraction) was detoxified over activated carbon [46]. Detoxification was performed in 250 mL Erlenmeyer flasks that were loaded with 100 mL of liquid and 10 g of activated carbon powder (Neon, Suzano, Brazil), having a surface area of 507.9 $m^2 \cdot g^{-1}$ and an average pore size of 1.29 nm. The flasks were covered with Parafilm M to prevent evaporation and placed on a shaker incubator at 25 °C and 120 rpm for 10 min. Then, the suspension was centrifuged at 2500 rpm for 15 min, and aliquots were removed from the supernatant, filtered through a 0.22 µm PVDF filter (Millipore, Burlington, MA, EUA), and analyzed by high-performance liquid chromatography (HPLC) to quantify furfural, 5-HMF, acetic acid, formic acid, xylose, glucose, and arabinose, using the chromatographic conditions that are described in Section 2.4. Then, the supernatant was passed through a 0.1-µm ash-less quantitative filter paper (Whatman® (Maidstone, UK)) to remove any remaining suspended solids. The detoxification process was repeated three times to ensure that acetic acid, furfural, and 5-HMF were brought to non-inhibitory concentrations.

2.2. Enzymatic Hydrolysis

The commercial enzymes used for hydrolysis were provided by Novozymes Latin America (Araucária, SP, Brazil). Cellic CTec3 is a commercial cellulase preparation while Cellic HTec3 contains hemicellulase activity predominantly [20]. Enzyme loading was always based on the wet weight of the commercial enzyme preparation that was added to the reaction system for enzymatic hydrolysis.

Enzymatic hydrolyses of SEB-UW and SEB-WW substrates were performed at 50 °C and 150 rpm for 96 h in acetate buffer (50 mmol·L^{-1}, pH 5.2) using 250 mL Erlenmeyer flasks in a shaker incubator (Ecotron, Infors HT). The best condition was derived from a factorial design that was based on three independent variables in two levels (2^3): substrate TS (10 and 20 wt%), Cellic CTec3 loading (20 to 60 mg g^{-1} TS), and Cellic HTec3 addition in a 10% mass ratio (wet basis) to Cellic CTec3 (2 to 6 mg·g^{-1} TS) [20]. Four quadratic polynomial equations were obtained using the R Studio® 3.4.3 software to describe the mathematical relationship between glucose release (g·L^{-1}) and the selected process variables. The goodness-of-fit of the models was evaluated by determining their adjusted R^2. For yield calculations, aliquots were collected at different incubation times and analyzed using HPLC to quantify cellobiose, glucose, and xylose using the chromatographic conditions described in Section 2.4. Hydrolysis yields were determined in percentage by expressing the total glucose release (glucose equivalents) in relation to the total glucose content (quantified as glucans) of the pretreated solids. Xylose was not considered in yield calculations because it was always found in very low quantities.

Enough substrate hydrolysate (C6 stream) for fermentation was obtained by performing the best hydrolysis conditions from the factorial design in a 3.6 L Labfors bioreactors (Infors HT, Bottmingen-Basel, Switzerland). Enzymatic hydrolyses of SEB-UW and SEB-WW substrates were performed at 50 °C and 150 rpm for 72 h in acetate buffer (50 mmol·L^{-1}, pH 5.2) using 20 wt% TS and 60 mg g^{-1} TS of Cellic CTec3, with and without addition of Cellic HTec3 (6 mg·g^{-1} TS). The total volume of this reaction system was 1000 mL. Aliquots were collected once again at different incubation times and analyzed using HPLC, and hydrolysis yields were determined as described above.

2.3. Microorganism and Fermentation

B. coagulans DSM2314 was acquired as a freeze-dried stock from the Germany Collection of Microorganisms and Cell Cultures (DSMZ, Leibniz Institute, Germany). Cells were grown on Man, Rogosa, and Sharpe (MRS) agar medium (HiMedia, Mumbai, India) and transferred to 50 mL flasks of MRS medium to be cultured for 16 h at 50 °C. The media were pre-sterilized for 15 min at 121 °C. When the optical density measured at 660 nm reached two, the pre-culture was added as inoculum to the fermentation, which was carried out at 50 °C and 150 rpm for 24 h in Multifors 2 bioreactor (Infors HT) that were pre-sterilized empty for 20 min at 121 °C.

Both SHF and SHCF fermentation experiments were carried out in duplicate in the Multifors 2 bioreactor using a working volume of 300 mL for a total volumetric capacity of 500 mL. SCB hydrolysates (C5, C6 and C5/C6 mixtures) were transferred to the bioreactor vessel and mixed with 1% yeast extract (Kasvi®, Conda Laboratories, Madri, Spain) and 10% (v/v) inoculum (30 mL). Temperature and agitation were set at 50 °C and 150 rpm, respectively, and dilute NaOH (5 mol L^{-1}) was used to maintain the fermentation broth at pH 6.0 during the entire reaction course. Fermentation ran for 24 h under anaerobic conditions using continuous N_2 purging. Aliquots were obtained at different times and analyzed using HPLC for carbohydrates and LA as described below. LA yields were determined as percentage in relation to the theoretical amount of LA that could have been produced from the fermentable sugars available in substrate hydrolysates.

2.4. Chromatographic Analysis

C5 (from pretreatment) and C6 (from enzymatic hydrolysis) streams and fermentations broths were analyzed at 65 °C using a Shimadzu HPLC, LC-20AD series, and a Rezex RHM column (Phenomenex, 300 × 7.8 mm) that was preceded by a Carbo H guard column (300 × 7.8 mm). The column was eluted with 5 mmol L^{-1} H_2SO_4 at a flow rate of 0.6 mL min^{-1}. Sample injection (20 μL) was performed using a Shimadzu SIL-10AF autosampler. Quantitative analyses were carried out by external calibration using differential refractometry (Shimadzu RID-10A) for carbohydrates and organic acids, while UV spectrophotometry (Shimadzu SPD-M10AVP) at 280 nm was used to quantify furfural and 5-HMF. HPLC calibration curves were based on analyzing six independent primary standard solutions, and the corresponding linear regression coefficients (R^2) were always around 0.99.

2.5. Statistical Analysis

The Tukey's Test ($p \leq 0.05$) was applied to evaluate the statistical significance of the experimental data, and the experimental design was validated with analysis of variance (ANOVA) using the R Studio® 3.4.3 software [47]. Hydrolysis and fermentation yields were expressed as averages with their corresponding standard deviations for experiments carried out in two or three replicates.

3. Results

3.1. SCB Pretreatment and Characterization

The chemical composition of SCB before and after pretreatment is shown in Table 1. The untreated material had glucans (mainly cellulose), hemicelluloses (mostly xylans), total lignin, total extractives, and ash contents like those already reported elsewhere [20,21,27,31]. Variations in SCB chemical composition are attributed to its source and maturation stage upon harvesting, as well as the edaphoclimatic conditions used for cultivation and the technology used for its industrial processing [48].

Table 1. Chemical composition (%) of untreated and water-insoluble steam-exploded SCB and the corresponding mass recovery (%) of the main SCB components after pretreatment.

Component	Untreated	195 °C, 7.5 min		195 °C, 15 min	
		Content	Recovery	Content	Recovery
Glucans [1]	37.8 ± 0.7	54.7 ± 2.4	89.7 ± 1.8	55.5 ± 0.7	82.5 ± 1.1
Xylans [2]	21.0 ± 0.8	3.0 ± 0.4	5.9 ± 0.1	1.2 ± 0.1	3.8 ± 0.1
Arabinosyl residues [2]	1.3 ± 0.3	bdl [3]	-	-	-
Acetyl groups [2]	2.6 ± 0.1	bdl [3]	-	bdl [3]	-
Hexoses identified as 5-HMF [4]	1.1 ± 0.2	bdl [3]	-	bdl [3]	-
Pentoses identified as furfural [4]	0.6 ± 0.1	bdl [3]	-	bdl [3]	-
Total lignin [5]	22.8 ± 0.7	30.5 ± 0.8	92.2 ± 1.9	31.2 ± 0.2	96.6 ± 0.3
Acid-soluble lignin	5.0 ± 0.1	5.7 ± 0.2	70.4 ± 1.5	5.9 ± 0.2	78.5 ± 0.2
Acid-insoluble lignin	19.1 ± 0.6	27.8 ± 0.8	96.3 ± 0.9	28.9 ± 0.1	101.4 ± 0.3
Ash	4.0 ± 0.1	6.1 ± 1.5	93.7 ± 1.9	6.8 ± 1.1	113.8 ± 0.3
Total	98.2	94.3		94.7	

[1] Present as β-(1→4)-D-glucans (cellulose). [2] Present as heteroxylan components (hemicelluloses). [3] bdl, below the detection limit of the method. [4] 5-HMF and furfural are released in pretreatment liquors by hexose and pentose dehydration, respectively. [5] Summation of acid-soluble and ash-free acid-insoluble lignin.

Steam explosion at 195 °C for 7.5 and 15 min reduced the SCB hemicellulose content by 85.7 and 94.3%, respectively, with a corresponding rise in both glucans and total lignin content (Table 1) [49]. Hemicelluloses were almost entirely depleted of their arabinosyl residues and acetyl groups because HPLC did not detect arabinose and acetic acid in sulfuric acid hydrolysates. Pentoses and hexoses were partly detected as furfural and 5-HMF due to dehydration, but in both situations, the reported values were not added to the corresponding polysaccharide quantification because their actual source was not elucidated. Hence, SCB hemicellulose content would be the summation of xylans, arabynosyl residues, and pentoses identified as furfural and acetyl groups, totaling 25.5%. Likewise, the total glucan content in Table 2 should be estimated at 38.9%, even though some 5-HMF may have come from hemicelluloses as well. The 5-HMF formation was higher in sulfuric acid hydrolysates because furfural was partially involved in side-reactions producing humins, while furans were not formed after acid hydrolysis of steam-treated materials because their hemicellulose content was very low [50,51].

Table 2. Chemical composition of pretreatment liquors before and after detoxification.

Component ($g{\cdot}L^{-1}$)	SEB7.5 [1]		SEB15 [1]	
	Untreated	**Detoxified**	**Untreated**	**Detoxified**
Glucose	2.2 ± 0.1	2.2 ± 0.1	2.2 ± 0.1	2.1 ± 0.1
Xylose	11.2 ± 0.2	11.1 ± 0.2	4.3 ± 0.1	4.3 ± 0.2
Arabinose	0.6 ± 0.1	0.60 ± 0.04	0.5 ± 0.1	0.4 ± 0.1
HMF [2]	1.6 ± 0.2	0.02 ± 0.02	1.6 ± 0.1	0.03 ± 0.01
Furfural [2]	0.8 ± 0.1	0.04 ± 0.02	1.4 ± 0.3	0.03 ± 0.01
Acetic acid [3]	9.7 ± 0.3	4.4 ± 0.5	12.1 ± 0.5	5.2 ± 0.2

[1] SEB7.5 and SEB15, steam explosion of cane bagasse (SEB) at 195 °C for 7.5 and 15 min; [2] 5-HMF and furfural are released in pretreatment liquors by hexose and pentose dehydration; [3] Acetic acid is released in pretreatment liquors by hemicellulose deacetylation.

Table 1 reveals the total recovery of SCB components in water-washed steam-exploded materials, in which the mass recovery of pretreated solids was also considered. Glucan recovery was around 90% when steam explosion was carried out at 195 °C for 7.5 min but doubling the pretreatment time to 15 min decreased this value by roughly 9%. The hemicellulose content (mostly xylans) in steam-exploded materials was very low and some lignin condensation may have occurred mainly at the highest pretreatment severity, in which lignin recovery was above 100%. Finally, ash recoveries above 100% may have indicated partial corrosion or abrasion of reactor walls [21,27].

3.2. *Chemical Composition of Pretreatment Liquors before and after Detoxification*

Pretreatment liquors were characterized using HPLC for their carbohydrate, acetic acid, 5-HMF, and furfural contents before and after detoxification with activated carbon powder (Table 2). Post-hydrolysis of these fractions with dilute sulfuric acid revealed xylose concentrations of 13.2 ± 0.4 and 10.5 ± 0.5 $g{\cdot}L^{-1}$ for pretreatments carried out for 7.5 and 15 min, respectively. The presence of acetic acid in pretreatment liquors was due to hemicellulose deacetylation, and its concentration was higher when pretreatment was carried out at more drastic conditions ($p < 0.05$). Acetic acid coming from *O*-acetyl groups is partially responsible for the auto-hydrolysis effect, which converts hemicelluloses to mono and oligosaccharides even without an exogenous acid catalyst. Furfural and 5-HMF derived from pentose and hexose dehydration, respectively, were also detected in pretreatment liquors in concentrations that increased with pretreatment severity. Accumulating organic acids, furans, and phenolic compounds in pretreatment liquors is undesirable due to their inhibitory effect on hydrolysis and fermentation [52,53].

Controlling microbial inhibition is essential for maximizing biomass conversion. Du et al. [54] identified and quantified 40 potentially toxic compounds after pretreating three different lignocellulosic materials with eight different pretreatment techniques. Fockink et al. [20] demonstrated that the autocatalytic steam explosion of SCB at 205 °C induced the accumulation of formic, levulinic, and acetic acid in the C5 fraction. Concerning furan compounds, furfural was present in higher concentration, while the accumulation of

5-HMF and 5-methyl-furan (5-MF) was higher at the highest pretreatment temperatures. This clearly indicates that a detoxification stage must be carried out for pretreatment liquors in which the presence of fermentation inhibitors is unavoidable [20,21].

The data presented in Table 2 showed that, by treating SCB pretreatment liquors with activated carbon powder, furfural and 5-HMF concentrations were reduced by 97–98% ($p < 0.05$), regardless of conditions used for pretreatment. Acetic acid removals were 55.2 and 57.2% for pretreatments carried out at 195 °C for 7.5 and 15 min, respectively. By contrast, no sugar losses were observed after adsorption on activated carbon powder at ambient temperature.

In our work, detoxification with activated carbon was highly efficient at room temperature, while other studies required heating to be effective. Lu, Dong, and Yang [55] reported using 2% ($m \cdot v^{-1}$) commercial activated carbon to remove 80% and 87.9% of furfural and 5-HMF present in wood chips acid hydrolysates, respectively. Adsorption was performed at 90 °C for 30 min under constant stirring (150 rpm). Miura, Suzuki, and Aoyama [56] detoxified wood acid hydrolysates using adsorption on 10% ($m \cdot v^{-1}$) activated carbon powder at 30 °C for 1 h. Around 83% of furfural and other low molar mass phenolic compounds were removed, while carbohydrate and acetic acid concentrations remained practically unaltered. Better detoxification efficiencies may have been due to the better textural properties of the activated carbon used in our studies.

3.3. *Enzymatic Hydrolysis*

Enzymatic hydrolysis of unwashed (SEB-UW) and water-washed (SEB-WW) substrates was investigated using an experimental design that was based on the following variables: substrate TS, Cellic CTec3 loading, and Cellic HTec3 supplementation. The latter variable was introduced because pretreatment liquors contained oligosaccharides that are known to inhibit total cellulase activity [57,58]. The results obtained after 96 h of hydrolysis were subjected to multiple linear regression analyses to generate mathematical models that could describe trends to the response function, which corresponded to the release of glucose equivalents in the reaction environment (in $g \cdot L^{-1}$). This was the only response function treated statistically because of its relevance for LA production since high concentrations of fermentable sugars are desirable to achieve high fermentation yields. Table 3 shows the mathematical equations that were generated to fit the experimental data for both SEB-UW and SEB-WW enzymatic hydrolyses, whereas Table 4 presents their corresponding analysis of variance (ANOVA). The quadratic models developed to adjust the enzymatic hydrolysis data were generally adequate, with adjusted regression coefficients (R^2) always above 0.99.

Table 3. Mathematical models for glucose release (g/L) from enzymatic hydrolysis experiments performed in shake flasks for 96 h.

Substrate	Equation	R^2
SEB7.5-WW	$Glc = -17.36(x1)^2 + 11.35x1 + 4.24x2 + 3.32x3 + 2.36x1x2 + 0.79x1x3 + 2.27x2x3 + 52.32$	0.9986
SEB15-WW	$Glc = -10.84(x1)^2 + 17.84x1 + 10.95x2 + 2.29x3 + 5.79x1x2 - 0.48x1x3 + 2.02x2x3 + 57.65$	0.9982
SEB7.5-UW	$Glc = -18.36(x1)^2 + 11.54x1 + 4.43x2 + 2.86x3 + 4.22x1x2 + 0.17x1x3 + 3.32x2x3 + 50.73$	0.9969
SEB15-UW	$Glc = -0.30(x1)^2 + 16.78x1 + 10.08x2 + 4.86x3 + 6.30x1x2 - 1.47x1x3 + 3.04x2x3 + 36.58$	0.9998

Table 4. Analysis of variance (ANOVA) of the mathematical models presented in Table 3 at a confidence level of 95% ($p < 0.05$).

Substrate	Conditions	Degrees of Freedom	RSR *	Adjusted R^2	F-Value	*p*-Value
SEB-WW	195 °C, 7.5 min	10	0.5219	0.9986	1702	1.459×10^{-14}
	195 °C, 15 min	10	0.8164	0.9969	779	7.208×10^{-13}
SEB-UW	195 °C, 7.5 min	10	0.9194	0.9982	1337	4.867×10^{-14}
	195 °C, 15 min	10	0.9317	0.9980	1207	8.156×10^{-14}

* Residual standard error.

The highest glucan conversion (74.4%) was obtained from SEB15-WW in the presence of Cellic HTec3 for a final glucose concentration of 84.5 g·L^{-1} after 96 h or hydrolysis (Table 5). This value was 25% higher than that obtained with the same substrate in the absence of Cellic HTec3 (67.2% glucan conversion for a 78.25 g·L^{-1} glucose concentration). In fact, Cellic HTec3 improved the hydrolysis performance of Cellic CTec3 in all reaction configurations. Therefore, residual xylans that were retained in steam-exploded materials seemed to have a role in limiting cellulose accessibility. Also, for SEB-UW substrates, additional hemicellulase activity helped converting water-soluble xylo-oligomers to fermentable xylose, justifying the achievement of slightly higher xylose recoveries. However, it is worth noticing that Cellic HTec3 contains some residual cellulase activity (~5 FPU g^{-1}), which may have been partially responsible for its boosting effect over Cellic CTec3.

Table 5. Glucose and xylose release after enzymatic hydrolysis of both SEB-UW and SEB-WW in shake flasks for 96 h using Cellic CTec3 in the absence and presence of 10% Cellic HTec3 (wet basis).

Parameter	SEB-WW [1]		SEB-UW [2]	
	CTec3	CTec3/HTec3	CTec3	CTec3/HTec3
Steam explosion at 195 °C for 7.5 min				
Glucose concentration (g·L^{-1})	46.17 ± 0.13	59.65 ± 0.33	45.61 ± 0.30	59.50 ± 0.37
Cellobiose concentration (g·L^{-1})	1.50 ± 0.12	2.52 ± 0.04	1.53 ± 0.12	2.25 ± 0.16
Glucan conversion (%)	39.26 ± 0.07	51.23 ± 0.19	38.27 ± 0.17	50.15 ± 0.22
Xylose concentration (g·L^{-1})	2.59 ± 0.15	3.00 ± 0.09	6.24 ± 0.12	6.89 ± 0.11
Xylan conversion	40.52 ± 0.32	46.9 ± 0.19	97.5 ± 0.16	97.7 ± 0.25
Steam explosion at 195 °C for 15 min				
Glucose concentration (g·L^{-1})	78.25 ± 0.28	84.52 ± 0.31	63.72 ± 0.13	75.14 ± 0.11
Cellobiose concentration (g·L^{-1})	3.09 ± 0.12	3.40 ± 0.06	1.51 ± 0.11	2.63 ± 0.12
Glucan conversion (%)	67.02 ± 0.16	74.44 ± 0.20	52.93 ± 0.08	63.15 ± 0.10
Xylose concentration (g·L^{-1})	1.78 ± 0.09	2.16 ± 0.12	2.88 ± 0.07	3.30 ± 0.02
Xylan conversion	27.8 ± 0.17	33.7 ± 0.11	45.0 ± 0.31	51.6 ± 0.47

[1] Water-washed steam-exploded sugarcane bagasse. [2] Unwashed steam-exploded sugarcane bagasse.

SEB-WW was better than SEB-UW for enzymatic hydrolysis, but the difference was minor for substrates produced at the lowest severity (195 °C, 7.5 min). Comparing the total processing time and the resulting glucose concentration in Table 5, water washing seems to be dispensable when pretreatment was carried out for 7.5 min, but mandatory when pretreatment was carried out for 15 min. This suggests that, by doubling the reaction time, the accumulation of potential hydrolysis inhibitors in the C5 stream became critical and this could not be attributed to XOS because these oligos tend to be almost completely hydrolyzed under more drastic pretreatment conditions. On the other hand, in both scenarios, better substrates for hydrolysis were produced when steam explosion was carried out for a longer reaction time or higher pretreatment severity.

Enzymatic hydrolysis of pretreated cellulosic materials was subsequently scaled-up by a factor of 10. Selected hydrolysis conditions (20 wt% TS and 60 mg g^{-1} TS of Cellic CTec3 plus 6 mg g^{-1} TS of Cellic HTec3) were used in the Infors-HT bioreactor to produce enough substrate hydrolysate for fermentation [20,22,59]. Figure 2 shows the hydrolysis profile of both SEB-UW and SEB-WW, with the results given in glucose release in g·L^{-1}. The effects of pretreatment severity, water washing, and Cellic HTec3 supplementation on hydrolysis efficiency were the same as those observed in shake flasks. Cellobiose was always within 1 and 3% of the total glucose release, meaning that the b-glucosidase activity of Cellic CTec3 was high enough to keep it below inhibitory levels. Xylan conversion was higher for SEB7.5-UW, which may have been caused by the xylose concentration of the C5 stream that was retained in the steam-exploded material after centrifugation.

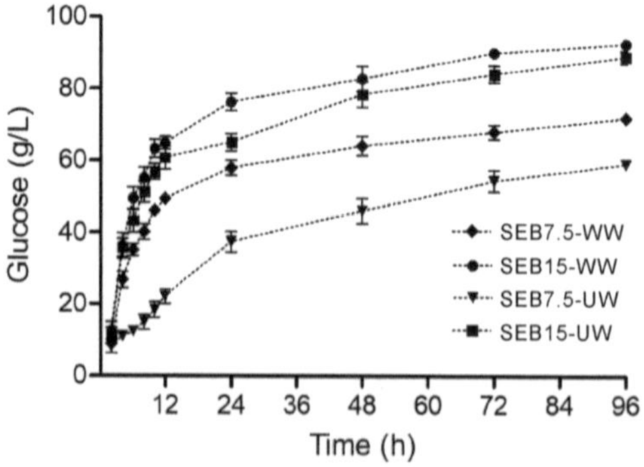

Figure 2. Glucose released (in g·L^{-1}) during the enzymatic hydrolysis of steam-exploded substrates for experiments that were carried out in Infors HT bioreactors.

3.4. Fermentation of SEB Pretreatment Liquors

SEB pretreatment liquors and enzymatic hydrolysates were fermented either alone or in combination using SHF and SHCF protocols. Initially, both fractions were fermented separately to evaluate the influence of detoxification on fermentation yields. There was a visual difference in the turbidity of the media (increase in cell biomass) and a significant difference in LA production between undetoxified and detoxified fractions ($p < 0.05$). Table 6 compares the fermentation profiles of SEB7.5 and SEB15 pretreatment liquors.

Table 6. Lactic acid fermentation of the pretreatment liquors before and after detoxification with a commercial activated carbon powder.

Parameter	SEB7.5 Liquor		SEB15 Liquor	
	Non-Detoxified	Detoxified	Non-Detoxified	Detoxified
Initial Xyl (g·L^{-1}) [1]	11.2 ± 0.2	11.1 ± 0.2	4.3 ± 0.2	4.3 ± 0.1
Initial Glc (g·L^{-1}) [1]	2.2 ± 0.2	2.3 ± 0.1	2.2 ± 0.1	2.1 ± 0.2
Initial Ara (g·L^{-1}) [1]	0.5 ± 0.1	0.5 ± 0.1	0.5 ± 0.2	0.4 ± 0.1
Lactic acid (g·L^{-1}) [2]	2.3 ± 0.6	13.4 ± 0.5	1.7 ± 0.4	6.6 ± 0.6
$Y_{p/s}$ (g·g^{-1}) [3]	0.17 ± 0.03	0.96 ± 0.02	0.25 ± 0.10	0.97 ± 0.12
$Y_{x/s}$ (g·g^{-1}) [4]	0.19 ± 0.03	0.16 ± 0.04	0.09 ± 0.05	0.11 ± 0.07
P_{LA} (g·L^{-1}·h^{-1}) [5]	<0.01	0.54 ± 0.02	<0.01	0.24 ± 0.03
OD_{600} [6]	1.0 ± 0.3	5.9 ± 0.1	1.6 ± 0.4	3.8 ± 0.6

[1] Xyl, Glc, Ara—xylose, glucose, and arabinose. [2] Lactic acid produced by fermentation. [3] Lactic acid yield (gram of product per gram of substrate). [4] Biomass yield (gram of biomass per gram of substrate). [5] Lactic acid productivity (gram of product per liter per hour). [6] Optical densities at 600 nm.

After fermentation, LA contents in non-detoxified pretreatment liquors were lower than those obtained from detoxified fractions for both pretreatment conditions. LA production from non-detoxified media was 2.3 and 1.7 g·L^{-1} for SEB7.5 and SEB15, respectively. These values are statistically equal at a 95% confidence level, meaning that pretreatment liquors without detoxification have similar fermentation performances. By contrast, LA contents from detoxified samples were 13.4 and 6.6 g·L^{-1}, respectively. Lower residence times into the steam reactor led to a lower level of carbohydrate degradation and lower release of fermentation inhibitors in pretreatment hydrolysates. By contrast, the poor fermentation performance of non-detoxified samples was attributed to the presence of inhibitory compounds such as furans, organic acids, and phenolic compounds.

Oliveira et al. [60] used a synthetic media containing xylose and glucose to produce LA, as well as SCB hemicellulose hydrolysates that were obtained by acid hydrolysis at 10% TS using 0.5% (v/v) HCl at 140 °C for 15 min. This acid hydrolysate, containing 48 g·L^{-1} xylose, 7.86 g·L^{-1} glucose, 0.08 g·L^{-1} 5-HMF, and 0.01 g·L^{-1} furfural, was supplemented with 20 $g{\cdot}L^{-1}$ yeast extract and fermented with *B. coagulans* 14–300 to produce 56 $g{\cdot}L^{-1}$ lactic acid for a theoretical yield of 87%. However, an evaporation step was added to concentrate the acid hydrolysate, which may have reduced its already low concentration of inhibitory compounds [21,27].

Ahorsu et al. [61] used *B. coagulans* DSM2314 to ferment nutshell hydrolysates obtained by microwave-assisted autohydrolysis. The experiments were carried out at 190 °C for up to 25 min in an equipment configuration that allowed five reactions to occur simultaneously. An LA productivity of 0.2 $g{\cdot}L^{-1}{\cdot}h^{-1}$ was obtained, which corresponded to a 93% xylose conversion (6 $g{\cdot}L^{-1}$ of LA). Furthermore, 0.45 $g{\cdot}L^{-1}$ furfural and 2.42 $g{\cdot}L^{-1}$ acetic acid did not affect the *B. coagulans* DSM2314 fermentation performance. However, the presence of xylo-oligosaccharides was not investigated, although probably present in pretreatment liquors, while fermentation times were up to 48 h.

van der Pol et al. [62] pre-cultured *B. coagulans* DSM2314 in the presence of fermentation inhibitors using glucose and xylose as carbon source. Inhibitors were found in the following concentrations: 1.6 $g{\cdot}L^{-1}$ furfural, 0.2 $g{\cdot}L^{-1}$ HMF, and 3.1 $g{\cdot}L^{-1}$ acetic acid. Although an increase in LA production was observed in the presence of furfural, fermentation trials were only carried out in synthetic media that simulated the carbohydrate composition of hemicellulose hydrolysates. Furthermore, the observed increase in productivity was not directly associated with furfural consumption or conversion.

A possible explanation for the albeit small formation of LA in the non-detoxified environment (Table 6, Figure 2) would be that, despite the toxicity of furfural being mainly caused by the formation of reactive oxygen species [63,64], several *B. coagulans* genes are known to encode for enzymes such as superoxide dismutase and catalase that can reduce these reactive species to less inhibitory compounds such as 2-furoic acid [65]. Since the DSM2314 strain is catalase-positive, this mechanism may be involved in its tolerance to the presence of furfural [62]. Ye et al. [66] used *B. coagulans* JI12 to ferment cellulosic and hemicellulosic hydrolysates from oil palm empty fruit bunches that contained glucose and xylose in different proportions (1:10 and 1:1). The strain was able to ferment both hydrolysates in the presence of 4 $g{\cdot}L^{-1}$ furfural, which was partially metabolized to 2-furoic acid.

3.5. Fermentation of SEB-UW and SEB-WW Enzymatic Hydrolysates

Table 7 presents the fermentation profile of both SEB-UW and SEB-WW enzymatic hydrolysates. C6 (glucose) fermentation predominated in both systems, but the former involved more C5/C6 co-fermentation because the substrate retained part of the pretreatment liquor (C5 stream) after centrifugation. Carbohydrates-to-LA conversions ($Y_{p/s}$) of 0.88 and 0.93 $g{\cdot}g^{-1}$ were achieved for SEB7.5-WW and SEB7.5-UW, while these values decreased to 0.61 and 0.78 $g{\cdot}g^{-1}$ for SEB15-WW and SEB15-UW, respectively. Therefore, UW hydrolysates were not inhibitory to LA fermentation and microbial growth (see OD_{600} values in Table 7). Also, *B. coagulans* DSM2314 consumed both glucose and xylose indistinctively, as demonstrated by the percentages of residual sugar detected after fermentation. On the other hand, lower $Y_{p/s}$ values for both SEB15-WW and SEB15-UW hydrolysates were probably due to the higher osmotic stress caused by applying higher initial glucose concentrations in the fermentation media.

For SHCF co-fermentation experiments, the C5/C6 ratio was based on van der Pol et al. [62], which was achieved by combining pretreatment liquors (C5 stream) and SEB-WW enzymatic hydrolysates to achieve 10% C5 and 90% C6. For this, pretreatment liquors were used without any further treatment (e.g., filtration or rotary evaporation) or after physical adsorption on activated carbon power to eliminate most fermentation inhibitors (detoxification). Table 8 presents the fermentation profile for SEB-WW enzymatic hydrolysates to which non-detoxified or detoxified C5 streams were added. Adding non-detoxified pretreatment liquor to the corresponding SEB7.5-WW enzymatic hydrolysate decreased fermentation efficiency ($Y_{p/s}$) by ~90%, from 0.88 ± 0.01 to 0.07 ± 0.02 $g{\cdot}g^{-1}$. This fact was observed by the low conversion of carbohydrates in the medium and the low cell density at the end of fermentation. Likewise, $Y_{p/s}$ decreased from 0.93 ± 0.01 to 0.07 ± 0.01 g·g^{-1} after adding the untreated C5 stream to SEB7.5-WW hydrolysates. Nevertheless, *B. coagulans* DSM 2314 produced 4.6 g·L^{-1} LA in the presence of fermentation inhibitors, demonstrating its ability to adapt to relatively high concentrations of furanic compounds such as furfural. It is important to notice that a possible elimination of the washing stage decreased LA production from 64.2 g·L^{-1} to 48.8 g·L^{-1} LA for SEB7.5, nearly 24% of reduction. SEB15 also showed a decrease in LA, from 56.9 g·L^{-1} to 52.4 g·L^{-1}, nearly 8%. In fact, these results can be explained by the superior hydrolysis performance of water-washed substrates, which also impacted the subsequent fermentation yields.

Table 7. Lactic acid fermentation of enzymatic hydrolysates derived from water-washed (SEB-WW) and unwashed (SEB-UW) pretreatment solids.

Parameter	SEB7.5-WW	SEB7.5-UW	SEB15-WW	SEB15-UW
Initial Xyl (g·L^{-1}) [1]	2.1 ± 0.1	6.4 ± 0.3	0.9 ± 0.7	4.9 ± 0.9
Initial Glc (g·L^{-1}) [1]	71.0 ± 0.4	59.8 ± 0.3	93.0 ± 1.8	89.9 ± 1.6
Residual Xyl (g·L^{-1}) [1]	0.2 ± 0.1	2.1 ± 0.3	0.13 ± 0.07	2.4 ± 0.8
Residual Glc (g·L^{-1}) [1]	17.1 ± 1.2	14.3 ± 1.6	34.6 ± 1.7	42.0 ± 1.1
Lactic acid (g·L^{-1}) [2]	64.2 ± 1.3	48.8 ± 0.5	56.9 ± 1.2	52.4 ± 0.8
$Y_{p/s}$ (g·g^{-1}) [3]	0.88 ± 0.01	0.74 ± 0.01	0.61 ± 0.01	0.55 ± 0.02
$Y_{x/s}$ (g·g^{-1}) [4]	0.09 ± 0.02	0.09 ± 0.01	0.08 ± 0.01	0.08 ± 0.3
P_{LA} (g·L^{-1}·h^{-1}) [5]	2.68 ± 0.08	2.03 ± 0.03	2.37 ± 0.07	2.19 ± 0.05
OD_{600} [6]	14.6 ± 0.8	15.3 ± 0.9	14.1 ± 0.3	15.9 ± 0.4

[1] Xyl, Glc —xylose and glucose. [2] Lactic acid produced by fermentation. [3] Lactic acid yield (gram of product per gram of substrate). [4] Biomass yield (gram of biomass per gram of substrate). [5] Lactic acid productivity (gram of product per liter per hour). [6] Optical densities at 600 nm.

Table 8. Lactic acid fermentation of enzymatic hydrolysates derived from water-washed (SEB7.5-WW) pretreatment solids with the addition of non-detoxified and detoxified pretreatment liquors (C5 stream) for a C5/C6 ratio of 1:10.

Parameter	SEB7.5-WW Hydrolysates Containing:		SEB15-WW Hydrolysates Containing:	
	Non-Detoxified C5	Detoxified C5	Non-Detoxified C5	Detoxified C5
Initial Xyl (g·L^{-1}) [1]	6.9 ± 0.7	6.4 ± 0.3	8.9 ± 0.9	5.5 ± 0.6
Initial Glc (g·L^{-1}) [1]	54.4 ± 0.9	59.8 ± 0.3	87.2 ± 0.9	91.0 ± 1.4
Residual Xyl (g·L^{-1}) [1]	4.4 ± 0.9	1.1 ± 0.5	6.2 ± 0.6	1.2 ± 0.3
Residual Glc (g·L^{-1}) [1]	51.9 ± 0.3	3.8 ± 0.7	85.7 ± 0.4	26.4 ± 0.2
Lactic acid (g·L^{-1}) [2]	4.6 ± 0.3	61.4 ± 1.8	4.1 ± 0.5	76.7 ± 1.4
$Y_{p/s}$ (g·g^{-1}) [3]	0.07 ± 0.02	0.93 ± 0.02	0.04 ± 0.02	0.78 ± 0.02
$Y_{x/s}$ (g·g^{-1}) [4]	0.07 ± 0.01	0.07 ± 0.02	0.07 ± 0.01	0.08 ± 0.01
P_{LA} (g·L^{-1}·h^{-1}) [5]	0.19 ± 0.06	2.55 ± 0.11	0.17 ± 0.03	3.20 ± 0.08
OD_{600} [6]	3.7 ± 0.2	16.2 ± 0.4	3.3 ± 0.2	16.0 ± 0.7

[1–6] See legends in Table 6.

Organic acids and phenolic acids are toxic to the bacteria because they can cross the cell membrane, decreasing the intracellular pH and causing damage to cell functions. Upon inhibition, the metabolic energy is spent to maintain homeostasis instead of being used for cell growth. Protein denaturation, metabolism inhibition, and cell death may also occur in the presence of potent inhibitory compounds [67–69].

Cubas-Cano et al. [34] studied the effect of inhibitory compounds on lactic acid fermentation using *B. coagulans* A162 and DSM2314 strains. Acid hydrolysates from garden plant waste acid-catalyzed steam explosion (180 °C for 10 min plus 60 mg·g^{-1} H_2SO_4) were used as carbon source. After pretreatment, the hemicellulose hydrolysate presented the following composition: 5 g L^{-1} glucose, 15 g L^{-1} xylose, 3.4 g L^{-1} arabinose, 0.20 g L^{-1} furfural, 0.23 g L^{-1} 5-HMF, 1.15 g L^{-1} acetic acid, and 0.23 g L^{-1} formic acid. The highest

LA productivity of 2.4 g·L^{-1}·h^{-1} was attained by the A162 strain, which was tolerant to the presence of fermentation inhibitors at these concentrations.

van der Pol et al. [70] applied pSSF for LA production using H_2SO_4-impregnated steam-exploded SCB (170 °C for 15 min using 0.72 vol% H_2SO_4 in relation to SCB dry mass). The substrate was pre-saccharified for 4 to 6 h with an enzyme cocktail containing both xylanase and cellulase activities (15 FPU·g^{-1} Genencor GC220 in relation to the substrate dry mass), when an inoculum (5%, *v*/*v*) that was pre-cultured in the presence of furfural was added to the medium. LA production reached 70.4 g·L^{-1} in 68 h, representing 89.7% carbohydrate conversion and an estimated productivity of 0.98 g·L^{-1}·h^{-1}. LA concentrations above 90 g·L^{-1} have already been reported in other studies for fermentation times up to 50 h [71].

4. Conclusions

Sugarcane bagasse steam explosion was successfully used to produce second generation LA by SHF and SHCF under anaerobic conditions. The pretreatment liquor (C5 stream) was shown to be highly inhibitory to *B. coagulans* DSM2314. This problem was circumvented by adsorption of fermentation inhibitors on activated carbon powder without causing noticeable sugar losses. Furan compounds such as furfural and 5-HMF were almost completely removed, while acetic acid was decreased by nearly 50%. The presence of xylo-oligosaccharides in substrate hydrolysates reduced hydrolysis efficiency, but their effect on fermentation was negligible. SHF and SHCF produced 64.2 and 61.4 g·L^{-1} LA from materials pretreated for 7.5 min, while samples pretreated for 15 min produced 56.9 and 76.7 g·L^{-1}, respectively. However, lactic acid yields were better for pretreatment at 195 °C for 7.5 min, reaching $Y_{p/s}$ values of 0.88 and 0.93 g·L^{-1} for SHCF with and without adding the detoxified C5 stream. Doubling the pretreatment time caused a decrease in $Y_{p/s}$ to 0.78 g·g^{-1}, but the corresponding LA productivity from SHCF with the detoxified C5 stream reached 3.20 g·L^{-1}·h^{-1}. For pretreatment at 195 °C for 7.5 min, elimination of substrate water washing seemed feasible, while the use of longer pretreatment times made it mandatory to eliminate fermentation inhibitors.

Author Contributions: Conceptualization, W.R.A., T.A.d.S. and L.P.R.; methodology, W.R.A., T.A.d.S. and A.Z.F.; validation, W.R.A., T.A.d.S. and L.P.R.; formal analysis, L.P.R. and A.Z.F.; investigation, W.R.A.; resources, L.P.R.; writing—original draft preparation, W.R.A., T.A.d.S. and L.P.R.; writing—review and editing, L.P.R.; supervision, T.A.d.S. and L.P.R.; project administration, L.P.R.; funding acquisition, L.P.R. All authors have read and agreed to the published version of the manuscript.

Funding: This research was funded by the National Council of Technological and Scientific Development (CNPq, Brazil—Grants 309506/2017-4 and 315930/2021-7) and by the Coordination for the Improvement of Higher Education Personnel (CAPES—Financial Code 001).

Institutional Review Board Statement: Not applicable.

Informed Consent Statement: Not applicable.

Data Availability Statement: No new data were created or analyzed in this study. Data sharing is not applicable to this article.

Acknowledgments: The authors are grateful to Raízen (Piracicaba, SP, Brazil) and Novozymes Latin America (Araucária, PR, Brazil) for donating batches of sugarcane bagasse and enzymes for hydrolysis, respectively.

Conflicts of Interest: The authors declare no conflict of interest.

References

1. Alsaheb, R.A.A.; Aladdin, A.; Othman, N.Z.; Malek, R.A.; Leng, O.M.; Aziz, R.; Enshasy, H.A.E. Lactic Acid Applications in Pharmaceutical and Cosmeceutical Industries. *J. Chem. Pharm. Res.* **2015**, *7*, 729–735.
2. Abedi, E.; Hashemi, S.M.B. Lactic Acid Production–Producing Microorganisms and Substrates Sources-State of Art. *Heliyon* **2020**, *6*, e04974. [CrossRef]

3. Grand View Research. *Lactic Acid Market Size, Share & Trends Analysis Report by Raw Material (Sugarcane, Corn, Cassava), by Application (PLA, Food & Beverages, Personal Care, Pharmaceuticals), by Region, and Segment Forecasts 2022–2030*; Grand View Research: San Francisco, CA, USA, 2021.
4. Singhvi, M.; Joshi, D.; Adsul, M.; Varma, A.; Gokhale, D. D-(−)-Lactic Acid Production from Cellobiose and Cellulose by *Lactobacillus lactis* Mutant RM2-24. *Green Chem.* **2010**, *12*, 1106–1109. [CrossRef]
5. Eş, I.; Khaneghah, A.M.; Barba, F.J.; Saraiva, J.A.; Sant'Ana, A.S.; Hashemi, S.M.B. Recent Advancements in Lactic Acid Production-a Review. *Food Res. Int.* **2018**, *107*, 763–770. [CrossRef] [PubMed]
6. Santoro, M.; Shah, S.R.; Walker, J.L.; Mikos, A.G. Poly(lactic acid) nanofibrous scaffolds for tissue engineering. *Adv. Drug Deliv. Rev.* **2016**, *107*, 206–212. [CrossRef]
7. Tsuji, H.; Fukui, I. Enhanced Thermal Stability of Poly(Lactide)s in the Melt by Enantiomeric Polymer Blending. *Polymer* **2003**, *44*, 2891–2896. [CrossRef]
8. Komesu, A.; de Oliveira, J.A.R.; Martins, L.H.d.S.; Maciel, M.R.W.; Filho, R.M. Lactic Acid Production to Purification: A Review. *BioResources* **2017**, *12*, 4364–4383. [CrossRef]
9. Reddy, L.V.; Park, J.; Wee, Y. Homofermentative Production of Optically Pure L -Lactic Acid from Sucrose and Mixed Sugars by Batch Fermentation of *Enterococcus faecalis* RKY1. *Biotechnol. Bioprocess Eng.* **2015**, *1105*, 1099–1105. [CrossRef]
10. Nwamba, M.C.; Sun, F.; Mukasekuru, M.R.; Song, G.; Harindintwali, J.D.; Boyi, S.A.; Sun, H. Trends and Hassles in the Microbial Production of Lactic Acid from Lignocellulosic Biomass. *Environ. Technol. Innov.* **2021**, *21*, 101337. [CrossRef]
11. Pleissner, D.; Demichelis, F.; Mariano, S.; Fiore, S.; Gutiérrez, I.M.N.; Schneider, R.; Venus, J. Direct production of lactic acid based on simultaneous saccharification and fermentation of mixed restaurant food waste. *J. Clean. Prod.* **2011**, *143*, 615–623. [CrossRef]
12. Wang, Y.; Meng, H.; Cai, D.; Wang, B.; Qin, P.; Wang, Z.; Tan, T. Improvement of L-Lactic Acid Productivity from Sweet Sorghum Juice by Repeated Batch Fermentation Coupled with Membrane Separation. *Bioresour. Technol.* **2016**, *211*, 291–297. [CrossRef]
13. Krishna, B.S.; Sai, S.; Gantala, N.; Tarun, B. Industrial Production of Lactic Acid and Its Applications. *Int. J. Biotech Res.* **2018**, *1*, 42–54.
14. Wang, Q.; Sun, H.; Wu, S.; Pan, S.; Cui, D.; Wu, D.; Xu, F.; Wamg, Z. Production of biomass-based carbon materials in hydrothermal media: A review of process parameters, activation treatments and practical applications. *J. Energy Inst.* **2023**, *110*, 101357. [CrossRef]
15. Bondesson, P.-M.; Galbe, M.; Zacchi, G. Ethanol and Biogas Production after Steam Pretreatment of Corn Stover with or without the Addition of Sulphuric Acid. *Biotechnol. Biofuels* **2013**, *6*, 11. [CrossRef] [PubMed]
16. Carrasco, C.; Baudel, H.M.; Sendelius, J.; Modig, T.; Roslander, C.; Galbe, M.; Hahn-Hägerdal, B.; Zacchi, G.; Lidén, G. SO_2-Catalyzed Steam Pretreatment and Fermentation of Enzymatically Hydrolyzed Sugarcane Bagasse. *Enzym. Microb. Technol.* **2010**, *46*, 64–73. [CrossRef]
17. Xia, J.; Jiang, S.; Liu, J.; Yang, W.; Oui, Z.; Liu, X.; He, A.; Li, D.; Xu, J. Efficient reduction of 5-hydroxymethylfurfural to 2, 5-bis (hydroxymethyl) furan by *Bacillus subtilis* HA70 whole cells. *Mol. Catal.* **2023**, *542*, 113122. [CrossRef]
18. Schneiderman, S.J.; Johnson, R.W.; Menkhaus, T.J.; Gilcrease, P.C. Quantifying second generation ethanol inhibition: Design of Experiments approach and kinetic model development. *Bioresour. Technol.* **2015**, *179*, 219–226. [CrossRef]
19. Wang, L.-Q.; Cai, L.-Y.; Ma, Y.-L. Study on Inhibitors from Acid Pretreatment of Corn Stalk on Ethanol Fermentation by Alcohol Yeast. *RSC Adv.* **2020**, *10*, 38409–38415. [CrossRef]
20. Fockink, D.H.; Urio, M.B.; Sa, J.H.; Ramos, L.P. Enzymatic Hydrolysis of Steam-Treated Sugarcane Bagasse: Effect of Enzyme Loading and Substrate Total Solids on Its Fractal Kinetic Modeling and Rheological Properties. *Energy Fuels* **2017**, *31*, 6211–6220. [CrossRef]
21. Neves, P.V.; Pitarelo, A.P.; Ramos, L.P. Production of Cellulosic Ethanol from Sugarcane Bagasse by Steam Explosion: Effect of Extractives Content, Acid Catalysis and Different Fermentation Technologies. *Bioresour. Technol.* **2016**, *208*, 184–194. [CrossRef]
22. Xue, S.; Uppugundla, N.; Bowman, M.J.; Cavalier, D.; Da, L.; Sousa, C.; Dale, B.E.; Balan, V. Sugar Loss and Enzyme Inhibition Due to Oligosaccharide Accumulation during High Solids-Loading Enzymatic Hydrolysis. *Biotechnol. Biofuels* **2015**, *8*, 195. [CrossRef]
23. Fockink, D.H.; Sánchez, J.H.; Ramos, L.P. Comprehensive Analysis of Sugarcane Bagasse Steam Explosion Using Autocatalysis and Dilute Acid Hydrolysis (H_3PO_4 and H_2SO_4) at Equivalent Combined Severity Factors. *Ind. Crop. Prod.* **2018**, *123*, 563–572. [CrossRef]
24. Horváth, I.S.; Sjöde, A.; Nilvebrant, N.O.; Zagorodni, A.; Jönsson, L.J. Selection of Anion Exchangers for Detoxification of Dilute-Acid Hydrolysates from Spruce. *Appl. Biochem. Biotechnol.* **2004**, *114*, 525–538. [CrossRef]
25. Cantarella, M.; Cantarella, L.; Gallifuoco, A.; Spera, A.; Alfani, F. Comparison of Different Detoxification Methods for Steam-Exploded Poplar Wood as a Substrate for the Bioproduction of Ethanol in SHF and SSF. *Process. Biochem.* **2004**, *39*, 1533–1542. [CrossRef]
26. Larsson, S.; Reimann, A.; Nilvebrant, N.-O.; Jönsson, L.J. Comparison of Different Methods for the Detoxification of Lignocellulose Hydrolyzates of Spruce. *Appl. Biochem. Biotechnol.* **1999**, *77*, 91–104. [CrossRef]
27. Pitarelo, A.P.; Fonseca, C.S.; Chiarello, L.M.; Gírio, F.M.; Ramos, L.P. Ethanol Production from Sugarcane Bagasse Using Phosphoric Acid-Catalyzed Steam Explosion. *J. Braz. Chem. Soc.* **2016**, *27*, 1889–1898. [CrossRef]
28. Jurado, M.; Prieto, A.; Martínez-Alcalá, Á.; Martínez, Á.T.; Martínez, M.J. Laccase Detoxification of Steam-Exploded Wheat Straw for Second Generation Bioethanol. *Bioresour. Technol.* **2009**, *100*, 6378–6384. [CrossRef] [PubMed]

29. Zhang, Y.; Xia, C.; Lu, M.; Tu, M. Effect of overliming and activated carbon detoxification on inhibitors removal and butanol fermentation of poplar prehydrolysates. *Biotechnol. Biofuels* **2018**, *11*, 1–14. [CrossRef]
30. Abdel-Rahman, M.A.; Tashiro, Y.; Sonomoto, K. Lactic acid production from lignocellulose-derived sugars using lactic acid bacteria: Overview and limits. *J. Biotechnol.* **2011**, *156*, 286–301. [CrossRef]
31. Zhang, Y.; Chen, X.; Qi, B.; Luo, J.; Shen, F.; Su, Y.; Khan, R.; Wan, Y. Improving Lactic Acid Productivity from Wheat Straw Hydrolysates by Membrane Integrated Repeated Batch Fermentation under Non-Sterilized Conditions. *Bioresour. Technol.* **2014**, *163*, 160–166. [CrossRef]
32. Ye, L.; Hudari, M.S.B.; Li, Z.; Wu, J.C. Simultaneous Detoxification, Saccharification and Co-Fermentation of Oil Palm Empty Fruit Bunch Hydrolysate for L-Lactic Acid Production by *Bacillus coagulans* JI12. *Biochem. Eng. J.* **2014**, *83*, 16–21. [CrossRef]
33. Malacara-Becerra, A.; Melchor-Martínez, E.M.; Sosa-Hernández, J.E.; Riquelme-Jiménez, L.M.; Mansouri, S.S.; Iqbal, H.M.N.; Parra-Saldívar, R. Bioconversion of Corn Crop Residues: Lactic Acid Production through Simultaneous Saccharification and Fermentation. *Sustainability* **2022**, *14*, 11799. [CrossRef]
34. Cubas-Cano, E.; Venus, J.; González-Fernández, C.; Tomás-Pejó, E. Assessment of different *Bacillus coagulans* strains for L-lactic acid production from defined media and gardening hydrolysates: Effect of lignocellulosic inhibitors. *J. Biotechnol.* **2020**, *323*, 9–16. [CrossRef]
35. Murariu, M.; Dubois, P. PLA composites: From production to properties. *Adv. Drug Deliv. Rev.* **2016**, *107*, 17–46. [CrossRef] [PubMed]
36. Lassalle, V.; Ferreira, M.L. PLA nano- and microparticles for drug delivery: An overview of the methods of preparation. *Macromol. Biosci.* **2007**, *7*, 767–783. [CrossRef] [PubMed]
37. Zhang, Z.; Tsapekos, P.; Alvarado-Morales, M.; Zhu, X.; Zervas, A.; Jacobsen, C.S.; Angelidaki, I. Enhanced fermentative lactic acid production from source-sorted organic household waste: Focusing on low-pH microbial adaptation and bio-augmentation strategy. *Sci. Total Environ.* **2022**, *808*, 152129. [CrossRef] [PubMed]
38. Kawaguchi, H.; Hasunuma, T.; Ogino, C.; Kondo, A. Bioprocessing of bio-based chemicals produced from lignocellulosic feedstocks. *Curr. Opin. Biotechnol.* **2016**, *42*, 30–39. [CrossRef]
39. Carrillo-Nieves, D.; Alanís, M.J.R.; Quiroz, R.d.l.C.; Ruiz, H.A.; Iqbal, H.M.; Parra-Saldívar, R. Current status and future trends of bioethanol production from agro-industrial wastes in Mexico. *Renew. Sustain. Energy Rev.* **2019**, *102*, 63–74. [CrossRef]
40. Wahono, S.K.; Rosyida, V.T.; Darsih, C.; Pratiwi, D.; Frediansyah, A. Optimization of simultaneous saccharification and fermentation incubation time using cellulose enzyme for sugarcane bagasse on the second-generation bioethanol production technology. *Energy Procedia* **2015**, *65*, 331–336. [CrossRef]
41. Michelson, T.; Kask, K.; Jõgi, E.; Talpsep, E.; Suitso, I.; Nurk, A.L. (+)-Lactic acid producer *Bacillus coagulans* SIM-7 DSM 14043 and its comparison with *Lactobacillus delbrueckii* ssp. *lactis* DSM 20073. *Enzym. Microb. Technol.* **2006**, *39*, 861–867. [CrossRef]
42. Sluiter, A.; Hames, B.; Hyman, D.; Payne, C.; Ruiz, R.; Scarlata, C.; Sluiter, J.; Templeton, D.; Wolfe, J. *Determination of Total Solids in Biomass and Total Dissolved Solids in Liquid Process Samples, Laboratory Analytical Procedure (LAP)*; Technical Report NREL/TP-510-42621; National Renewable Energy Laboratory: Golden, CO, USA, 2008.
43. Sluiter, A.; Hames, B.; Ruiz, R.; Scarlata, C.; Sluiter, J.; Templeton, D. *Determination of Ash in Biomass, Laboratory Analytical Procedure (LAP)*; Technical Report NREL/TP-510-42622; National Renewable Energy Laboratory: Golden, CO, USA, 2008.
44. Sluiter, A.; Ruiz, R.; Scarlata, C.; Sluiter, J.; Templeton, D. *Determination of Extractives in Biomass, Laboratory Analytical Procedure (LAP)*; Technical Report NREL/TP-510-42619; National Renewable Energy Laboratory: Golden, CO, USA, 2008.
45. Sluiter, A.; Hames, B.; Ruiz, R.; Scarlata, C.; Sluiter, J.; Templeton, D.; Crocker, D. *Determination of Structural Carbohydrates and Lignin in Biomass, Laboratory Analytical Procedure (LAP)*; Technical Report NREL/TP-510-42618; National Renewable Energy Laboratory: Golden, CO, USA, 2008.
46. Candido, J.P.; Claro, E.M.T.; de Paula, C.B.C.; Shimizu, F.L.; de Oliveria Leite, D.A.N.; Brienzo, M.; de Angelis, D.D.F. Detoxification of sugarcane bagasse hydrolysate with different adsorbents to improve the fermentative process. *World J. Microbiol. Biotechnol.* **2020**, *36*, 43. [CrossRef]
47. RStudio Team. *RStudio: Integrated Development for R*; RStudio, Inc.: Boston, MA, USA, 2015; Available online: http://www.rstudio.com/.2015 (accessed on 1 June 2023).
48. Andrade, L.P.; Crespim, E.; de Oliveira, N.; de Campos, R.C.; Teodoro, J.C.; Galvão, C.M.A.; Filho, R.M. Influence of sugarcane bagasse variability on sugar recovery for cellulosic ethanol production. *Bioresour. Technol.* **2017**, *241*, 75–81. [CrossRef]
49. Wang, G.; Chen, H. Carbohydrate elimination of alkaline-extracted lignin liquor by steam explosion and its methylolation for substitution of phenolic adhesive. *Ind. Crop. Prod.* **2014**, *53*, 93–101. [CrossRef]
50. Antal, M.J., Jr.; Mok, W.S.; Richards, G.N. Mechanism of formation of 5-(hydroxymethyl)-2-furaldehyde from D-fructose and sucrose. *Carbohydr. Res.* **1990**, *199*, 91–109. [CrossRef]
51. Brown, D.W.; Floyd, A.J.; Kinsman, R.G.; Roshan-Ali, Y. Dehydration reactions of fructose in non-aqueous media. *J. Chem. Technol. Biotechnol.* **1982**, *32*, 920–924. [CrossRef]
52. El-Sheshtawy, H.S.; Fahim, I.; Hosny, M.; El-Badry, M.A. Optimization of lactic acid production from agro-industrial wastes produced by Kosakonia cowanii. *Curr. Res. Green Sustain. Chem.* **2022**, *5*, 100228. [CrossRef]
53. Ramos, L.P. The chemistry involved in the pretreatment of lignocellulosic materials. *Química. Nova.* **2003**, *26*, 863–871. [CrossRef]

54. Du, B.; Sharma, L.N.; Becker, C.; Chen, S.F.; Mowery, R.A.; van Walsum, G.P.; Chambliss, C.K. Effect of varying feedstock–pretreatment chemistry combinations on the formation and accumulation of potentially inhibitory degradation products in biomass hydrolysates. *Biotechnol. Bioeng.* **2010**, *107*, 430–440. [CrossRef]
55. Lu, C.; Dong, J.; Yang, S.T. Butanol Production from wood pulping hydrolysate in an integrated fermentation-gas stripping process. *Bioresour. Technol.* **2013**, *143*, 467–475. [CrossRef]
56. Miura, M.; Suzuki, T.; Aoyama, M. Detoxification of Japanese white birch wood hemicellulose hydrolysate with a carbonaceous sorbent prepared from birch wood hydrolysis residue. *Cellul. Chem Tech.* **2016**, *50*, 265–268.
57. Yang, J.; Yue, H.R.; Pan, L.Y.; Feng, J.X.; Zhao, S.; Suwannarangsee, S.; Zhao, X.Q. Fungal strain improvement for efficient cellulase production and lignocellulosic biorefinery: Current status and future prospects. *Bioresour. Technol.* **2023**, *385*, 129449. [CrossRef] [PubMed]
58. Mathew, G.M.; Sukumaran, R.K.; Singhania, R.R.; Pandey, A. Progress in research on fungal cellulases for lignocellulose degradation. *J. Sci. Ind. Res.* **2008**, *67*, 898–907. Available online: http://nopr.niscpr.res.in/handle/123456789/2417 (accessed on 1 June 2023).
59. Wang, L.; Feng, X.; Zhang, Y.; Chen, H. Lignocellulose particle size and rheological properties changes in periodic peristalsis enzymatic hydrolysis at high solids. *Biochem. Eng. J.* **2022**, *178*, 108284. [CrossRef]
60. Oliveira, R.A.; Schneider, R.; Rossel, C.E.V.; Maciel Filho, R.; Venus, J. Polymer grade L-lactic acid production from sugarcane bagasse hemicellulosic hydrolysate using *Bacillus coagulans*. *Bioresour. Technol. Rep.* **2019**, *6*, 26–31. [CrossRef]
61. Ahorsu, R.; Cintorrino, G.; Medina, F.; Constantí, M. Microwave processes: A viable technology for obtaining xylose from walnut shell to produce lactic acid by *Bacillus coagulans*. *J. Clean. Prod.* **2019**, *231*, 1171–1181. [CrossRef]
62. van der Pol, E.; Springer, J.; Vriesendorp, B.; Weusthuis, R.; Eggink, G. Precultivation of *Bacillus coagulans* DSM2314 in the presence of furfural decreases inhibitory effects of lignocellulosic by-products during l(+)-lactic acid fermentation. *Appl. Microbiol. Biotechnol.* **2016**, *100*, 10307–10319. [CrossRef]
63. Feron, V.J.; Til, H.P.; De Vrijer, F.; Woutersen, R.A.; Cassee, F.R.; Van Bladeren, P.J. Aldehydes: Occurrence, carcinogenic potential, mechanism of action and risk assessment. *Mutat. Res.* **1991**, *259*, 363–385. [CrossRef]
64. Allen, S.A.; Clark, W.; McCaffery, J.M.; Cai, Z.; Lanctot, A.; Slininger, P.J.; Lewis Liu, Z.; Gorsich, S.W. Furfural induces reactive oxygen species accumulation and cellular damage in *Saccharomyces cerevisae*. *Biotechnol. Biofuels* **2010**, *2*, 2–10. [CrossRef]
65. Cabiscol, E.; Tamarit, J.; Ros, J. Oxidative stress in bacteria and protein damage by reactive oxygen species. *Int. Microbiol.* **2000**, *3*, 3–8. Available online: http://hdl.handle.net/10459.1/56751 (accessed on 1 June 2023).
66. Ye, L.; Hudari, M.S.B.; Zhou, X.; Zhang, D.; Li, Z.; Wu, J.C. Conversion of acid hydrolysate of oil palm empty fruit bunch to L-lactic acid by newly isolated *Bacillus coagulans* JI12. *App. Microbiol. Biotechnol.* **2013**, *97*, 4831–4838. [CrossRef]
67. Trček, J.; Mira, N.P.; Jarboe, L.R. Adaptation and tolerance of bacteria against acetic acid. *Appl. Microbiol. Biotechnol.* **2015**, *99*, 6215–6229. [CrossRef] [PubMed]
68. Abdel-Rahman, M.A.; Tashiro, Y.; Sonomoto, K. Recent advances in lactic acid production by microbial fermentation processes. *Biotechnol. Adv.* **2013**, *31*, 877–902. [CrossRef]
69. Othman, M.; Ariff, A.B.; Rios-Solis, L.; Halim, M. Extractive fermentation of lactic acid in lactic acid bacteria cultivation: A review. *Front. Microbiol.* **2017**, *8*, 2285. [CrossRef]
70. van der Pol, E.C.; Eggink, G.; Weusthuis, R.A. Production of L(+)-lactic acid from pretreated sugarcane bagasse using *Bacillus coagulans* DSM2314 in a simultaneous saccharification and fermentation strategy. *Biotechnol. Biofuels* **2016**, *9*, 248. [CrossRef] [PubMed]
71. Zhang, F.; Liu, J.; Han, X.; Gao, C.; Ma, C.; Tao, F.; Xu, P. Kinetic characteristics of long-term repeated fed-batch (LtRFb) l-lactic acid fermentation by a *Bacillus coagulans* strain. *Eng. Life Sci.* **2020**, *20*, 562–570. [CrossRef] [PubMed]

fermentation

MDPI

Article

Microbial Fuel Cell Using a Novel Ionic Liquid-Type Membrane–Cathode Assembly for Animal Slurry Treatment and Fertilizer Production

Eduardo Iniesta-López [1], Adrián Hernández-Fernández [1], Yolanda Garrido [1], Ioannis A. Ieropoulos [2] and Francisco José Hernández-Fernández [1,*]

[1] Department of Chemical Engineering, Faculty of Chemistry, University of Murcia (UMU), Campus de Espinardo, E-30100 Murcia, Spain; eduardo.iniestal@um.es (E.I.-L.); adrian.h.f@um.es (A.H.-F.); ygh46006@um.es (Y.G.)

[2] Civil, Maritime & Environmental Engineering Department, University of Southampton, Bolderwood Campus, Southampton SO16 7QF, UK; i.ieropoulos@soton.ac.uk

* Correspondence: fjhernan@um.es; Tel.: +34-868889758

Abstract: The implementation of a microbial fuel cell for wastewater treatment and bioenergy production requires a cost reduction, especially when it comes to the ion exchange membrane part and the catalysts needed for this purpose. Ionic liquids in their immobilized phase in proton exchange membranes and non-noble catalysts, as alternatives to conventional systems, have been intensively investigated in recent years. In the present study, a new microbial fuel cell technology, based on an ionic liquid membrane assembly for CoCu mixed oxide catalysts, is proposed to treat animal slurry. The new low-cost membrane–cathode system is prepared in one single step, thus simplifying the manufacturing process of a membrane–cathode system. The novel MFCs based on the new low-cost membrane–cathode system achieved up to 51% of the power reached when platinum was used as a catalyst. Furthermore, the removal of organic matter in suspension after 12 days was higher than that achieved with a conventional system based on the use of platinum catalysts. Moreover, struvite, a precipitate consisting of ammonium, magnesium, and phosphate, which could be used as a fertilizer, was recovered using this membrane–cathode system.

Keywords: microbial fuel cell; ionic liquid membrane; slurry treatment; fertilizer production

Citation: Iniesta-López, E.; Hernández-Fernández, A.; Garrido, Y.; Ieropoulos, I.A.; Hernández-Fernández, F.J. Microbial Fuel Cell Using a Novel Ionic Liquid-Type Membrane–Cathode Assembly for Animal Slurry Treatment and Fertilizer Production. *Fermentation* **2023**, *9*, 844. https://doi.org/10.3390/fermentation9090844

Academic Editors: Miguel Ladero, Victoria E. Santos and Christian Kennes

Received: 21 July 2023
Revised: 31 August 2023
Accepted: 6 September 2023
Published: 14 September 2023

1. Introduction

Microbial fuel cells (MFCs) form part of an emerging technology that makes it possible to convert the chemicals and biowastes contained in wastewater into electrical energy and other high-value products. Most publications are related to laboratory-scale applications, although there are also pilot- or semi-industrial-scale publications. The change of scale requires the search for cheaper and more efficient materials than conventional materials [1–6]. A microbial fuel cell consists of a cathode, an anode, and an ion exchange membrane. Platinum (Pt) is widely used as a catalyst in cathode systems to combat slow reactions, but it is expensive and prevents commercialization progress; Pt is also used in the ion exchange membrane, which is usually a perfluorinated organic polymer [7,8]. The research in this area in recent years has focused on the development of new nanostructured chemical catalysts, which are comparable with platinum in terms of reaction velocity, by using low-cost catalysts based on non-noble metals, in the hope of developing technologies that can be industrially implemented. Specifically, the catalysts described below were developed and tested in MFCs: (i) iron streptomycin [9], (ii) MnO_2 [10,11], (iii) iron aminoantipyrine [12], (iv) ferroelectric materials such as $LiTaO_3$ [13], (v) activated carbon [14], and (vi) biocathodes [15–17].

Regarding the proton exchange membrane, ionic liquids are used in proton exchange membranes as the active phase being used in the innovative phase in proton exchange membranes. Ionic liquids (ILs) are salts that exist in a liquid state at or near room temperature. Ionic liquids are composed of organic cations, the most common of which are tetraalkylammonium, tetraalkylphosphonium, N-alkylpyridine, and N,N′-dialkylimidazolium, and anions, such as hexafluorophosphate, tetrafluoroborate, triflimide, and triflate anions. Due to their unique properties, such as high ionic conductivity, a wide electrochemical stability window, and thermal stability, they were extensively studied as alternative electrolytes for various electrochemical devices, including batteries, capacitors, sensors, and electroplating [18]. Furthermore, as was recently demonstrated, they allow for the selective transport of ions [19]. Three generations of proton exchange membranes composed of ionic liquids were used in MFCs: (i) supported ionic liquid membranes [20]; (ii) polymeric ionic liquid membranes, which are also called ionogel, based on ionic liquids [21] (the advantage of the latter over the former being that they allow for a much larger quantity of ionic liquids to be immobilized, while increasing stability); and (iii) polymeric ionic liquid inclusion membranes, based on ionic liquids assembled onto the catalyst [22]. For these last membranes, the catalyst was sprayed on a carbon cloth, and the ionogel was created on the carbon cloth, which allows for close contact between the membrane and the cathode, with a consequent reduction in electrical resistance. In this work, a new simple procedure was assayed, in which the catalyst is located within the ionogel, based on the ionic liquid phase, which also favors contact between the proton exchange membrane and catalyst. These new membranes were assayed for slurry treatment.

Livestock farms are one of the sectors with great economic activity in Europe. Increased demand for meat products led to the development of intensive farming, which results in large amounts of slurry, contaminating both our atmosphere and soils due to the nitrogenous content. This made conventional treatment for slurry infeasible and, therefore, motivated the scientific community and business sector to seek new and sustainable solutions for the treatment of slurry. Slurry is a mixture of animal excrement, unconsumed feed material, and water, and it is commonly used in agriculture as a fertilizer and soil conditioner. However, if not managed properly, slurry can cause environmental problems. The high concentration of nitrogen and phosphorus in slurry can contribute to eutrophication in watercourses and reservoirs, leading to ecological imbalances in aquatic ecosystems. Furthermore, slurry has high values of biochemical and chemical oxygen demand (BOD and COD), which make slurry treatment by conventional methods difficult. The high COD and BOD levels can lead to oxygen depletion in water bodies, causing harm to aquatic life [23].

In the present study, slurry at a high COD concentration was used as a fuel in a microbial fuel cell technology based on a new type of ionic-liquid-type membrane–cathode assembly. The methodology proposed in this study requires only one step, i.e., the impregnation of the diffusion layer with a mixture of the ionic liquid and catalyst. This novel approach simplifies the fabrication of the membrane–catalyst system, thus reducing costs, especially with a view of industrialization. The bioenergy produced by microbial fuel cells, as well as the reduction in COD, BOD, and other physical–chemical parameters, was analyzed. The chemical composition and the morphology of the membrane–catalyst system were extensively studied using SEM–EDX techniques.

2. Materials and Methods

2.1. Fuel and Chemicals

Livestock slurry from the livestock farm of the Veterinary Faculty of the Universidad de Murcia was used as fuel. The wastewater also acted as the inoculum for the formation of an anodic bacterial community. The wastewater's soluble chemical oxygen demand (COD) was found to be 2540 mg/L.

Polyvinylidene chloride and ionic liquids, which were used for preparing the ionic liquid membranes, were purchased from Sigma-Aldrich-Fluka, Kawasaki, Japan.

2.2. Synthesis of Copper and Cobalt Mixed Valence Oxides

Thermal decomposition method was used to synthesize $(Cu_{0.3}Co_{0.7})Co_2O_4$ [9,23]. Previous studies prepared copper–cobalt oxides at varying atomic ratios of Cu/Co and found that $(Cu_{0.3}Co_{0.7})Co_2O_4$ had the highest power output [9]. This catalyst also showed up to three Co oxidation states in its structure [24]. To prepare the co-precipitates of copper and cobalt hydroxides ($CoOH_2$, $CuOH_2$), an excess quantity of 3 M NaOH solution was slowly added over a mixture of $CuCl_2$ and $CoCl_2$, stirred for 7 h at 25 °C, and filtered. The resulting precipitates ($(Cu_{0.3}Co_{0.7})Co_2O_4$) were washed with deionized hot water and dried at 60 °C for 24 h. The dried powder was heated between 350 and 400 °C for 8 h.

The crystallographic structure and purity of the oxides were characterized using X-ray diffraction (XRD) with a Bruker D8 Advance diffractometer (Bruker; Karlsruhe, Germany) with a generator of X-ray, Kristalloflex K 7608-80 F, with a copper anode. Particle size was measured with a JEOL JEM2100 transmission electron microscopy (TEM).

The TEM images showed copper–cobalt oxides nanoparticle with dimensions ranging between 17 and 45 nm. The XRD patterns of the catalyst revealed that the oxides have spinel lattice packed in cubic structure and purity of 100% (semi-quantitative analysis), with no other phases appearing in the samples analyzed.

2.3. Preparation of New Proton Exchange Membranes with Catalytic Activity Based on Ionic Liquids

The embedded membrane–cathode assembly was a polymeric inclusion membrane type. Polymer inclusion membranes based on ionic liquids were obtained by casting methods using PVC polymer and methyltrioctyl ammonium chloride ($[MTOA^+][Cl^-]$). The amount of ionic liquid was 70% *w*/*w* of the PVC/IL mixture. Two new types of embedded membrane–cathode assemblies were prepared and tested in MFCs. In the first preparation (P1), the IL and PVC base polymer were dissolved in THF, and 10 mg of the catalyst was added to the solution, which was then stirred for 10 min. The resulting suspension was poured into a Fluka glass ring (28 mm inner diameter, 30 mm height) on a glass plate and left to settle overnight until complete evaporation of THF. After evaporation, a thin plastic membrane was obtained, which was carefully peeled off the glass plate. Finally, the membrane and a piece of 3 cm diameter carbon cloth (Fuel Cell Earth, EEUU) were fixed with a round joint clip. In the second procedure (P2), the mixture of the IL methyltrioctylammonium chloride, ($[MTOA^+][Cl^-]$), with PVC (polyvinyl chloride), THF (tetrahydrofuran), and 10 mg of catalyst were poured directly onto a piece of 3 cm diameter carbon cloth cathodes placed between a Fluka glass ring and a Fluka glass plate. This protocol produces a composite material comprising proton exchange membrane, catalyst, and diffusion layer.

As a control (C), a PILIM based on $[MTOA^+][Cl^-]$ and PVC 70% *w*/*w* and a platinum cathode was used. The membrane was manufactured by dissolving the IL and PVC in THF, before pouring the solution into a Fluka glass ring (Sigma-Aldrich International GmbH; Buchs, Switzerland) and evaporating the THF [25]. The cathode, based on platinum on carbon cloth, was fixed together with the ionic liquid membrane with a round join clip. The cathode was a mixture of 10.5 mg of a Pt/C mixture (nominally 60% platinum on a high surface area advance carbon support, Thermo Scientific, Waltham, MA, USA), water, isopropanol, and PTFE sprayed onto a 3 cm diameter piece of 10% water-proof carbon cloth (Fuel Cell Earth, Sacramento, CA, USA). The final load of platinum on the carbon cloth was 0.5 mg Pt cm^{-2}.

2.4. MFC Studies

Newly embedded membrane–cathode assemblies were employed in one-chamber MFCs for slurry treatment and evaluated in terms of power output, COD, and BOD reduction (Figure 1).

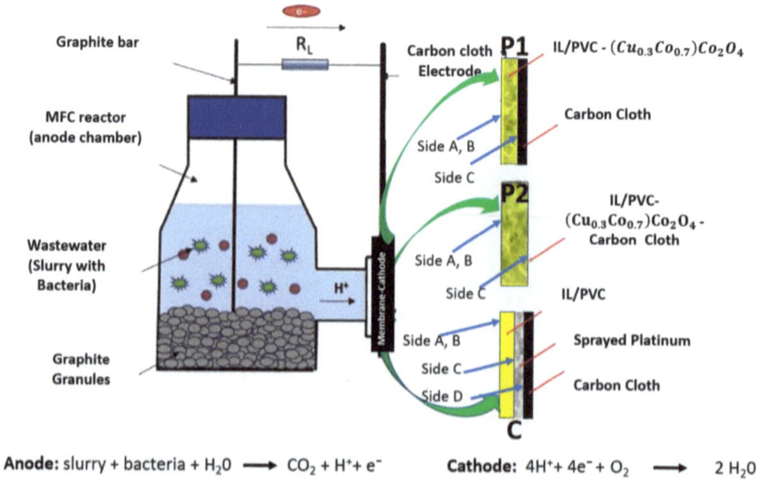

Figure 1. Schematic diagram of the reactor used in the experiments. P1 is the catalyst suspended in the mixture IL/PVC, and the synthesized catalytic ionogel is placed on carbon cloth. P2 is the catalytic ionogel synthesized in the carbon cloth. C is a conventional method in which the catalyst is sprayed on carbon cloth and then the proton exchange membrane base on ionic liquid is placed on the carbon cloth.

The experimental setup consisted of reactors made from modified 250 mL glass bottles with cylindrical flanges, where the temperature was maintained at 25 °C. The proton exchange membrane–cation system was identified by the labels P1, P2, and C, as mentioned earlier. The cathode was connected to the anode using a 1 kΩ resistor. The anode was composed of 100 g of graphite granules with a diameter of 3–5 mm and a graphite rod with a diameter of 3.18 mm. Anode chambers contained 200 mL of feed and were sealed with a lid equipped with a sampling port, ensuring anaerobic conditions throughout the experiment. The membrane–cathode system was securely fastened to the reactor flange using a round joint clip (see Figure 1). All tests were conducted in batch mode using wastewater as the sole source of microorganisms and fuel.

To establish biofilm attachment and stabilization, an initial period was required, after which polarization experiment was carried out at 192 h. Anodes and membrane–cathode systems were replaced for each trial, and daily sampling was performed by extracting a 5 mL aliquot with a syringe, which was then filtered through a 0.45 μm pore diameter membrane filter. A 2 mL aliquot was retained for COD analysis, which was carried out throughout the experiment. To ensure accuracy, three MFC replicates were run, and the data presented are an average of three reactor runs for each system.

2.5. Analytical Methods

2.5.1. Chemical Analysis

During treatment, the wastewater was characterized in terms of COD (chemical oxygen demand), BOD_5 (biochemical oxygen demand at 5 days), and UV254 (absorbance at 254 nm). Wastewater samples were collected regularly, using 0.45 nm nylon syringe filters (Fisher brand of Fisher Scientific, Waltham, MA, USA). When necessary, the samples were digested in a TR 420 thermoreactor of Spectroquant, from Merck Millipore, Burlington, MA, USA. The chemical composition of the samples was determined with spectrophotometry using Spectroquant 300 prove equipment from Merck Millipore (Danvers, MA, USA) using kits. The methods used were as follows:

- COD (chemical oxygen demand): test conducted in COD 145541 Supelco cuvettes (Sigma-Aldrich). Procedure was according to DIN ISO 15705 and approved by the

USEPA for wastewater. The relative standard deviation of the assay was found to be less than 3%, indicating a high degree of repeatability.

- Biochemical oxygen demand at 5 days (BOD_5): A system of six Velp Scientifica DBO sensors was used for the manometric determination of BOD, using dicyanamide as an inhibitor for nitrification and sodium hydroxide as alkali to capture CO_2.
- Absorbance at 254 nm (UV254): as described previously [26], there is a correlation between dissolved organic matter and absorbance at 254 nm. The reduction in UV254 has also been related to the removal of organic micropollutants [27].
- Other parameters, including pH, conductivity, and temperature, were determined with a digital multimeter (sensION + MM150 from Hach Company; Loveland, CO, USA).

2.5.2. Electrochemical Analysis

To analyze the performance of the new fuel material (wastewater) in terms of electricity generation, polarization, internal resistance, and Coulombic efficiency were calculated.

Polarization Test

The voltage was continuously monitored every minute by a PCI 6010 data acquisition system (National Instruments, Austin, TX, USA). The voltage was also read intermittently off-line using a DVM891 digital multimeter (HQ Power, Berlin, Germany). Polarization was measured using a variable resistor box (5.77 MΩ, 953 kΩ, 486 kΩ, 96.5 kΩ, 50 kΩ, 11 kΩ, 6 kΩ, 1.1 kΩ, 561 Ω, 94.5 Ω, and 1.5 Ω) after 192 h of operation.

The voltage measurement was performed when the cell had reached a pseudo steady state under a specific resistor value. The time taken for the cell to reach the pseudo steady state was approximately 1 min. To ensure the reliability of the results, each measurement was conducted three times, and the mean value was reported. The relative standard deviation of the assay was found to be less than 10%, indicating a high degree of repeatability.

Internal Resistance

The internal resistance R_{int} (Ω) of each MFC was determined from Equation (1) [28]:

$$R_{int}(\Omega) = \frac{OCV\ (V)}{I(A)} - R_{ext}(\Omega) \quad (1)$$

where OCV is the open circuit voltage, I is the current density at maximum power, and R_{ext} is the external resistance at maximum power.

Coulombic Efficiency

The concept of selectivity is used to evaluate Coulombic efficiency in microbial fuel cells, which measures the electric charge accumulated during substrate removal. Coulombic efficiency is determined by calculating the ratio of the number of coulombs transferred to the anode from the substrate to the maximum number of coulombs that would be transferred in the theorical case such that the entire substrate is transformed to current [28]:

$$Y_Q = \frac{\text{coulombs transferred}}{\text{total theoretical coulombs produced}} \cdot 100 \quad (2)$$

$$Y_Q = \frac{M_m \int_0^t i(t)dt}{F\Delta DQObV} \cdot 100 \quad (3)$$

where M_m is the molar mass of oxygen (32 $gmol^{-1}$), i(t) is the current intensity at a given time t, F is Faraday constant (96,485 $Cmole^{-1}$), ΔCOD is the variation of COD in (mg/L), b is the number of moles of electrons produced per mole of oxygen (b = 4), and V is the volume of liquid in the anodic chamber (0.2 L).

2.6. SEM-EDX Characterization

To study the morphological characteristics, chemical composition, and distribution of elements in the membranes, a SEM equipped with an energy-dispersive X-ray (EDX) analyzer (APREO S and a JEOL-6100, from Jeol Ltd., Tokyo, Japan) was employed. The PILIMs were characterized using SEM-EDX immediately after preparation (referred to as fresh membranes) and after MFC operation (referred to as test membranes).

3. Results and Discussion

3.1. Electrochemical Analysis

Initially, 200 mL of slurry was added to form a biofilm around the graphite granules. The MFC voltage was continuously monitored for more than 280 h with a 1 kΩ external resistor, for different membrane–cathode systems. During the first 72 h (maturation period), the voltage increased until a plateau was reached. For the tests, fresh fuel was added, waiting 192 h for the voltage to reach a pseudo-steady state, after which the polarization tests were carried out. Polarization curves were obtained by varying the resistance between both electrodes and determining at which resistance the maximum current density is produced.

Figure 2 shows the polarization curves with different embedded membrane–cathode assemblies (P1, P2, and C). Figure 2 shows the maximum value of OCV for C at 500 mV. P1 and P2 OCV´s values were around 430 mV. At low intensities, data represent the activation overpotential of the electrodes, the linear part then represents the ohmic losses of the system and finally mass transfer losses can be determined at high current concentrations. P1 and P2 show two different slopes corresponding to activation and ohmic losses, with limiting current densities reached at 653 mA m^{-2} and 1478 mA m^{-2} for P1 and P2, respectively. In the case of membrane–cathode system C, polarization behavior was linear throughout, and with limiting current densities of around 1753 mA m^{-2}.

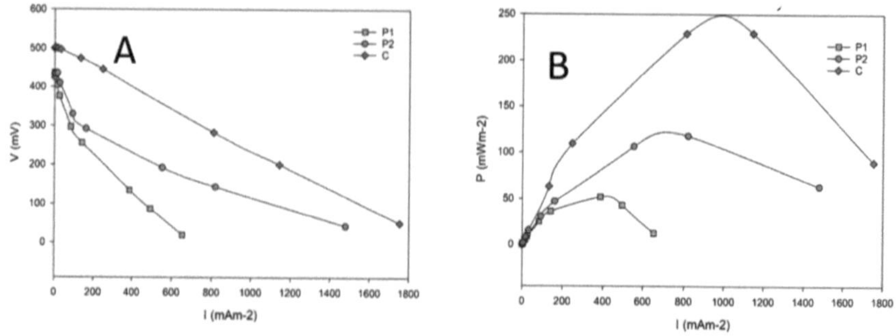

Figure 2. Polarization (**A**) and power (**B**) curves of MFCs with different embedded membrane–cathode assembly systems (P1, P2, and C). Calculations are based on the surface of the membrane–cathode assembly.

The power output curves (Figure 2B) showed that the highest power densities were obtained with the control membrane–cathode system based on platinum, with a maximum value of 230 mWm^{-2}. The power density for the P2 membrane–cathode system was 118 mWm^{-2} (51% the value of C), and for P1, the maximum recorded power was 52 mWm^{-2} (23% the value of C) (see Figure 2B). Kiely et al. [29] used an air cathode microbial fuel cell where the cathode is constructed by applying platinum (0.5 mg/cm^2) and four diffusion layers (PTFE) to 30% wet-proofed carbon cloth. They used dairy manure wastewater with a soluble chemical oxygen demand of 450 mg/L. The maximum power density was 189 mW/m^2.

Polarization curves provide information on the electrochemical behavior of system and help determine the internal resistance (see Figure 2). The different embedded membrane–cathode assembly systems showed different internal resistances due to their composition and the way in which they are manufactured. The control (C), which is composed of platinum, showed the lowest internal resistance (833 Ω), and consequently, the greatest maximum power. The resistance of P2 (1093 Ω) was similar to that of the MFC with platinum and lower than a half of the resistance of P1 (2465 Ω). The greater power produced by P2 compared with P1 could be due to the membrane–cathode system of P2 being manufactured in one step. For P1, the catalytic membrane is placed on a carbon cloth after the ionogel is prepared, which would increase the internal resistance compared with P2.

In all cases, the Coulombic efficiency increased with time, which could be due to a more efficient oxygen reduction reaction and, consequently, a better utilization of the chemical energy contained in organic matter. In general, the Coulombic efficiency of P2 was always higher than that of P1 (Figure 3). This could be explained by the fact that the new composite material P2 offers less ionic resistance since it is prepared in one single step.

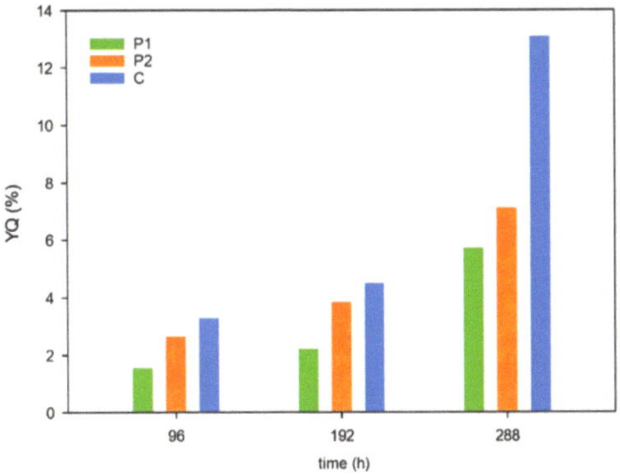

Figure 3. Profile of the Coulombic efficiency (YQ) of the MFC with different membrane–cathode systems (P1, P2, and C).

The most common method for preparing the cathode involves spraying Pt/C, water, isopropanol, and PTFE on a diffusion layer and used a proton exchange membrane on the diffusion layer [9]. The proposed method is simpler, as it is based on mixing the ionic liquid phase, which also acts as a membrane, and the catalyst. The new preparation technique is also malleable if intended for industrial purposes.

3.2. Slurry Wastewater Treatment Using Microbial Fuel Cell

The initial COD of the slurry wastewater sample was found to be 2540 mg/L, and its behavior was assessed over the test time (288 h) for the three types of microbial fuel cells (P1, P2, and C). The data obtained provided the COD reduction percentage for each MFC (Figure 4).

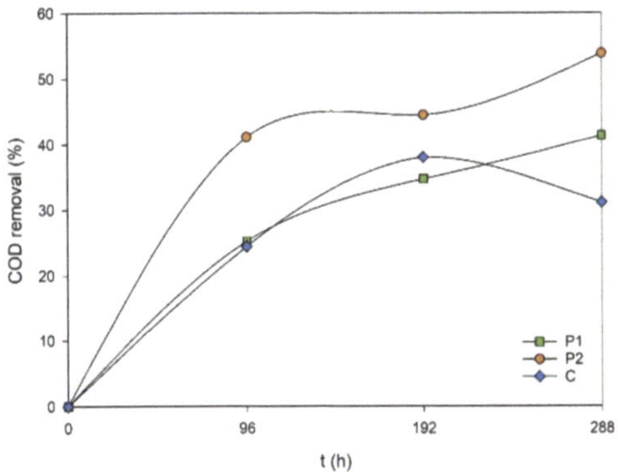

Figure 4. COD reduction (%) in MFC for each membrane–cathode system (P1, P2, and C).

The rate of COD reduction was nearly constant during the first 96 h with a 41% reduction obtained for P2, and around 25% for P1 and C. After this time, the rate of COD reduction decreased, reaching a COD reduction during the last phase (192–288 h) of 54% for the P2 system, 41% for the P1 system, and 31% for C. Considering the removal of COD at initial time (until 96 h), the COD removal rate values were 10 mg L^{-1} h^{-1} for P1 and 6.3 mg L^{-1} h^{-1} for P2 and C. Furthermore, taking into account the higher Coulombic efficiency of platinum compared with P1 and P2, the higher values of COD reduction for P1 and P2 could be explained due to the organic matter oxidation in the anodic compartment. The formulation of the platinum cathodes could affect the activity of microorganisms in the anodic compartment decreasing their natural activity towards the degradation of organic matter. The results of the analysis of dissolved organic compound (DOC) based on the absorbance at 254 nm are shown in Table 1.

Table 1. % DOC reduction in MFC based on membrane–cathode systems of P1, P2, and C.

	Absorbance at 254 nm (Dilution 1:10)		
MFC	$[UV254]_0$	$[UV254]_{t=288\ h}$	$[\%DOC]_{t=288\ h}$
P1	0.847	0.3465	59.09
P2	0.847	0.269	68.24
C	0.847	0.393	53.60

A similar trend was observed for %COD and %DOC removal, since higher values were observed for P2 and lower values for P1. Indeed, these two parameters correlated well with the literature. The initial value of biodegradable organic matter was 1167 mg L^{-1}. The % of BOD reduction reached at the end of the experiment was 89%, 77%, and 63% for P1, P2, and C, respectively.

In the literature, there are very few research works that analyze the use of slurry as fuel in microbial fuel cells. Furthermore, other works usually consider slurry with considerably less COD than the one used in this work. Yokohama et al. [30] studied the purification of slurry containing the feces and urine of Holstein cows using cathode-in-air microbial fuel cells. The cathode contained 0.5 mg/cm^2 of platinum catalyst, attached to a Nafion-115 proton exchange membrane. The initial COD values were 1010 mg/L. In this work, the BOD reduction was around 80% at 70 h, and the COD reduction was around 70%. The maximum power density was around 0.34 mW/m^2. Similar BOD reduction values were found in both research works; however, higher COD reduction was found in the Yokoyama's work [30]. The lower COD reduction could be due to the higher ammonium concentration in our sample. The initial ammonium concentration in our wastewater was around 620 mg/L and in the previous work was around 38.9 mg/L. The high values of COD could involve a high value of ammonium, which is toxic and hinders the degradability of the COD. However, the BOD values are relatively high, and they are similar to those obtained in the literature.

3.3. *Characterization of the Membrane–Catalyst Assembly*

The membrane–catalyst assembly systems (P1, P2, and C) were characterized using SEM-EDX before and after use in the MFCs. Figure 5 shows the SEM micrographs corresponding to the P1 membrane on the side in contact with the anodic solution before testing and after >280 h of testing. The air side after operation is also shown in Figure 5.

Figure 5A shows a relatively homogeneous surface, with some roughness and grooves. After using the catalytic membranes (Figure 5B), the surface becomes more homogeneous, and deposits appear on it, which are assumed to be salts from the anodic solution as are discussed below in Table 2. Figure 5C shows the SEM micrograph of the side in contact with the carbon cloth after 280 h of operation. This side presents a more homogeneous appearance than the opposite face in contact with the anodic solution (Figure 5B), with no deposits on its surface.

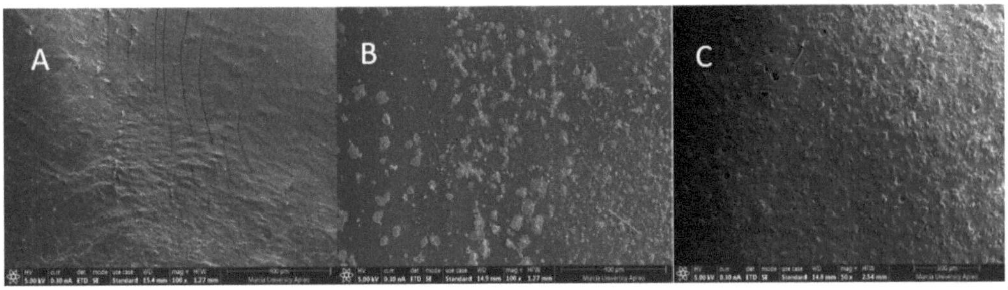

Figure 5. Scanning electron micrographs of PEM–catalyst assembly (P1) based on [MTOA$^+$][Cl$^-$] ionic liquids, 30% PVC, and ($Cu_{0.3}Co_{0.7}$)Co_2O_4. (**A**) Anodic solution side before operation; (**B**) anodic solution side after operation; and (**C**) air side after operation.

Table 2. Peak element vs. weight (%) and atomic (%) for EDX spectra of PEM–catalyst assembly (P1) based on [MTOA$^+$][Cl$^-$] ionic liquids, 30% PVC, and ($Cu_{0.3}Co_{0.7}$)Co_2O_4. (A) Anodic solution side before operation; (B) anodic solution side after 280 h operation; (B*) analysis of the surface deposits observed on the anodic solution side after 280 h operation; and (C) air side after 280 h operation.

PEM–Catalyst Assembly (P1) ([MTOA$^+$][Cl$^-$]-PVC-($Cu_{0.3}Co_{0.7}$)Co_2O_4)								
Peak Element	**Weight % A**	**Atomic % A**	**Weight % B**	**Atomic % B**	**Weight % B***	**Atomic % B***	**Weight % C**	**Atomic % C**
C K	56.25	74.46	53.87	75.30	16.69	33.55	40.16	65.15
N K	1.57	1.78	−0.43	−0.51	3.11	5.36	2.45	3.40
O K	8.23	8.18	3.62	3.81	-	-	4.38	5.34
Na K	1.82	1.26	4.54	3.32	22.75	23.89	-	-
S K	-	-	0.79	0.41	0.49	0.37	1.64	1.00
Cl K	31.65	14.20	34.22	16.22	51.03	34.76	37.88	20.82
K K	-	-	0.80	0.34	0.44	0.27	-	-
Ca K	-	-	2.13	0.89	0.84	0.51	2.16	2.05
Co K	0.36	0.10	0.38	0.11	0.37	0.15	6.49	2.15
Cu K	0.13	0.03	0.07	0.02	0.07	0.03	0.62	0.9
Zr K	-	-	-	-	4.22	1.12	4.22	0.9

Table 2 shows the results of the EDX technique applied to the surface of the catalytic membrane of P1 before and after use in a single-chamber microbial fuel cell for 280 h. Note that the EDX spectra are from a sample of up to a few micrometers thick/deep so surface catalytic membrane is analyzed.

The EDX spectrum of PVC presented the characteristic peaks assigned to the C and N K lines corresponding to the chemical formulation of PVC. Hydrogen and other light elements were not detected using EDX. The EDX study of the membrane–cathode assembly was based on the selection of characteristic elements: N and Cl for [MTOA$^+$][Cl$^-$], Cl for PVC, and Co and Cu for the metallic catalyst ($Cu_{0.3}Co_{0.7}$)Co_2O_4. The relative peak heights of identical elements in the different compounds were roughly related with their respective concentrations.

The nitrogen peaks are mainly due to the presence of the ionic liquid [MTOA$^+$][Cl$^-$]. Nitrogen is present in fresh (unused) P1 (Table 2, column A), which disappears when EDX is applied to the entire surface in contact with the anodic solution for 280 h (Table 2, column B). However, when the EDX was carried out for the deposits observed on the surface (see Figure 2B), traces of nitrogen were evident (Table 2, column B*). On the face in contact with the carbon cloth (external face), the presence of nitrogen was still evident after 280 h of use (Table 2, column C). As previously mentioned, EDX provides information on the composition of the first few micrometers of the membrane. Taking this into account and from the results discussed above, we can assume that the ionic liquid has been washed, to a certain extent, on the outermost surface of the face of P1 in contact with the anodic solution meanwhile the ionic liquid remained on the outer face (in contact with the carbon cloth). The selection of the ionic liquid ([MTOA$^+$][Cl$^-$]) was made taking into account the low solubility of this ionic liquid in water (<0.02% (v/v)), in order to increase the stability of the membrane in aqueous solutions. Furthermore, the inclusion method with PVC has been demonstrated to increase the stability of the liquid membrane in the face of aqueous media [31].

The deposits found in the membranes after use (Figure 2B), besides the ionic liquid, included other elements (Table 2, column B*) such as sodium, sulfur, potassium, and calcium, probably from salts contained in the anodic solution.

However, the proportion of catalytic metals, Co and Cu, remained practically constant before and after the test in the fuel cell (Table 2, columns A and B), which demonstrated the stability of the catalyst in the membrane during the test time (280 h).

The morphology of the P2 (Figure A1) type membrane–cathode assembly differs from that of P1. In P2, the ionic liquid and PVC solution, together with the catalyst, are poured onto a carbon cloth that acts as a diffusion layer and electrical conductor. The fiber-like appearance is due to the carbon cloth. P2 has a larger active surface than P1, which may be important for the rate of the chemical reactions taking place in it and, therefore, in the power of the fuel cell. SEM micrographs corresponding to the side in contact with the anodic solution before the test (A), the same side after 280 h of testing (B), and the side that is in contact with air after 280 h (C) (Figure 1, Appendix A). There is no great morphological difference between them, although, in micrograph C, a uniform layer is evident on the carbon cloth. This morphology may be explained by the fact that this side was the one that remained in contact with a glass sheet (upside down) when P2 was formed and, consequently, the ionic liquid and PVC mixture accumulated on this face. Table 3 shows the results of the EDX technique applied to the surface of the catalytic membrane of P2 before use and after use in a single-chamber microbial fuel cell for 280 h.

Table 3. Peak element vs. weight (%) and atomic (%) for EDX spectra of PEM–catalyst assembly (P2) based on $[MTOA^+][Cl^-]$ ionic liquids, 30% PVC, and $(Cu_{0.3}Co_{0.7})Co_2O_4$. (A) Anodic solution side before operation; (B) anodic solution side after 280 h operation; and (C) air side after 280 h operation.

PEM–Catalyst Assembly (P2) ($[MTOA^+][Cl^-]$-PVC-$(Cu_{0.3}Co_{0.7})Co_2O_4$)-Carbon Cloth						
Peak Element	**Weight % A**	**Atomic % A**	**Weight % B**	**Atomic % B**	**Weight % C**	**Atomic % C**
C K	68.46	86.56	74.92	90.76	52.08	75.34
N K	1.88	2.04	−3.17	−3.29	0.46	0.57
F K	-	-	5.85	4.48	-	-
O K	-	-	-	-	2.84	3.09
Si K	-	-	-	-	0.62	0.38
S K	-	-	0.29	0.13	1.10	0.59
Cl K	23.51	10.07	15.41	6.32	39.43	19.39
Ca K	-	-	2.22	0.80	-	-
Co K	3.02	0.78	0.69	0.17	0.47	0.14
Cu K	0.41	0.10	0.27	0.06	−0.10	−0.03
Zr K	2.72	0.45	3.53	0.56	3.09	0.59

As a first observation, it should be noted that the % weight of the carbon element is greater in P2 than in P1, which would be related with the composition of P2 based on carbon cloth. The nitrogen element, related with the composition of the ionic liquid, $[MTOA^+][Cl^-]$, disappears from the side in contact with the anode compartment due to the partial washing of the first micrometers of depth of the catalytic membrane (column B). However, it does not disappear from the side of the catalytic membrane in contact with air (column C). Furthermore, both Co and Cu catalysts remain on the membrane after use (columns C and B). Other elements present in the sample such as calcium, silicon, or fluorine may come from the residual water or impurities of the raw materials.

In the case of PEM–catalyst assembly control system (C) (Figure A2), the fresh membrane (A) had a smooth surface that began to show deposits after 280 h of operation (B). The side of the membrane in contact with the carbon cloth (C) had less deposits than B after 280 h of operation. The morphology of the carbon cloth in contact with the membrane–catalyst assembly after 280 h of operation (D) showed a conventional crossed-fiber appearance. The side of the carbon cloth in contact with the air after operation (data not shown) was similar in appearance to the side in contact with the membrane–catalyst assembly (D).

The behavior of the control system (composed of ionic liquid-based proton exchange membrane and diffusion layer impregnated with Pt with Teflon) (Table 4) was similar to that found in previous studies (P1 and P2) (Tables 2 and 3). The side in contact with the anodic dilution lost ionic liquid from the surface after 280 h of operation (column B), while the liquid remained on the membrane side in contact with the diffusion layer (carbon cloth) because the N element appeared in the EDX spectrum (column C). This surface also contained Pt since it was in contact with the diffusion layer, which was impregnated with Pt. The diffusion layer in contact with the ionic liquid-based PEM also contained Pt and F (column D). F element was used in the protocol to support Pt on the carbon cloth. In regards the SEM-EDX analysis, it is important to point out that crystals of struvite were found on the anodic side of the membrane–catalyst assembly systems (P1, P2, and C) (see Figure A3). Struvite (magnesium ammonium phosphate) is a phosphate mineral with the formula $NH_4MgPO_4{\cdot}6H_2O$, which crystallizes with an orthorhombic geometry. Struvite is used as an agricultural fertilizer since it contains P and N, two of the three major plant macronutrients, along with Mg as a minor macronutrient. SEM micrographs showed orthorhombic crystals on the surface of the membrane–catalyst systems. The EDX spectrum of struvite crystal showed two big peaks, which corresponded to Mg K and P K.

Table 4. Peak element vs. weight (%) and atomic (%) for EDX spectra of PEM–catalyst assembly (C-control) based on a PEM ($[MTOA^+][Cl^-]$/PVC, 70/30) and $(Cu_{0.3}Co_{0.7})Co_2O_4$ supported on carbon cloth. (A) Anodic solution side before operation; (B) anodic solution side after 280 h operation; and (C) air side after 280 h operation.

Control System (C) $[MTOA^+][Cl^-]$-PVC-Pt-Carbon Cloth								
Peak Element	**Weight % A**	**Atomic % A**	**Weight % B**	**Atomic % B**	**Weight % C**	**Atomic % C**	**Weight % D**	**Atomic % D**
C K	64.53	83.05	46.31	74.12	55.82	77.85	71.91	80.45
N K	2.22	2.44	−3.98	−5.47	1.17	1.40	0.73	0.70
F K	-	-	-	-	-	-	26.54	18.77
OK	-	-	-	-	4.26	4.46	-	-
S K	-	-	2.87	1.72	2.10	1.10	-	-
Cl K	33.25	14.50	53.09	28.79	31.15	14.42	0.06	0.02
K K	-	-	1.72	0.85	-	-	-	
Pt M	-	-	-	-	5.51	4.47	0.75	0.05

Recently, there has been a great deal of interest in the production of struvite in microbial fuel cells. Furthermore, previous work reported on the accelerated recovery of struvite with the addition of magnesium [32]. The effluent from this process was used to obtain bioenergy and to reduce its COD more efficiently.

When the concentrations of Mg^{2+}, NH_4^+, and PO_4^{3-} exceed the limit for struvite formation, it leads to the precipitation of struvite crystals. The solubility of struvite decreases as the pH increases, and it is also influenced by the ionic strength of the solution [33]. The decrease in pH and the reduction in ionic strength favor the formation of struvite. In this work, it is observed that struvite precipitation occurred on the membrane–catalyst assembly, indicating that the pH near the cathode should increase.

During oxygen reduction reactions, protons are consumed at the electrode surface, and an increase in pH in the vicinity of the electrode surface is expected [34]. Furthermore, the effect of the ionic liquid of the membrane in the cathode microenvironment also needs to be considered. Similar results were found by Ichihashi and Hirooka [35] when they treated swine wastewater by using microbial fuel cell technology. According to their proposal, struvite crystal formation was triggered by the cathodic reaction and resulted from the rise in pH near the cathode.

Although we have worked in an experimental one-chamber lab system, for an industrial application, we propose to work with a stack made up of different membrane–electrode assemblies (MEAs). During the MEA cleaning process, the struvite could be detached and recovered for fertilizer application since struvite is precipitated as microcrystals, which are easy to remove from the MEA. Also, it is important to point out that in this study, the lack of any biofilms observed in the membranes based on ionic liquids was due to the antibiofilm activity of ionic liquids at high concentration as it seen in ionic liquid membranes.

4. Conclusions

In this work, microbial fuel cell technology based on a new type of membrane–cathode systems was studied for slurry purification. In these new membrane–cathode systems, the non-noble catalyst copper and cobalt mixed valence oxides are suspended in the ionic liquid phase and not sprayed on the diffusion layer as it is conventionally conducted. In the P2 membrane–cathode assembly, the internal resistance was significantly reduced, and the maximum power reached was 120 mWm^{-2}, which corresponds to 51.3% of the value obtained with the control using a platinum-based cathode. This membrane–cathode system P2 attained >50% organic matter removal at 288 h, which was higher than that was achieved with the platinum based-cathode. This novel membrane–cathode system allows easier manufacturing and production of cheaper membranes and catalysts than that conventionally used in microbial fuel cells. This would reduce the capital and the processing costs involved in the case of the industrial manufacture of membrane–cathode systems. Furthermore, struvite fertilizer was obtained as a precipitate in all the membrane–cathode assemblies. The proposed MFC permits the optimal use of slurry wastewater for bioenergy and fertilizer production, accompanied by water decontamination.

Author Contributions: Conceptualization, F.J.H.-F. and I.A.I.; methodology, E.I.-L., A.H.-F. and Y.G.; validation, E.I.-L., A.H.-F. and Y.G.; investigation, E.I.-L., A.H.-F. and Y.G.; data curation, F.J.H.-F. and I.A.I.; writing—original draft preparation, F.J.H.-F. and I.A.I.; writing—review and editing, F.J.H.-F. and I.A.I.; supervision, F.J.H.-F.; project administration, F.J.H.-F.; funding acquisition, F.J.H.-F. All authors have read and agreed to the published version of the manuscript.

Funding: The authors wish to acknowledge the financial support of the Ministry of Science, Innovation, and Universities (MICINN) ref. RTI2018-099011-B-I00 and the Seneca Foundation Science and Technology Agency of the Region of Murcia ref. 20957/PI/18. The work was produced with the support of a 2021 Leonardo Grant for Researchers and Cultural Creators, BBVA Foundation ref. BBVA LEONARDO IN[21]_ING_0137. Adrián Hernández-Fernández has a grant 21817/FPI/22 from Seneca Foundation Science and Technology Agency of the Region of Murcia. The Foundation takes no responsibility for the opinions, statements, and contents of this project, which are entirely the responsibility of its authors.

Institutional Review Board Statement: Not applicable.

Informed Consent Statement: Not applicable.

Data Availability Statement: The datasets supporting reported results during the current study are available from the corresponding author on reasonable request.

Conflicts of Interest: The authors declare no conflict of interest.

Appendix A

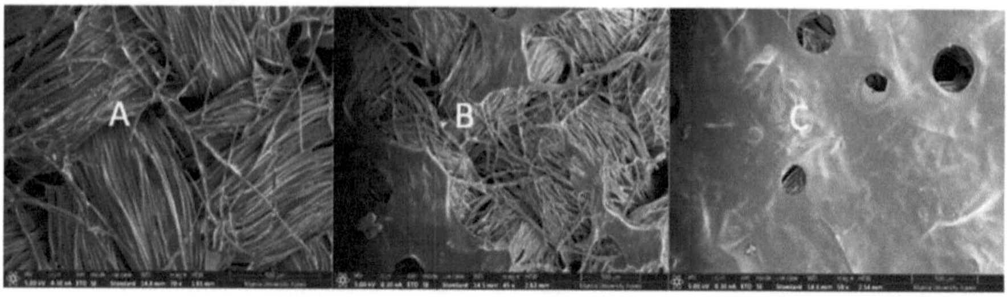

Figure A1. Scanning electron micrographs of PEM–catalyst assembly (P2) based on $[MTOA^+][Cl^-]$ ionic liquids, 30% PVC, $(Cu_{0.3}Co_{0.7})Co_2O_4$ and carbon cloth. (**A**) Anodic solution side before operation; (**B**) anodic solution side after 280 h operation; and (**C**) air side after 280 h operation.

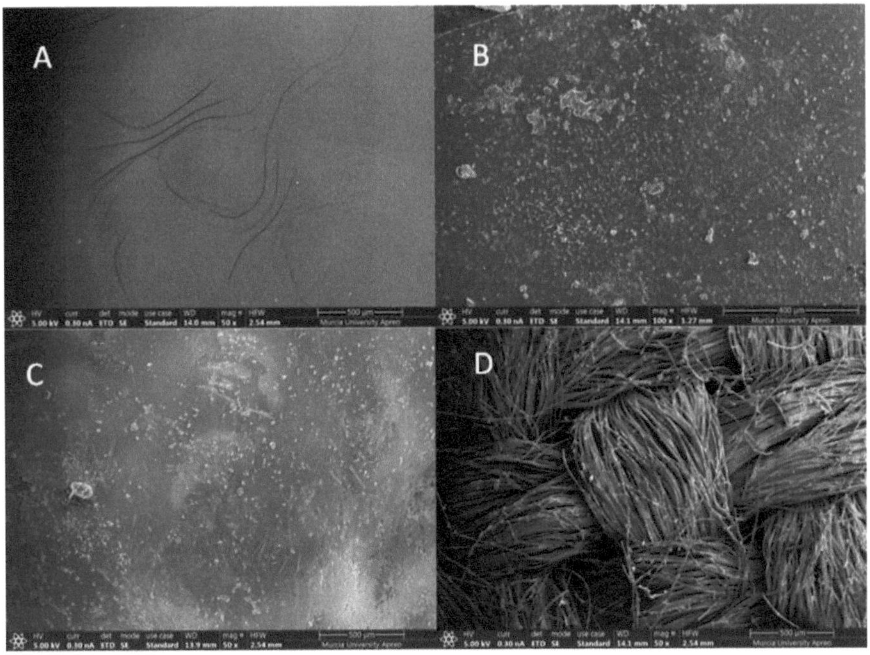

Figure A2. Scanning electron micrographs of PEM–catalyst assembly C—control (C-control) based on PEM ([MTOA$^+$][Cl$^-$]/PVC, 70/30) and $(Cu_{0.3}Co_{0.7})Co_2O_4$ supported on carbon cloth. (**A**) Anodic solution side before operation; (**B**) anodic solution side after 280 h operation; (**C**) membrane catalyst assembly in contact with carbon cloth; and (**D**) carbon cloth side in contact with membrane catalyst assembly.

Figure A3. SEM details of orthorhombic crystal of struvite on anodic side of PEM–catalyst systems P1, P2, and C.

References

1. Rossi, R.; Hur, A.Y.; Page, M.A.; Thomas, B.; Butkiewicz, J.J.; Jones, D.W.; Baek, G.; Saikaly, P.E.; Cropek, D.M.; Logan, B.E. Pilot scale microbial fuel cells using air cathodes for producing electricity while treating wastewater. *Water Res.* **2022**, *215*, 118208. [CrossRef]
2. Walter, X.A.; Madrid, E.; Gajda, I.; Greenman, J.; Ieropoulos, I. Microbial fuel cell scale-up options: Performance evaluation of membrane (c-MFC) and membrane-less (s-MFC) systems under different feeding regimes. *J. Power Sources* **2022**, *520*, 230875. [CrossRef]
3. Sevda, S.; Dominguez-Benetton, X.; Vanbroekhoven, K.; Wever, H.D.; Sreekrishnan, T.R.; Pant, D. High strength wastewater treatment accompanied by power generation using air cathode microbial fuel cell. *Appl. Energy* **2013**, *105*, 194–206. [CrossRef]
4. Schmidt, A.; Sturm, G.; Lapp, C.J.; Siebert, D.; Saravia, F.; Horn, H.; Ravi, P.P.; Lemmer, A.; Gescher, J. Development of a production chain from vegetable biowaste to platform chemicals, Microb. *Cell Factories* **2018**, *17*, 90. [CrossRef] [PubMed]
5. Galai, S.; de los Ríos, A.P.; Hernández-Fernández, F.J.; Kacem, S.H.; Mateo-Ramírez, F.; Quesada-Medina, J. Microbial Fuel Cell Application for Azoic Dye Decolorization with Simultaneous Bioenergy Production Using *Stenotrophomonas* sp. *Chem. Eng. Technol.* **2015**, *38*, 1511–1518. [CrossRef]

6. Hernández-Fernández, F.J.; de los Ríos, A.P.; Salar-García, M.J.; Ortiz-Martínez, V.M.; Godínez, L.J.L.C.; Tomás-Alonso, F.; Quesada-Medina, J. Recent progress and perspectives in microbial fuel cells for bioenergy generation and wastewater treatment. *Fuel Process. Technol.* **2015**, *138*, 284–297. [CrossRef]
7. Banerjee, A.; Calay, R.K.; Eregno, F.E. Role and Important Properties of a Membrane with Its Recent Advancement in a Microbial Fuel Cell. *Energies* **2022**, *15*, 444. [CrossRef]
8. Lozano, L.J.; Godínez, C.; de los Ríos, A.P.; Hernández-Fernández, F.J.; Segado, S.S.; Alguacil, F.J. Recent advances in supported ionic liquid membrane technology. *J. Membr. Sci.* **2011**, *376*, 1–14. [CrossRef]
9. Salar-García, M.J.; Santoroa, C.; Kodali, M.; Serovc, A.; Artyushkova, K.; Atanassov, P.; Ieropoulosa, I. Iron-streptomycin derived catalyst for efficient oxygen reduction reaction in ceramic microbial fuel cells operating with urine. *J. Power Sources* **2019**, *425*, 50–59. [CrossRef]
10. Roche, I.; Katuri, K.; Scotthamber, K. A microbial fuel cell using manganese oxide oxygen reduction catalysts. *J. Appl. Electrochem.* **2010**, *40*, 13–21. [CrossRef]
11. Touach, N.; Ortiz-Martínez, V.M.; Salar-García, M.J.; Benzaouak, A.; Hernández-Fernández, F.J.; de los Ríos, A.P.; Labjar, N.; Louk, S.; El Mahi, M.; Lotfi, E.M. Influence of the preparation method of MnO_2-based cathodes on the performance of single-chamber MFCs using wastewater. *Sep. Purif. Technol.* **2016**, *171*, 174–181. [CrossRef]
12. Gajda, I.; Greenman, J.; Santoro, C.; Serov, A.; Melhuish, C.; Atanassov, P.; Ieropoulos, I.A. Improved power and long term performance of microbial fuel cell with Fe-N-C catalyst in air-breathing cathode. *Energy* **2018**, *144*, 1073–1079. [CrossRef]
13. Benzaouak, A.; Touach, N.; Ortiz-Martínez, V.M.; Salar-García, M.J.; Hernández-Fernández, F.J.; de los Ríos, A.P.; El Mahi, M.; Lotfi, E.M. Ferroelectric LiTaO3 as novel photo-electrocatalyst in microbial fuel cells, Environ. *Prog. Sustain.* **2017**, *36*, 1568–1574. [CrossRef]
14. Santoro, C.; Artyushkova, K.; Babanova, S.; Atanassov, P.; Ieropoulos, I.; Grattieri, M.; Cristiani, P.; Trasatti, S.; Li, B.; Schuler, A.J. Parameters characterization and optimization of activated carbon (AC) cathodes for microbial fuel cell application. *Bioresour. Technol.* **2014**, *163*, 54–63. [CrossRef]
15. Kacem, S.H.; Galai, S.; de los Ríos, A.P.; Fernández, F.J.H.; Smaali, I. New efficient laccase immobilization strategy using ionic liquids for biocatalysis and microbial fuel cells applications. *J. Chem. Technol. Biotechnol.* **2018**, *93*, 174–183. [CrossRef]
16. Blázquez, E.; Gabriel, D.; Baeza, J.A.; Guisasola, A.; Ledezma, P.; Freguia, S. Implementation of a sulfide–air fuel cell coupled to a sulfate-reducing biocathode for elemental sulfur recovery. *Inter. J. Environ. Res. Public Health* **2021**, *18*, 5571. [CrossRef] [PubMed]
17. Zhang, G.; Zhang, H.; Zhang, C.; Zhang, G.; Yang, F.; Yuan, G.; Gao, F. Simultaneous nitrogen and carbon removal in a single chamber microbial fuel cell with a rotating biocathode. *Process Biochem.* **2013**, *48*, 893–900. [CrossRef]
18. Gancarz, P.; Zorębski, E.; Dzida, M. Influence of experimental conditions on the electrochemical window. Case study on bis(trifluoromethylsulfonyl)imide-based ionic liquids. *Electrochem. Commun.* **2021**, *130*, 107107. [CrossRef]
19. Baicha, Z.; Salar-García, M.J.; Ortiz-Martínez, V.M.; Hernández-Fernández, F.J.; de los Ríos, A.P.; Marín, D.P.M.; Collado, J.A.; Tomás-Alonso, F.; El Mahi, M. On the selective transport of nutrients through polymer inclusion membranes based on ionic liquids. *Processes* **2019**, *7*, 544. [CrossRef]
20. Hernández-Fernández, F.J.; de los Ríos, A.P.; Mateo-Ramírez, F.; Godinez, C.; Lozano-Blanco, L.J.; Moreno, J.I.; Tomas-Alonso, F. New application of supported ionic liquids membranes as proton exchange membranes in microbial fuel cell for wastewater treatment. *J. Chem. Eng.* **2015**, *279*, 115–119. [CrossRef]
21. Hernández-Fernández, F.J.; de los Ríos, A.P.; Mateo-Ramírez, F.; Juárez, M.D.; Lozano-Blanco, L.J.; Godinez, C. New application of polymer inclusion membrane based on ionic liquids as proton exchange membrane in microbial fuel cell. *Sep. Purif. Technol.* **2016**, *160*, 51–58. [CrossRef]
22. Ortiz-Martínez, V.M.; Salar-García, M.J.; Hernández-Fernández, F.J.; de los Ríos, A.P. Development and characterization of a new embedded ionic liquid based membrane-cathode assembly for its application in single chamber microbial fuel cells. *Energy* **2015**, *93*, 1748–1757. [CrossRef]
23. Marszałek, M.; Kowalski, Z.; Makara, A. Physicochemical and microbiological characteristics of pig slurry. *Tech. Trans. Chemical.* **2014**, *111*, 81–91.
24. Cong, H.N.; El Abbassi, K.; Chartier, P. Electrocatalysis of Oxygen Reduction on Polypyrrole/Mixed Valence Spinel Oxide Nanoparticles. *J. Electrochem. Soc.* **2002**, *149*, A525. [CrossRef]
25. Mateo-Ramírez, F.; Addi, H.; Hernández-Fernández, F.J.; Godínez, C.; de los Rios, A.P.; El Lotfi, M.; El Mahi, M.; Blanco, L.J.L. Air breathing cathode-microbial fuel cell with separator based on ionic liquid applied to slaughterhouse wastewater treatment and bio-energy production. *J. Chem.Technol. Biotechnol.* **2017**, *92*, 642–648. [CrossRef]
26. Garrido, Y.; Tudela, J.A.; Marín, A.; Allende, A.; Gil, M.I. Reconditioning of Wash Water for the Fresh-Cut Industry. *Adv. Food Process Technol.* **2020**, *3*, 126.
27. Fu, J.; Lee, W.N.; Coleman, C.; Nowack, K.; Carter, J.; Huang, C.H. Removal of disinfection byproduct (DBP) precursors in water by two-stage biofiltration treatment. *Water Res.* **2017**, *123*, 224–235. [CrossRef]
28. Logan, B.E.; Hamelers, B.; Rozendal, R.; Keller, U.S.J.; Freguia, S.; Aelterman, P.; Verstraete, W.; Rabaey, K. Microbial Fuel Cells: Methodology and Technology. *Environ. Sci. Technol.* **2006**, *40*, 5181–5191. [CrossRef]
29. Kiely, P.D.; Cusick, R.; Call, D.F.; Selembo, P.A.; Regan, J.M.; Logan, B.E. Anode microbial communities produced by changing from microbial fuel cell to microbial electrolysis cell operation using two different wastewaters. *Bioresour. Technol.* **2011**, *102*, 388–394. [CrossRef]

30. Yokoyama, H.; Ohmori, H.; Ishida, M.; Waki, M.; Tanaka, Y. Treatment of cow-waste slurry by a microbial fuel cell and the properties of the treated slurry as a liquid manure. *Anim. Sci. J.* **2006**, *77*, 634–638. [CrossRef]
31. Tomás-Alonso, F.; Rubio, A.M.; Giménez, A.; de los Ríos, A.P.; Salar-García, M.J.; Ortiz-Martínez, V.M.; Hernández-Fernández, F.J. Influence of ionic liquid composition on the stability of polyvinyl chloride-based ionic liquid inclusion membrnes in aqueous solution. *AIChE J.* **2017**, *63*, 770–780. [CrossRef]
32. You, J.; Greenman, J.; Melhuish, C.; Ieropoulos, I. Electricity generation and struvite recovery from human urine using microbial fuel cells. *J. Chem. Technol. Biotechnol.* **2016**, *91*, 647–654. [CrossRef]
33. Doyle, J.D.; Parsons, S.A. Struvite formation control and recovery. *Water Res.* **2002**, *36*, 3925–3940. [CrossRef] [PubMed]
34. Zhao, F.; Harnisch, F.; Schrder, U.; Scholz, F.; Bogdanoff, P.; Herrmann, I. Challenges and constraints of using oxygen cathodes in microbial fuel cells. *Environ. Sci. Technol.* **2006**, *40*, 5193–5199. [CrossRef] [PubMed]
35. Ichihashi, O.; Hirooka, K. Removal and recovery of phosphorus as struvite from swine wastewater using microbial fuel cell. *Bioresour. Technol.* **2012**, *114*, 303–307. [CrossRef]

fermentation

MDPI

Article

Maximizing Bioethanol Production from *Eucalyptus globulus* Using Steam Explosion Pretreatment: A Multifactorial Design and Fermenter Development for High Solid Loads

Eduardo Troncoso-Ortega [1], Roberto Valenzuela [2], Pablo Reyes-Contreras [3,4], Patricia Castaño-Rivera [5], L-Nicolás Schiappacasse [6] and Carolina Parra [1,7,*]

1 Renewable Resourses Laboratory, Biotechnology Center, University Campus, University of Concepcion, Concepcion 4030000, Chile; etroncoso@udec.cl
2 Innocon S.A., Concepción 4070304, Chile; rvalenzuela@innocon.cl
3 Center of Excellence in Nanotechnology (CEN), Leitat Chile, Santiago 8320000, Chile; preyes@leitat.cl
4 National Center of Excellence for the Wood Industry (CENAMAD), Pontifical Catholic University of Chile, Santiago 7820436, Chile
5 Technological Development Unit (UDT), University of Concepcion, Coronel 4190000, Chile; p.castano@udt.cl
6 Faculty of Engineering, Catholic University of Temuco, Temuco 4780000, Chile; lschiappacasse@uct.cl
7 ANID—Millennium Science Initiative Program-Millennium Nuclei on Catalytic Process towards Sustainable Chemistry (CSC), Santiago 8320000, Chile
* Correspondence: roparra@udec.cl; Tel.: +56-41-220730

Abstract: Steam explosion pretreatment is suitable for bioethanol production from *Eucalyptus globulus* wood. Multifactorial experiment designs were used to find the optimal temperature and residence time required to obtain the best glucose yield from the enzymatic hydrolysis of pretreated materials. The chemical composition, crystallinity index, morphology and polymerization degree of the pretreated materials were correlated with enzymatic accessibility. Simultaneous saccharification and fermentation (SSF) using a fed-batch strategy was applied to three different laboratory-scale fermenters. The optimization of the pretreatment was obtained at 208 °C and 11 min. However, the enzymatic hydrolysis performance did not show significant differences from the material obtained at 196 °C and 9.5 min, which was determined to be the real optimum, owing to its lower energy requirement. The vertical fermenter with type "G" blades and the horizontal fermenter with helical blades were both highly efficient for reaching ethanol yields close to 90% based on dry wood, and ethanol concentrations close to 9.0% *v*/*v*.

Keywords: high-concentration bioethanol; response surface; simultaneous saccharification and fermentation; fed batch; fermenters design; steam explosion pretreatment

Citation: Troncoso-Ortega, E.; Valenzuela, R.; Reyes-Contreras, P.; Castaño-Rivera, P.; Schiappacasse, L.-N.; Parra, C. Maximizing Bioethanol Production from *Eucalyptus globulus* Using Steam Explosion Pretreatment: A Multifactorial Design and Fermenter Development for High Solid Loads. *Fermentation* **2023**, *9*, 965. https://doi.org/10.3390/fermentation9110965

Academic Editors: Miguel Ladero and Victoria E. Santos

Received: 16 September 2023
Revised: 20 October 2023
Accepted: 20 October 2023
Published: 10 November 2023

1. Introduction

Lignocellulosic biomass (LCB) is considered to play an important role in the building of a sustainable society, as it has the potential to replace fossil fuels and chemicals. For the conversion and use of biomass in value-added products, the "biorefinery" concept has emerged, which can contribute to the development of the circular economy by recovering and recycling bio-based products [1]. Currently, one of the main products of biomass is biofuels, which are estimated at a world production of 137 billion liters in 2016 [2]. Currently, bioethanol is the most ideal bio-based fuel or fuel additive for use in motor vehicles as a partial substitute for fossil fuels. Compared to gasoline, bioethanol contains 34.7% oxygen, which enables a 15% higher combustion efficiency, thereby resulting in fewer emissions of particulate nitrogen oxides, which are harmful to the environment [3]. The production of second-generation bioethanol (from lignocellulosic materials) requires several stages: pretreatment, hydrolysis, fermentation and distillation, where each stage will have consequences for the quality and cost of bioethanol [4]. The objective of pretreatments is to increase the susceptibility of the material to obtain a reactive lignocellulosic substrate that is accessible to the enzyme. The efficiency of pretreatment is measured by the intensity at which hydrolysis becomes easier with minimal degradation and/or loss of carbohydrates, avoiding the formation of inhibitory compounds for the fermentative process. Among the different kinds of pretreatment methods, steam explosion (SE) is one of the most successful and widely used for fractionating lignocellulosic biomass [5,6]. This pretreatment has its origin in the Masonite process to produce fiber boards, described in 1930, where the high-pressure vapor penetrates the cell structure for a short time, and is then released, expanding rapidly when the reactor is depressurized, thereby causing the cell system to deconstruct (as a result of mechanical force by pressure drop). Before depressurization, a hydrolytic mechanism occurs that promotes the breakdown of hemicelluloses (autohydrolysis of the acetyl group), lignin modifications and cellulose exposure, promoting the formation of pores in the biomass and improving the efficiency of hydrolysis and saccharification during the bioethanol production process. It is known that the biomass depressurizes quickly in the expansion chamber. It is also known that the rapid depressurization of the biomass in the expansion chamber, causes a subito stress, which disrupts the glycosidic linkages and hydrogen bonds, facilitating enzymatic hydrolysis of the sugars (pentoses and hexoses) released from the hemicellulose and cellulose [7–9].

Additionally, another challenge for the viability of the bioethanol process and for bioethanol production is obtainment of highly concentrated ethanol during the fermentation or saccharification and simultaneous fermentation (SSF) stage, which would allow for an optimization of the energy consumed during distillation. Approximately 37% of the energy required by the bioethanol production plant is consumed during distillation [10].

To achieve a higher ethanol concentration, high pretreated substrate loading is crucial for the economy of the SSF process [11–13]. Working with the highest solid load has a significant effect on the capital and production cost of the process due to the reduction in size of the required equipment, such as tanks and distillation columns, as well as the reduced amount of energy utilized during distillation and the reduced production of wastewater [14]. For bioethanol produced from lignocellulosic material to be economically feasible on an industrial scale, the ethanol produced must be above 4% (w/v) [15]. High substrate loadings mean that the process is taking place at pretreated solid levels where there are not significant initial amounts of free water [16]. Less water causes an increase in the viscosity of the substrate matrix and creates rheological problems when the mass transfer rate in the substrate matrix is obstructed, thereby increasing the time required to liquefy the matrix and to perform efficient hydrolysis by the enzymes using classical batch-mode SSF [12], [17]. Thus, it has been demonstrated that the glucose and ethanol yields in enzymatic hydrolysis and SSF, respectively, decreased linearly with increasing solid concentration [18,19]. However, this statement has been subject to debate taking into consideration the efficiency and amount of enzyme used during mixing [20].

It is well known that the enzymatic activity of the cellulases complex decreases the viscosity of the lignocellulosic substrate during enzymatic hydrolysis. Some authors have proposed a gradual increment in the substrate concentration, in the fed-batch mode, instead of adding all of the substrate initially in the batch mode [21]. This strategy helps to maintain the viscosity at relatively low levels, thereby avoiding mixing and heat and mass transfer problems [13,22]. Several works using the fed-batch SSF approach have been performed using different raw materials and pretreatments, demonstrating the feasibility of this strategy, not only due to the improvement of the ethanol concentration obtained, but also due to the product yields achieved [17,23]. Guigou et al. [24] published a comparison of ethanol yields and concentrations achieved using different species of Eucalyptus, with initial solid percentages ranging from 4%wt to 27%wt and concentrations between 5.4 g/L to 30.7 g/L. The highest ethanol concentration was achieved using the presaccharification and simultaneous fermentation (PSSF) strategy for *Eucalyptus grandis* pretreated by steam explosion and *S. cerevisiae* PE-2 as the fermentative microorganism.

This work was focused on the optimization of the SE pretreatment for *Eucalyptus globulus*, a most important commercial hardwood species in the world and the second most important in Chile [25,26]. A multifactorial experimental design was used. The optimal parameters of pretreatment were obtained by using a response surface methodology, followed by a physicochemical characterization of pretreated materials to different severities. The physicochemical characteristics of these materials were correlated with the efficiency of enzymatic hydrolysis. The best-pretreated materials, with high enzymatic accessibility, were evaluated in bioethanol production, showing high yields and high concentrations when using fermenters designed for a SSF process with high solid loads.

2. Materials and Methods

2.1. Raw Material

Woodchips from 10- to 12-year-old *E. globulus* trees were donated by CMPC Cellulose S.A., Chile. The woodchips were selected to be a specific size of 2.5 cm × 1.5 cm × 0.3 cm to obtain a homogeneous sample. They were air-dried to a moisture content of 10% and stored under dry conditions until they were used.

2.2. Chemical Characterization of Samples

The woodchips were milled using a knife mill (Maestranza Proinco S.A., Concepción, Chile) and then classified using an ASTM sieve (45–60 mesh). The milled wood was extracted with a 90% acetone solution for 16 h in a Soxhlet extractor to measure the amount of extractives present. The steam-exploded samples were dried until they reached a humidity level of 10% and then they were milled and sieved. However, the sample from pretreatment process not were extracted with acetone. To hydrolyze the milled samples, 3 mL of 72% sulfuric acid was added to 300 mg of each sample and incubated at 30 °C for 1 h. The acid was then diluted to 4% by the addition of 84 mL of water, and the mixture was heated at 121 °C (1 atm) for 1 h. Finally, the resulting residual material was cooled and filtered through a number 4 porous glass filter. The solids fractions were dried at 105 °C until a constant weight was achieved, and then determined to be insoluble lignin. The concentration of soluble lignin in the filtrate was determined by measuring the absorbance at 205 nm, using an absorption coefficient of 110 L/g cm. Glucose, xylose, and acetyl groups were analyzed using a HPLC (Merck Hitachi, Tokyo, Japan) equipped with a refractive index detector and an Aminex HPX-87H (Bio-Rad, CA, USA) column at 45 °C, eluted at a flow rate of 0.6 mL/min with 5 mM H_2SO_4 [27]. To convert sugar monomers to anhydromonomers, 0.90 for glucose (reported as glucan) and 0.88 for xylose (reported as xylans) were used. These factors were calculated based on the water added to polysaccharides during acid hydrolysis, taking into account the molar mass of each original anhydromonomer in the polysaccharide. This resulted in a 10% increase for glucose (from 162 g/mol in the anhydromonomer to 180 g/mol in glucose) and a 12% increase for xylose (from 132 g/mol in the anhydromonomer to 150 g/mol in xylose). The acetyl content was calculated by multiplying the acetic acid content by 0.7. The percentages of reaction products (glucan, xylan, and lignin) in the solid and liquid fractions were calculated based on the initial amount of each component loaded into the SER (*w*/*w* dry basis). The moisture content of the solid fraction was determined using a moisture analyzer (Sartorius MA35).

2.3. Experimental Design to Optimize Steam Explosion Pretreatment

The range of variables for this study was defined based on previous work and the safety restrictions of the reactor and its boiler. Response surface methodology (RSM) was used to determine the optimal pretreatment condition that yields the highest glucose (Y_{EH}) in enzymatic hydrolysis. The influence and optimization of the variables, temperature and residence time were determined using Central Composite Circumscribed design (CCC) with Modde 7.0 software (Umetrics, Sweden). A two-level factorial design was used for each parameter, with levels of -1 and +1, star points with levels of $-\sqrt{2}$ and $+\sqrt{2}$ and a central point (0, 0) which is performed in triplicate to obtain the experimental standard deviation, as shown in Table 1. The temperature range evaluated was 180–213 °C while the residence time range was 4–15 min. The response variable, Y_{EH}, was obtained through enzymatic hydrolysis for 72 h. The data were analyzed using Modde 7.0 software to obtain the polynomials related to the reaction system and the response surface plot. The model was statistically validated using analysis of variance (ANOVA) with a confidence level of 95%, using the same software. Variables such as wood particle size (3 × 3 cm^2), and impregnation reagent (2 L of water for 200 g of dry base wood) were kept constant.

Table 1. Experimental design results and optimal conditions for steam explosion pretreatment of *E. globulus*. Solid yield, raw material and pretreated materials chemical composition, glucose yields obtained in enzymatic hydrolysis.

Pretreatment Conditions				**Raw and Pretreated Material Composition (% dwb) [a]**				**%Y_{EH} [e] (dwb) (72 h)**
Exp N°	**Temperature (°C)**	**Time (min)**	**Log S_0**	**Solid Yield**	**Glucans**	**Xylans**	**Lignin**	**Glucose Yield**
E.g. [b]	--	--	--	--	45.5	15.3	23.5	--
1	180 (−1)	4 (−1)	2.96	92.0	47.0	16.9	28.0	6.4
2	213 (+1)	4 (−1)	3.93	71.9	64.9	3.2	35.3	71.2
3	180 (−1)	15 (+1)	3.53	85.1	53.1	12.1	31.4	28.5
4	213 (+1)	15 (+1)	4.50	68.1	68.4	1.6	41.3	73.1
5	173 (−1.4)	9.5 (0)	3.13	93.0	48.2	18.3	26.0	8.1
6	220 (+1.4)	9.5 (0)	4.51	66.6	66.8	1.5	40.3	76.9
7	196 (0)	1.7 (−1.4)	3.06	74.3	50.6	15.3	29.3	19.8
8	196 (0)	17 (−1.4)	4.06	68.6	65.0	3.2	36.6	75.9
9 [c]	196 (0)	9.5 (0)	3.80	70.9 ± 3.0	63.2 ± 1.3	4.4 ± 0.4	35.1 ± 1.0	73.3 ± 4.8
10 [d]	208	11	4.22	67.5 ± 0.9	58.3 ± 1.4	2.7 ± 0.1	31.8 ± 0.7	74.5 ± 2.3

[a] Dry wood basis. [b] *Eucalyptus globulus* chips. [c] Exp N° 9 represents the average of triplicate experiments at the central point. [d] Average values obtained from triplicate pretreated material under optimized conditions. [e] Yield enzymatic hydrolysis.

2.4. Steam Explosion Pretreatment

Wood chips were pretreated in a steam explosion reactor (SER) (Maestranza Proinco S.A., Concepción, Chile) consisting of a steam generator, 5 L vessel reactor and an expansion chamber. The steam generator has a volumetric capacity of 150 L and a 25 kW electric heater, allowing it to generate saturated steam between 220 °C and 230 °C. The expansion chamber with a capacity of 120 L is equipped with an automated fast open valve. The reaction vessel is equipped with temperature and pressure sensors. The reaction time and temperature are controlled using a console (control keypad), as shown in Figure 1. For each experiment, the reactor was loaded with 200 g of *E. globulus* woodchips, based on dry wood basis (dwb). Saturated steam at various temperatures (ranging from 173 to 220 °C) was then introduced for different residence times (ranging from 1.7 to 17 min), following the multivariate experimental design outlined in Table 1. Finally, the reactor was depressurized to atmospheric pressure. The slurry was collected and filtered, and the liquid was then stored a temperature of 4 °C for further determine of glucose, xylose and arabinose contents. In order to determine the solids yield (%), the pretreated wood chips were left to dry in air for approximately 24 h. After this period, the wood chips were weighed, and the moisture content was measure. These two data were utilized to calculate amount of dry mass obtained after the pretreatment process. The severity factor (S_0) was calculated using Equation (1):

$$(S_0) = \log(t \times \exp\left[\frac{T_H - T_R}{14.75}\right]) \quad (1)$$

where t is the reaction time in minutes, T_H is the hydrolysis temperature in °C and T_R is a reference temperature of 100 °C. The value of 14.75 represents a constant (ω), which is an empirical parameter related to the activation energy and temperature of the reaction.

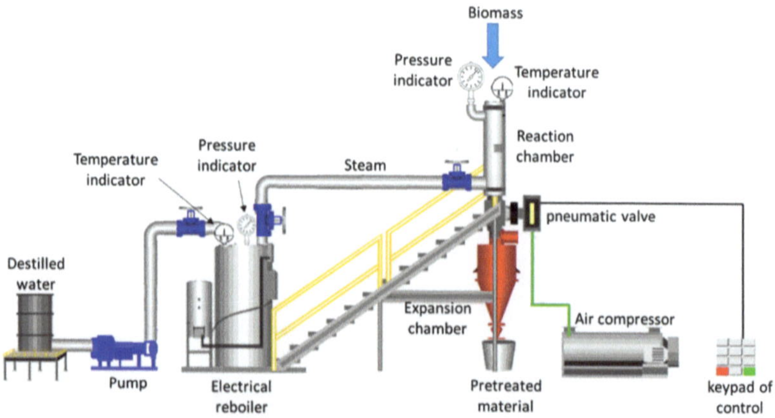

Figure 1. This is a schematic representation of the steam explosion reactor located at the Renewable Resources Laboratory of the University of Concepción.

2.5. *Xylose Recovery from Liquid Fraction Post Steam Explosion Pretreatment*

The liquid fraction was processed according the method outlined by Castro et al. [28] to measure the amount of xylose present in the liquid fraction. In this method, a sample of containing primarily xylose and xylooligomers is subjected to hydrolysis using 0.6% H_2SO_4 at 121 °C for 60 min in an autoclave. After hydrolysis, the sample is filtered through 0.22 μm porosity membranes. The total xylose is determined using HPLC described above for the quantification of carbohydrates. The proportion of xylooligomers in the liquor is calculated by subtracting the xylose content in hydrolyzed liquor from the xylose content in the non-hydrolyzed liquor.

2.6. *Characterization of Cellulose*

2.6.1. Cellulose Extraction

The cellulose isolation of pretreated pulps with a wood-like texture at low severity (2.96; 3.53; 3.13 and 3.06) was conducted using the following procedure [29]: 2 g of the sample were treated with an aqueous solution consisting of 80 mL of water, 2.50 g $NaClO_3$ and 2.0 mL of acetic acid at 80 °C for 90 min. The resulting insoluble residue (holocellulose = cellulose + hemicellulose) was cooled, filtered through a glass filter number 2 and washed with distilled water and acetone. The holocellulose was then dried at 40 °C, weighed, and 0.5 g it was treated with 12.5 mL of 0.5% KOH at 20 °C for 5 min. The obtained cellulose was filtered through a glass filter number 4, washed with distilled water, a 1 M solution of acetic acid and water again. Finally, it was dried at 40 °C. The cellulose isolation of high severity pretreated materials with a pulp-like texture (3.93; 4.50; 4.51; 4.06 and 3.80) was performed as follows: 1 g of the sample was treated with an aqueous solution consisting of 40 mL of water, 1.25 g of $NaClO_2$ and 0.5 mL of acetic acid at 80 °C for 25 min. The resulting insoluble residue (holocellulose = cellulose + hemicellulose) was cooled, filtered through a sintered glass filter number 2 and washed with distilled water and acetone. The holocellulose was then dried at 40 °C, weighed, and 0.5 g it was treated with 12.5 mL of 0.5% KOH at 20 °C for 5 min. Subsequently, 12.5 mL of water was added, and the mixture was treated for additional 5 min at 20 °C. The obtained cellulose was filtered through a glass filter number 4, washed with distilled water, a 1 M solution of acetic acid and water again. Finally, it was dried at 40 °C.

2.6.2. Average Degree of Polymerization

The average degree of polymerization (DP) of isolated cellulose samples was calculated from its intrinsic viscosity [η] using the equation: $DP^{0.90} = 1.65\ [\eta]$, as described by Monrroy et al. [29]. Specifically, the viscosity of a 0.5% cellulose solution in 0.5 M cupper ethylenediamine was measured using a capillary viscometer (Cannon Fenske Viscometer, Cannon Instrument Co., State College, PA, USA).

2.6.3. The Cellulose Crystallinity Index by Spectroscopic FT-IR

The cellulose crystallinity index (CrI) was determined using FT-IR. FTIR spectra of isolated cellulose samples were measured by directly transmitting light through a KBr pellet. To prepare the pellet, 1.5 mg of dried sample was mixed with 200 mg of KBr (spectroscopy grade, Merck, Germany) in an agate mortar. The resulting mixture was then pressed at 5000 psi for 2 min. Spectra were recorded between 4000 and 400 cm^{-1}, using a Perkin Elmer System 2000 FT-IR (Perkin Elmer, Inc., Waltham, MA, USA) equipped with a DTGS detector. The background used for FT-IR measurements was a KBr pellet that did not contain any sample. All spectra were measured at a spectral resolution of 8 cm^{-1} and 64 scans were taken per sample. These spectra were normalized to the intensity of the band at 2900 cm^{-1} (C–H stretching vibration), which is a relatively constant band that does not change with the temperature of pretreatment (ranging from 30 to 250 °C). This band was used as a reference to compare the spectra from different samples and determine the cellulose crystallinity. The CrI was evaluated by calculating the ratio of the absorption bands H1372/H2900, which was determined from the averages of four measurements for each sample. The band at 1372 cm^{-1} is related to cellulose crystallinity. The ratio of peak heights at 1372 and 2900 cm^{-1} (H1372/H2900), which represents the ratio of C–H bending to C–H stretching, has been used in various studies to predict cellulose crystallinity [30].

2.7. Surface Morphology by Scanning Electronic Microscopy

Images were captured of the fiber surfaces both before and after pretreatment using a JSM- 6380LV SEM instrument (Jeol, MA, USA) under high vacuum conditions. The SEM is equipped with a secondary electron detector. The samples were dried at room temperature and then coated with conductive gold paint (particle size 500 Å) using an S150 Edwards Sputter Coater (Atlas Copco Group, Stockholm, Sweden). The imaging process was conducted at a beam-accelerating voltage of 20 kV with a tungsten filament serving as the electron source.

2.8. Enzymatic Hydrolysis

Enzymatic hydrolysis (EH) experiments were conducted in 125 mL Erlenmeyer flasks at 50 °C using an orbital shaking incubator (Labtech LSI-4018A, Sorisole, BG, Italy) set at 150 rpm for 72 h. Each experiment was performed in triplicate with a total volume of 50 mL containing 10% solid pretreated material (*w*/*v*). A commercial cellulase enzyme complex (NS-22128 CCN3128; 71 FPU mL/mL) supplemented with β-glucosidase (NS-22128 DCN00216; 370 CB mL/mL) and a pH 4.8 buffer of 0.05 M sodium citrate were used. The enzyme dosages employed were 20 FPU and 20 CBU of cellulase and β-glucosidase, respectively, per gram of dry material. The glucose content released during the enzymatic treatment was analyzed by HPLC [8]. Enzymatic digestibility was expressed as enzymatic hydrolysis yield ($\%Y_{EH}$) and was calculated using Equation (2):

$$\%Y_{EH} = \frac{\left(\frac{Gs}{\text{dry substrate loading (g)}}\right) \times \%SR}{Gi} \times 100 \quad (2)$$

where Gs represents the amount of glucose released in grams, the dry substrate loading of pretreated biomass in grams, the percentage of solid recovered after the pretreatment (%SR) and Gi denotes the initial amount of glucose in wood expressed as grams of monomer.

2.9. Description of the Horizontal Fermenter with Helical Blades

The horizontal fermenter (HF) is a stainless-steel container with a semi-cylindrical shape and an internal volume of 500 mL. To regulate the temperature, it was designed with a double jacket. The helical blade, made of 316 stainless-steel are responsible for agitating the pretreated material. The fermenter is equipped with an acrylic cover, allowing for observation of the agitation. To facilitate gas exchange during fermentation, a rubber stopper with a syringe needle is placed on the acrylic cap. Rubber seals are used to ensure a hermetic seal for the fermenter. The stirring speed is externally controlled at 50 rpm using motor.

2.10. Description of the Vertical Fermenter with Type "G" Blades

The vertical fermenter (VF) consisted of a glass container with an internal volume of 420 mL. It was provided by a plastic cover with a rubber stopper to which a syringe needle was connected to allow gaseous exchange during fermentation. The cover and blade assembly were designed using Teflon seals to remain airtight throughout the fermentation stage. The stirring speed was maintained at 50 rpm by using an external motor. The fermenter was immersed in thermostatic water bath to maintain a constant fermentation temperature. The blades are made of 316 stainless steels.

2.11. Simultaneous Saccharification and Fermentation

Thermotolerant *Saccharomyces cereviseae* IR2-9a [31] was used as the fermentative microorganism. *S. cereviseae* IR2-9a are maintained in agar plates with a medium composed of 20 g/L of glucose, 20 g/L of agar, 20 g/L of peptone and 10 g /L of yeast extract; in a grow oven at 40 °C. Prior to the simultaneous saccharification and fermentation (SSF) and fed-batch SSF (FB-SSF) process, an inoculum of yeast was prepared in a growth liquid medium; contained 100 g/L of glucose, 10 g/L of yeast extract, 10 g/L of peptone, 4.14 g/L NH_4Cl, 1.17 g/L of KH_2PO_4 and 0.36 g/L of $MgSO_4 \times 7H_2O$. They were then inoculated with colonies isolated from an agar plate. The mixture was incubated at 40 °C with shaking at 150 rpm for 24 h. Subsequently the cells were harvested by centrifugation at 2500× *g* rpm, washed four times with sterile 0.9% *w*/*v* NaCl and suspended in the same solution. The cell concentration was determined gravimetrically. SSF experiments were performed at two dry-substrate loadings, 10 and 20% *w*/*v*. The enzyme dosages employed were 20 FPU and 20 CBU per gram of pretreated material and initial inoculums of 12 g/L of *S. cereviseae* IR2-9a. Samples were collected after 24, 48 and 72 h, unless otherwise stated. FB-SSF experiments were performed a final dry-substrate loading of 20% *w*/*v*. The initial substrate loading was 12%, and every two hours, the substrate loading was increased by 2% until the final concentration was reached; this process took a period of 8 h. Samples were collected at 24, 48 and 72 h, unless otherwise. The enzyme dosages and initial inoculum of S. cerevisiae IR2-9a were the same as those used in the SSF. All SSF and FB-SSF experiences were performed in 50 mM citrate buffer (pH 4.8) supplemented with 10 g/L of yeast extract, 10 g/L of peptone, 4.14 g/L NH_4Cl, 1.17 g/L of KH_2PO_4 and 0.36 g/L of $MgSO_4 \times 7H_2O$.

The ethanol yield of the reactions was calculated on a dry wood basis ($\%Y_E$) according to Equation (3), considering the ethanol in grams produced, dry substrate loading in grams of pretreated biomass, percentage of solid recovered after the pretreatments (%SR) and potential ethanol from the stoichiometric conversion of potential glucose in the feedstock.

$$\%Y_E = \frac{\left(\frac{\text{ethanol (g)}}{\text{dry substrate loading (g)}}\right) \times \%SR}{\text{potential glucose in wood } \times 0.51} \times 100 \quad (3)$$

2.12. Ethanol Analysis

The ethanol concentration was determined using a Merck Hitachi (Tokyo, Japan) HPLC system equipped with a refractive index detector, a Aminex HPX 87H column (Bio-Rad, Hercules, CA, USA) at 45 °C and H_2SO_4 5.0 mM as mobile phase at a flow rate of 0.6

mL/min was using to determination of ethanol concentration. The ethanol standards were prepared using commercial grade ethanol.

3. Results and Discussion

The chemical composition of *E. globulus* woodchips and pretreated materials by SE is summarized in Table 1. The solid yields ranged from 66 to 93% (dry matter), depending on the severity of pretreatment, expressed as S_0.

3.1. Steam Explosion Pretreatment Optimizaztion by CCC Design

A set of 11 experiments were performed for the different pretreatment conditions, defined by the CCC design. The effects of the pretreatment condition on the pretreated materials and experimental $\%Y_{EH}$ are shown in Table 1. The responses were fitted to a second-order polynomial equation using Modde 7.0 software (Unimetrics), as shown in Equation (4). The model equation was validated through ANOVA with 95% confidence level. The predicted polynomial responses were close to the experimental responses with a correlation coefficient $r^2 = 0.95$.

$$\%Y_{EH} = 74.31 + 25.74\,T + 12.52\,t - 15.82\,T^2 - 13.42\,t^2 - 45.25\,Tt \quad (4)$$

where $\%Y_{EH}$ is the enzymatic hydrolysis yield predicted, (in Wood dry weight basis); T = temperature, °C and t = time, minutes.

The contour diagram describing the estimated response surface for $\%Y_{EH}$ on a wood dry weight basis of *E. globulus* pretreated with SE is shown in Figure 2. Through the response surface and applying least squares regression, the optimal pretreatment values for the temperature and reaction time predicted by mathematical model were 208 °C and 11 min. These parameters were used to validate the optimal condition predicted by polynomial. The chemical composition of the materials pretreated under optimal conditions is shown in Table 1, codified as Exp. N° 10 and represents the average of three experiments. The $\%Y_{EH}$ of the pretreated material under optimal predicted conditions reached a value of 74.5 ± 2.3% glucose on a dry wood basis with a S_0 of 4.22. Similar values were obtained for the materials obtained from experiments 2, 4, 6, 8 and 9, with $\%Y_{EH}$ ranging from 71 to 77% of glucose. Considering the energy involved during pretreatments (S_0), it was evident that the center point experiments (S_0 = 3.80) were more competitive than those obtained under the optimal predicted condition. To verify this, a statistical test (Student's *t*-test) allowed asseveration that there is no significant difference between $\%Y_{EH}$ reached by the material obtained at optimal conditions (208 °C for 11 min) compared to those obtained from the center point conditions (196 °C for 9.5 min). For the statistical test, a population of nine replicates per material was considered, resulting in t-calculated = 0.79, which was less than the t-critical (1.74); therefore, there was no significant difference at $p = 0.05$.

3.2. Analysis of Chemical Composition

The chemical composition of *E. globulus* used in this research is detailed in Table 1. The main component found were glucans with 45.5% followed by hemicellulose with 15.3% and lignin with 23.5%. In relation to the pretreated materials, it was observed that the chemical composition of each biomass samples pretreated at low severity, showed no greater removal of wood components towards the liquid fraction during the pretreat-ments. As S_0 increased above 3.8, the increase in dissolved materials became evident, and mainly hemicellulose and lignin were removed. At higher severity values, the degradation of xylan to xylose, which degrades products such a furfural and carboxylic acids, is favored. The curves of xylose and xylooligomer content in the liquid phase versus severity are shown in Figure 3a. It was possible to observe the depolymerization of hemicellulose to xylose up to a severity factor of 4.3. Above this value, xylose concentrations begin to decrease, probably due to its conversion into other compounds such as those mentioned above.

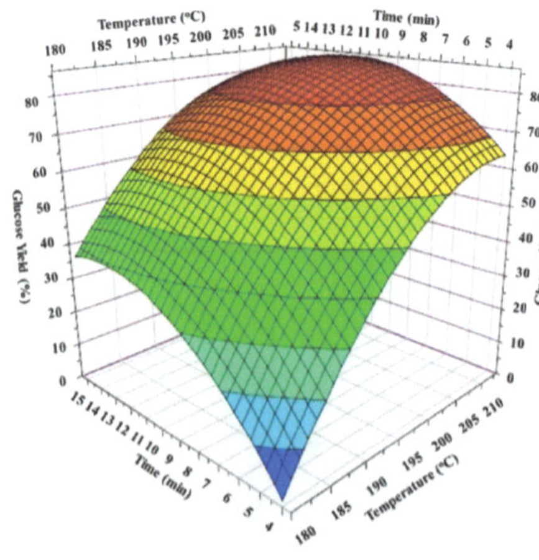

Figure 2. Estimated second-order polynomial response surface to enzymatic hydrolysis yield (%Y_{EH}) expressed in glucose released, in dry wood basis of the *Eucalyptus globulus* pretreated by steam explosion. Blue: lower glucose yields. Red: higher glucose yields.

In contrast, the amount of residual xylans obtained in the materials subjected to pretreatment between S_0 2.96 and 4.51 ranged from 18.9% to 1.5%, approximately 100% and 1.0%, respectively, of the original amount in the wood. The solubility of xylans depends on their molecular weight and the presence of side-chain substituents. Chen et al. [32] mention that acetyl groups, arabinose and uronic acid increase the solubility of xylans. Other authors have described SE pretreatment as an effective method for reducing LCB recalcitrance by removing hemicellulose, disrupting the lignin-hemicellulose matrix, and redistributing lignin in the cell wall layers [8,33] and thus allowing the preservation of glucans under controlled conditions. The increases in S_0 in the pretreatments, also a glucan preservation varied between 100% and 82.6%, while the lignin content in the same pretreated materials varied between 21.5% and 28.1% (Table 1), slightly higher than the lignin content in the raw material. As mentioned above, the reason for this increase could be the generation of pseudo-lignin during the SE pretreatment. Araya et al. [34] indicated that the increase in the lignin content of autohydrolysis pretreated materials at higher severity conditions is partly due to the concomitant loss of polysaccharides and the formation of condensed lignin products. This increase can also be attributed to the formation of lignin-like compounds from lignin and carbohydrate degradation.

3.3. Structural Changes in Cellulose in Pretreated Materials

The DP and crystallinity of cellulose are considered important factors in the recalcitrance of LCB [35]. The DP of cellulose in the raw material and pretreated materials was determined by the intrinsic viscosity method, while its crystallinity index (CrI) was determined by the ratio of bands of the infrared spectrum. *E. globulus* wood showed an initial DP of 3493, decreasing rapidly as the severity of the pretreatment increased, until reaching a value of 317 with the maximum severity (Figure 3b). The FTIR crystallinity index was calculated using the absorbance ratio of the 1372 and 2900 cm^{-1} bands. The CrI is between 0.45 and 0.38, observing a slight decrease as the severity increases (Figure 3b). Therefore, it is possible to attribute a good performance in enzymatic hydrolysis to materials that present a high removal of hemicellulose (xylan in *E. globulus*) and a lower DP of cellulose; however, it cannot be established that CrI plays an important role in the performance value of EH.

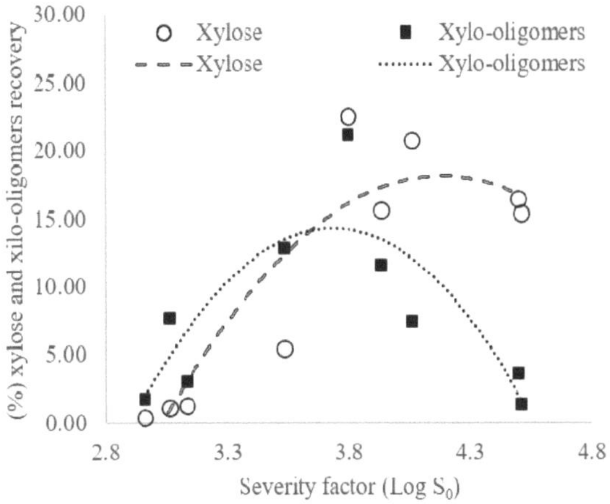

(**a**)

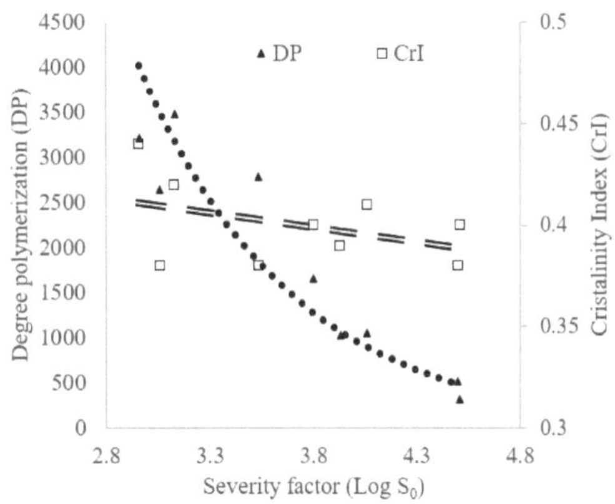

(**b**)

Figure 3. (**a**) Correlation between the recovery of xylose and xylo-oligomers in the liquid fraction with the severity of pretreatment and (**b**) relationship between the severity factor of the process with the degree polymerization and the crystallinity index of the cellulose isolated from pretreated material by steam explosion.

3.4. Scanning Electron Microscopy

Figure 4 shows images and scanning electron microscopy images for the most representative materials of the present study, that is, the pretreated material with the least severity (Exp. 1: 2.96), with the highest severity (Exp. 6: 4.51), central point of the design (Exp 9: 3.80) and material optimized by response surface (Exp. 10: 4.22). The first row shows a photograph of the materials pretreated by steam explosion, the second row shows scanning electron microscopy (SEM) images at a magnification of 130× to obtain a general idea of the fiber arrangement, and the third row shows SEM images at 1400× magnification to further characterize the surface of the fibers. A low severity does not manage to deconstruct the fibrillar structure of the wood, probably due to the high content of hemicellulose still present in the pretreated material. By applying a high severity of pretreatment, it is possible to completely deconstruct the structure of the wood, and even shorter fibers are observed, which is probably related to the drastic decrease in the DP of cellulose. At intermediate severities (Exp. 9 and Exp. 10), a mixture is observed between structures that have not yet been deconstructed and detached fibers, although they are longer than those present in Exp. 6. This can be explained for the sudden depressurization led to an "explosion" of the steam inside the lignocellulosic matrix, which promotes breakdown and defibrillation of its structure generating a solid fraction which a more open structure. The good-results were observed in the enzymatic hydrolysis of Exp. 9 and Exp. 12 indicate that no further deconstruction is necessary to achieve good glucose release if a significant amount of xylan is removed.

3.5. Simultaneous Sacharification and Fermentation

In Section 3.1 it was mentioned that the pretreated material obtained under the conditions of the central point of the experimental design (Exp. 9) presented a similar performance to the material obtained under the optimization conditions (Exp. 10), but with a lower energy expenditure from the pretreatment point of view (less severity). Severity it was used in subsequent studies. The fermentability of the Exp. 9 pretreated material were studied by SSF at two different substrate loads, 10% and 20% *w*/*v*. With a load of 20% solids, the challenge of achieving good agitation that allows the diffusion of the enzymes, liberating the glucose from the cellulose and then its fermentation to ethanol was verified. Agitation is an important parameter for overcome mass transfer related limitations. To overcome the agitation barrier, a fed-batch strategy was implemented (Figure 5), which allows the feeding of pretreated material to the SSF reactor as it is liquefied by the action of enzymes. This strategy was successfully implemented in three types of laboratory-scale fermenters.

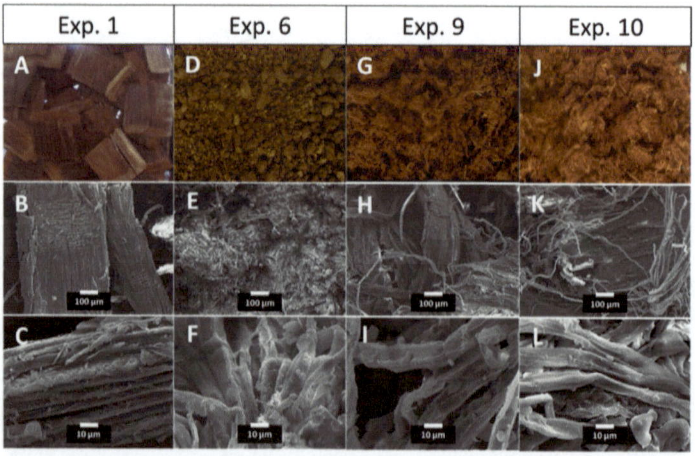

Figure 4. Macroscopic and microscopic images of pretreated materials selected according to severity. (**A**,**D**,**G**,**J**) correspond to the original texture of samples Exp. 1, Exp. 6, Exp. 9 and Exp. 10, respectively. (**B**,**E**,**H**,**K**) correspond to images by scanning electron microscopy with 130× magnification of samples Exp. 1, Exp. 6, Exp. 9 and Exp. 10, respectively. (**C**,**F**,**I**,**L**) correspond to images by scanning electron microscopy with 1400× magnification of samples Exp. 1, Exp. 6, Exp. 9 and Exp. 10, respectively.

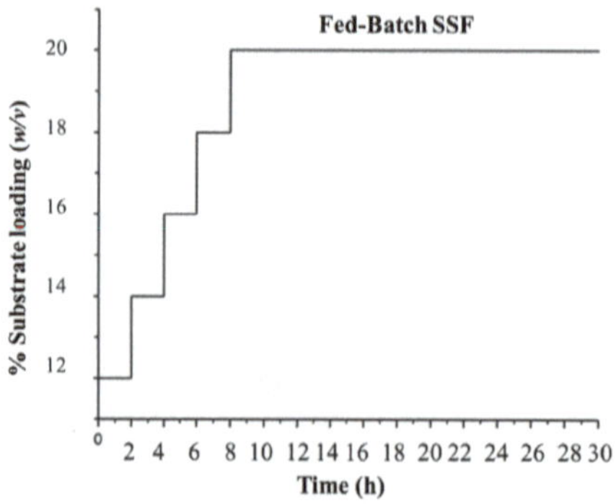

Figure 5. Fed-batch SSF strategy with 20 FPU/20 CBU and 12 g/L of *S. cereviseae* IR2-9a. Every two hours, the substrate loading was increased in 2% until to reach the final concentration, this process took a period of 8 h.

The first laboratory-scale fermenter was a traditional orbitally stirred Erlenmeyer flask. The second and third SSF fermenters are scalable prototypes of the original design with blade systems for high solid loads. The second fermenter corresponds to a horizontal design with helical-type blades, where its design can be seen in Figure 6A and the third fermenter is of a vertical design with "G-type" blades shown in Figure 6B.

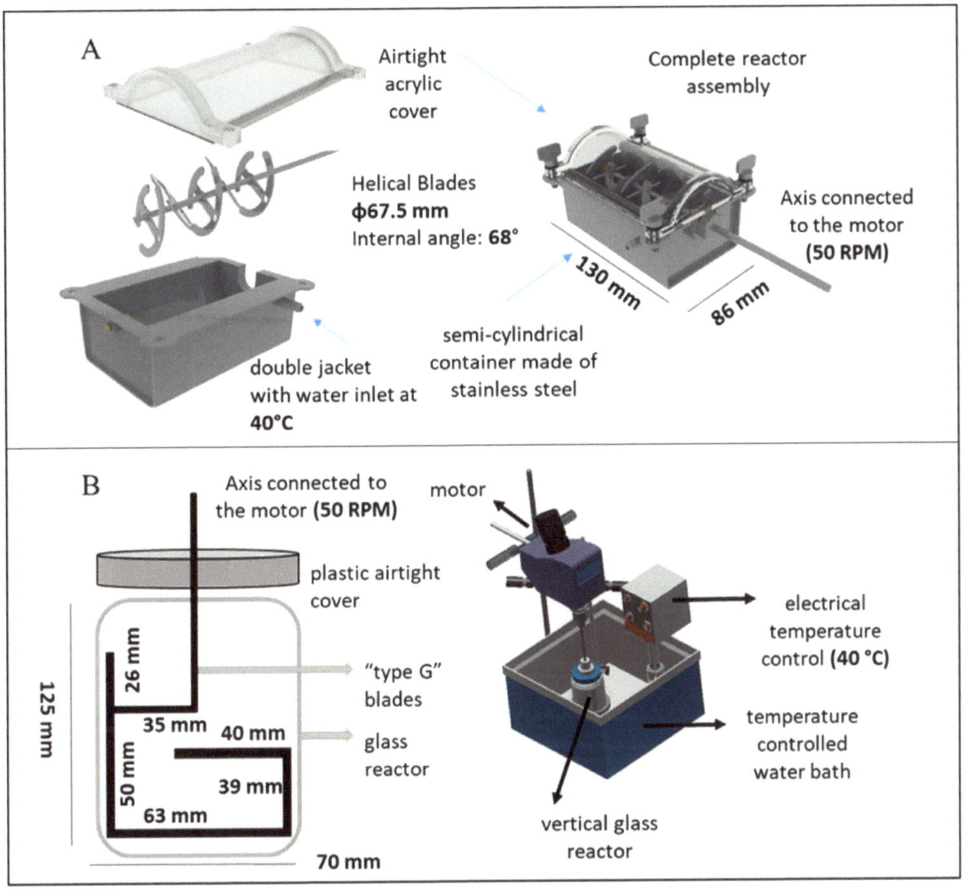

Figure 6. (**A**) Schematic representation of the horizontal fermenter (HF) with helical blades and (**B**) schematic representation of the vertical fermenter (VF) with "type G" blades.

In a traditional fermenter (Erlenmeyer flask), when the solid pretreated biomass was increased from 10% *w/v* to 20% *w/v*, the yield decreased from 61.9% to 55.0%. At high substrate loading, a slower rate of mass transfer of enzyme as well also soluble products. Diffusional limitation of enzymes increases with elevated substrate concentration and the same is responsible for slower mass transfer of enzymes [36]. However, when the fed-batch strategy was implemented, the glucose yield increased to 70.6%. SSF increases the yields because the glucose released by the enzymes is rapidly consumed by the yeast, reducing the inhibition per product. Table 2 shows a significant increase in the yield of FB-SSF with 20% *w/v* of pretreated material, which reached 78.0% at 72 h. However, the profitability and productivity of future bioethanol production plants remain very low. Very promising results were achieved with fermenter two and three, where we achieved a yield of 87.7% equivalent to 8.3% *v/v* of ethanol with the horizontal fermenter, and 92.5% equivalent to 9.1% *v/v* of ethanol with the vertical fermenter, both yields reached at 48 h of FB-SSF. Cost and energy studies are required to determine whether a vertical or horizontal fermenter is more viable for scaling up the bioethanol production.

Table 2. Glucose and ethanol yields (wood dry weight basis) obtained by EH, SSF and FB-SSF. Comparison between different fermenter designs.

Solids	SSF Erlenmeyer Flasks (Orbital Agitation)			SSF in Horizontal Fermenter (Helical Blades)			SSF in Vertical Fermenter (Type "G" Blades)		
	Ethanol Yields	Ethanol		Ethanol Yields	Ethanol		Ethanol Yields	Ethanol	
%	%	g/L	%v/v	%	g/L	%v/v	%	g/L	%v/v
10	63.8 ± 0.7 (72) [a]	24.7 ± 0.3	3.2 ± 0.0						
20	61.0 ± 0.1 (96)	44.6 ± 0.1	5.7± 0.0	78.6 ± 1.2 (72)	60.0 ± 1.0	7.7 ± 0.8	77.7 ± 2.3 (72)	57.0 ± 2.1	7.3 ± 2.0
20 [b]	78.0 ± 0.6 (72)	58.0 ± 0.4	7.4 ± 0.1	87.7 ± 0.9 (48)	65.0 ± 0.8	8.3 ± 0.6	92.5 ± 1.9 (48)	71.0 ± 1.5	9.1 ± 1.4

[a] Time in hours of the process at which the yields were reached. [b] Fed-batch strategy in first 8 h.

4. Conclusions

SE pretreatment of *E. globulus* was optimized using RSM. The response polynomial allowed us to obtain the optimal parameter combination of 208 °C and 11 min to maximize glucose release by EH. The study of samples generated during optimization allowed us to determine the physicochemical changes in a wide range of severities, where the removal of xylans and decrease in the DP of cellulose were the main effects caused by SE pretreatment. These effects correlated with the SEM images and Y_{EH}. Finally, a high Y_E was achieved, which is necessary to lower distillation costs through novel SSF fermenter designs and fed-batch strategies.

The combination of fermenters with blades designed for a high concentration of pretreated solids and their incorporation in the fed-batch mode allowed us to achieve a high ethanol concentration without sacrificing the potential yield. Furthermore, this strategy enabled us to achieve the highest ethanol concentration in a shorter time for SSF compared with the conventional orbital agitation method.

Author Contributions: E.T.-O., conceptualization, methodology, software, formal analysis, investigation, and writing—original draft. R.V., conceptualization, software, and formal analysis. P.R.-C., methodology, formal analysis, and writing—original draft. P.C.-R., methodology, investigation, and writing—original draft. L.-N.S., investigation, writing—review and editing, and funding acquisition. C.P., conceptualization, methodology, software, investigation, resources, writing—original draft, supervision, project administration, and funding acquisition. All authors have read and agreed to the published version of the manuscript.

Funding: This research was funded by the Innova Chile CORFO grant 208-7302 and the Innovation Found for Competitiveness of the Chilean Economic Development Agency (CORFO) under Grant no. 13CEI2-21839, FONDECYT-ANID 1231086 and ANID BASAL FB210015 CENAMAD.

Institutional Review Board Statement: Not applicable.

Informed Consent Statement: Not applicable.

Data Availability Statement: Data are contained within the article.

Acknowledgments: The authors are grateful to ANID—Millennium Science Initiative Program—NCN17_040. Eduardo Troncoso-Ortega thanks the ANID/2020-21202445 National Doctoral Scholarship.

Conflicts of Interest: Author Roberto Valenzuela is employed by the company Innocon S.A. But for purposes of this investigation, there was no financing relationship with the company; therefore, there are no conflicts of interest. The remaining authors declare that the research was conducted in the absence of any commercial or financial relationships that could be construed as a potential conflict of interest.

References

1. David, A.J.; Abinandan, S.; Vaidyanathan, V.K.; Xu, C.C.; Krishnamurthi, T. A critical review on current status and environmental sustainability of pre-treatment methods for bioethanol production from lignocellulose feedstocks. *3 Biotech* **2023**, *13*, 233. [CrossRef] [PubMed]
2. Galbe, M.; Wallberg, O. Pretreatment for biorefineries: A review of common methods for efficient utilisation of lignocellulosic materials. *Biotechnol. Biofuels* **2019**, *12*, 294. [CrossRef] [PubMed]
3. Mariano, A.P.B.; Unpaprom, Y.; Ponnusamy, V.K.; Ramaraj, R. Bioethanol production from coconut pulp residue using hydrothermal and postalkaline pretreatment. *Int. J. Energy Res.* **2021**, *45*, 8140–8150. [CrossRef]
4. Aditiya, H.B.; Mahlia, T.M.I.; Chong, W.T.; Nur, H.; Sebayang, A.H. Second generation bioethanol production: A critical review. *Renew. Sustain. Energy Rev.* **2016**, *66*, 631–653. [CrossRef]
5. Jeong, S.Y.; Lee, J.W. Hydrothermal Treatment. In *Pretreatment of Biomass: Processes and Technologies*; Elsevier: Amsterdam, The Netherlands, 2015; pp. 61–74. [CrossRef]
6. Troncoso, E.; Castillo, R.; Valenzuela, R.; Reyes, P.; Freer, J.; Norambuena, M.; Rodríguez, J.; Parra, C. Chemical and microstructural changes in Eucalyptus globulus fibers subjected to four different pretreatments and their influence on the enzymatic hydrolysis. *J. Chil. Chem. Soc.* **2017**, *62*, 3442–3446. [CrossRef]
7. Ramos, L.P. The chemistry involved in the steam treatment of lignocellulosic materials. *Quim. Nova* **2003**, *26*, 863–871. [CrossRef]
8. Reyes, P.; Márquez, N.; Troncoso, E.; Parra, C.; Teixeira Mendonça, R.; Rodríguez, J. Evaluation of Combined Dilute Acid-Kraft and Steam Explosion-Kraft Processes as Pretreatment for Enzymatic Hydrolysis of *Pinus radiata* Wood Chips. *BioResources* **2015**, *11*, 612–625. [CrossRef]
9. Das, N.; Jena, P.K.; Padhi, D.; Kumar Mohanty, M.; Sahoo, G. A comprehensive review of characterization, pretreatment and its applications on different lignocellulosic biomass for bioethanol production. *Biomass Convers. Biorefinery* **2023**, *13*, 1503–1527. [CrossRef]
10. Hill, C. Benchmarking and best practices. *Annu. Qual. Congr.* **2000**, *54*, 715–717.
11. Mithra, M.G.; Sajeev, M.S.; Padmaja, G. Comparison of SHF and SSF Processes under Fed Batch Mode on Ethanol Production from Pretreated Vegetable Processing Residues. *Eur. J. Sustain. Dev. Res.* **2018**, *3*, em0084. [CrossRef]
12. Kadhum, H.J.; Rajendran, K.; Murthy, G.S. Effect of solids loading on ethanol production: Experimental, economic and environmental analysis. *Bioresour. Technol.* **2017**, *244*, 108–116. [CrossRef] [PubMed]
13. Rudolf, A.; Alkasrawi, M.; Zacchi, G.; Lidén, G. A comparison between batch and fed-batch simultaneous saccharification and fermentation of steam pretreated spruce. *Enzyme Microb. Technol.* **2005**, *37*, 195–204. [CrossRef]
14. Jørgensen, H.; Vibe-Pedersen, J.; Larsen, J.; Felby, C. Liquefaction of lignocellulose at high-solids concentrations. *Biotechnol. Bioeng.* **2007**, *96*, 862–870. [CrossRef] [PubMed]
15. Wingren, A.; Galbe, M.; Zacchi, G. Techno-economic evaluation of producing ethanol from softwood: Comparison of SSF and SHF and identification of bottlenecks. *Biotechnol. Prog.* **2003**, *19*, 1109–1117. [CrossRef] [PubMed]
16. Hodge, D.B.; Karim, M.N.; Schell, D.J.; McMillan, J.D. Model-based fed-batch for high-solids enzymatic cellulose hydrolysis. *Appl. Biochem. Biotechnol.* **2009**, *152*, 88–107. [CrossRef]
17. Zhang, M.; Wang, F.; Su, R.; Qi, W.; He, Z. Ethanol production from high dry matter corncob using fed-batch simultaneous saccharification and fermentation after combined pretreatment. *Bioresour. Technol.* **2010**, *101*, 4959–4964. [CrossRef]
18. Cara, C.; Moya, M.; Ballesteros, I.; Negro, M.J.; González, A.; Ruiz, E. Influence of solid loading on enzymatic hydrolysis of steam exploded or liquid hot water pretreated olive tree biomass. *Process Biochem.* **2007**, *42*, 1003–1009. [CrossRef]
19. Hoyer, K.; Galbe, M.; Zacchi, G. Production of fuel ethanol from softwood by simultaneous saccharification and fermentation at high dry matter content. *J. Chem. Technol. Biotechnol.* **2009**, *84*, 570–577. [CrossRef]
20. Lu, Y.; Wang, Y.; Xu, G.; Chu, J.; Zhuang, Y.; Zhang, S. Influence of high solid concentration on enzymatic hydrolysis and fermentation of steam-exploded corn stover biomass. *Appl. Biochem. Biotechnol.* **2010**, *160*, 360–369. [CrossRef]
21. Amândio, M.S.T.; Rocha, J.M.S.; Xavier, A.M.R.B. Enzymatic Hydrolysis Strategies for Cellulosic Sugars Production to Obtain Bioethanol from Eucalyptus globulus Bark. *Fermentation* **2023**, *9*, 241. [CrossRef]
22. Chander Kuhad, R.; Mehta, G.; Gupta, R.; Sharma, K.K. Fed batch enzymatic saccharification of newspaper cellulosics improves the sugar content in the hydrolysates and eventually the ethanol fermentation by *Saccharomyces cerevisiae*. *Biomass Bioenergy* **2010**, *34*, 1189–1194. [CrossRef]
23. Valenzuela, R.; Priebe, X.; Troncoso, E.; Ortega, I.; Parra, C.; Freer, J. Fiber modifications by organosolv catalyzed with H_2SO_4 improves the SSF of Pinus radiata. *Ind. Crops Prod.* **2016**, *86*, 79–86. [CrossRef]
24. Guigou, M.; Guarino, J.; Chiarello, L.M.; Cabrera, M.N.; Vique, M.; Lareo, C.; Ferrari, M.D.; Ramos, L.P. Steam Explosion of *Eucalyptus grandis* Sawdust for Ethanol Production within a Biorefinery Approach. *Processes* **2023**, *11*, 2277. [CrossRef]
25. Bhalla, A.; Cai, C.M.; Xu, F.; Singh, S.K.; Bansal, N.; Phongpreecha, T.; Dutta, T.; Foster, C.E.; Kumar, R.; Simmons, B.A.; et al. Performance of three delignifying pretreatments on hardwoods: Hydrolysis yields, comprehensive mass balances, and lignin properties. *Biotechnol. Biofuels* **2019**, *12*, 213. [CrossRef]
26. Acuña, E.; Cancino, J.; Rubilar, R.; Parra, C. Potencial de bioetanol de culturas lenhosas conduzidas em alta densidade e curta rotação em terras marginais no Chile central. *Cerne* **2017**, *23*, 133–145. [CrossRef]
27. Mendonça, R.T.; Jara, J.F.; González, V.; Elissetche, J.P.; Freer, J. Evaluation of the white-rot fungi *Ganoderma australe* and *Ceriporiopsis subvermispora* in biotechnological applications. *J. Ind. Microbiol. Biotechnol.* **2008**, *35*, 1323–1330. [CrossRef]

28. Castro, J.F.; Parra, C.; Yáñez-S, M.; Rojas, J.; Teixeira Mendoncìa, R.; Baeza, J.; Freer, J. Optimal pretreatment of *Eucalyptus* by hydrothermolysis and alkaline extraction for microbial production of ethanol and xylitol. *Ind. Eng. Chem. Res.* 2 5713–5720. [CrossRef]
29. Monrroy, M.; Ortega, I.; Ramírez, M.; Baeza, J.; Freer, J. Structural change in wood by brown rot fungi and effect on er hydrolysis. *Enzym. Microb. Technol.* **2011**, *49*, 472–477. [CrossRef]
30. Monrroy, M.; Garcia, J.R.; Troncoso, E.; Freer, J. Fourier transformed near infrared (FT-NIR) spectroscopy for the estima parameters in pretreated lignocellulosic materials for bioethanol production. *J. Chem. Technol. Biotechnol.* **2015**, *90*, 1281 [CrossRef]
31. Edgardo, A.; Carolina, P.; Manuel, R.; Juanita, F.; Baeza, J. Selection of thermotolerant yeast strains *Saccharomyces cerevisiae* bioethanol production. *Enzym. Microb. Technol.* **2008**, *43*, 120–123. [CrossRef]
32. Chen, X.; Lawoko, M.; Heiningen, A. van Kinetics and mechanism of autohydrolysis of hardwoods. *Bioresour. Technol.* **2010**, *101*, 7812–7819. [CrossRef] [PubMed]
33. Leschinsky, M.; Sixta, H.; Patt, R. Detailed Mass Balances of the Autohydrolysis of eucalyptus globulus at 170 °C. *BioResources* **2009**, *4*, 687–703. [CrossRef]
34. Araya, F.; Troncoso, E.; Mendonça, R.T.; Freer, J. Condensed lignin structures and re-localization achieved at high severities in autohydrolysis of *Eucalyptus globulus* wood and their relationship with cellulose accessibility. *Biotechnol. Bioeng.* **2015**, *112*, 1783–1791. [CrossRef] [PubMed]
35. Yoo, C.G.; Meng, X.; Pu, Y.; Ragauskas, A.J. The critical role of lignin in lignocellulosic biomass conversion and recent pretreatment strategies: A comprehensive review. *Bioresour. Technol.* **2020**, *301*, 122784. [CrossRef] [PubMed]
36. Baksi, S.; Sarkar, U.; Villa, R.; Basu, D.; Sengupta, D. Conversion of biomass to biofuels through sugar platform: A review of enzymatic hydrolysis highlighting the trade-off between product and substrate inhibitions. *Sustain. Energy Technol. Assess.* **2023**, *55*, 102963. [CrossRef]

MDPI AG
Grosspeteranlage 5
4052 Basel
Switzerland
Tel.: +41 61 683 77 34

Fermentation Editorial Office
E-mail: fermentation@mdpi.com
www.mdpi.com/journal/fermentation